BIOLOGICAL SCIENCE
interaction of experiments and ideas

second edition

PRENTICE-HALL, INC., ENGLEWOOD CLIFFS, N.J.

BIOLOGICAL
SCIENCE

interaction of experiments and ideas   second edition

BIOLOGICAL SCIENCES CURRICULUM STUDY

BIOLOGICAL SCIENCE: Interaction of Experiments and Ideas, Second Edition
   Supplementary: Teachers Guide

Prepared by BIOLOGICAL SCIENCES CURRICULUM STUDY
Published by Prentice-Hall, Inc.
For permissions and other rights under this copyright,
please contact the Director, BSCS, P. O. Box 930, Boulder, Colorado 80302

Printed in the United States of America

10 9 8 7 6 5 4 3 2 1

ISBN 0-13-077008-6

Designed by Lynda de Victoria

BIOLOGICAL SCIENCE
interaction of experiments and ideas

THE WRITERS

NORMAN ABRAHAM: Supervisor, First Edition
Formerly of Yuba City Unified School
   District
Yuba City, California

ROBERT G. SCHROT: Supervisor, Second Edition
Yuba City High School
Yuba City, California

PATRICK BALCH
Biological Sciences Curriculum Study
Boulder, Colorado

DON E. BORRON
St. Stephens Episcopal School
Austin, Texas

FRANK ERK
State University of New York
Stony Brook, Long Island, N.Y.

WILLIAM KASTRINOS
Educational Testing Service
Princeton, New Jersey

JERRY P. LIGHTNER
National Association of Biology Teachers
Washington, D. C.

ROBERT MILLER
Western New Mexico University
Silver City, New Mexico

DONALD NEU
Flathead High School
Kalispell, Montana

GLEN E. PETERSON
Memphis State University
Memphis, Tennessee

LAWRENCE M. ROHRBAUGH
University of Oklahoma
Norman, Oklahoma

CONSUELO SAVIN
University of Mexico
Mexico, D. F.

GERALD SCHERBA
California State College
San Bernardino, California

PHILIP SNIDER
University of Houston
Houston, Texas

WILLIAM UTLEY
Yuba City High School
Yuba City, California

WAYNE UMBREIT
Rutgers, The State University
New Brunswick, New Jersey

BETTY WISLINSKY
San Francisco College for Women
San Francisco, California

THE BSCS STAFF

Chairman **ADDISON E. LEE**

Director **WILLIAM V. MAYER**

Associate Director **MANERT H. KENNEDY**

Assistant Director **GEORGE M. CLARK**

Staff Consultants **PATRICK BALCH**

THOMAS J. CLEAVER

JAMES T. ROBINSON

HAROLD A. RUPERT, JR.

RICHARD R. TOLMAN

Art Director **ROBERT F. WILSON**

Business Manager **KEITH L. BUMSTED**

Biological Sciences Curriculum Study Publications

BIOLOGICAL SCIENCE: Interaction of Experiments and Ideas, Second Edition
Prentice-Hall, Inc., Englewood Cliffs, New Jersey

REVISED QUARTERLY TESTS AND FINAL EXAMINATION (for above textbook)
Educational Programs Improvement Corporation (EPIC), Box 3406, Boulder, Colorado 80302
(First edition tests available from Prentice-Hall, Inc.)

BIOLOGICAL SCIENCE: An Inquiry into Life
Harcourt, Brace and World, Inc., New York

BIOLOGICAL SCIENCE: Molecules to Man
Houghton Mifflin Company, Boston

HIGH SCHOOL BIOLOGY: BSCS Green Version
Rand McNally and Company, Chicago

BSCS QUARTERLY TESTS AND FINAL EXAMS, 1968 editions (for above textbooks)
available from the version publishers

PROCESSES OF SCIENCE TEST (for all versions)
The Psychological Corporation, New York

BIOLOGICAL SCIENCE: Patterns and Processes, Second Edition (BSCS Special Materials)
Holt, Rinehart and Winston, Inc., New York

BSCS UNIT TESTS AND FINAL EXAMINATION FOR BIOLOGICAL SCIENCE: Patterns and Processes
Educational Programs Improvement Corporation (EPIC), Box 3406, Boulder, Colorado 80302 (First edition examinations available from the Psychological Corporation, New York)

BSCS LABORATORY BLOCKS (12 titles)
D.C. Heath and Company, Boston

TESTS AND TEACHER'S RESOURCE BOOK (for Laboratory Blocks)
D.C. Heath and Company, Boston

RADIATION AND ITS USE IN BIOLOGY (a Laboratory Block)
Educational Programs Improvement Corporation (EPIC), Box 3406, Boulder, Colorado, 80302

RESEARCH PROBLEMS IN BIOLOGY: Investigations for Students (Series 1–4)
Doubleday and Company, Inc., Garden City, New York

INNOVATIONS IN EQUIPMENT AND TECHNIQUES FOR THE BIOLOGY TEACHING LABORATORY
D.C. Heath and Company, Boston

BSCS SINGLE TOPIC INQUIRY FILMS (super 8-mm loop) (40 titles):
Harcourt, Brace and World, Inc., New York
Houghton Mifflin Company, Boston
Rand McNally and Company, Chicago

BSCS INQUIRY SLIDE SERIES (20 titles)
Harcourt, Brace and World., Inc., New York

POPULATION GENETICS, A Self-Instructional Program
General Learning Corporation, Silver Burdett Company Division, Morristown, New Jersey

STORY OF THE BSCS (information film),
BSCS, Boulder, Colorado

BSCS TEACHER PREP FILM, A BSCS SINGLE TOPIC FILM PRESENTATION (free on loan)
BSCS, Boulder, Colorado

BIOLOGY TEACHERS' HANDBOOK, Second Edition
John Wiley and Sons, Inc., New York

BSCS PAMPHLET SERIES (24 titles)
BSCS, Boulder, Colorado

BSCS PATTERNS OF LIFE SERIES (8 titles)
Rand McNally and Company, Chicago

BSCS BULLETIN SERIES (Nos. 1–3)
BSCS, Boulder, Colorado

BULLETIN NO. 4: THE CHANGING CLASSROOM: The Role of the Biological Sciences Curriculum Study by Arnold B. Grobman
Doubleday and Company, New York

BSCS SPECIAL PUBLICATIONS (Nos. 4–7 only)
BSCS, Boulder, Colorado

BSCS NEWSLETTER
BSCS, Boulder, Colorado

BSCS INTERNATIONAL NEWS NOTES
BSCS, Boulder, Colorado

FOREWORD

This course in biology presupposes certain biological knowledge on the part of the student and differs from most texts designed for a similar audience in being nonrepetitive. In this text the student will not again be taught the structure of the cell, mitosis, how to use the microscope, and other similar materials that, by their redundancy at various levels, do much to alienate the student while, at the same time, inducing the instructor to complain of the lack of time available to cover the field of biology. By building on previous knowledge with which the student easily may become reacquainted on his own, this program takes the student further into science as an enterprise than is common in such books. It involves him in the work of a scientist. There is no section on the "scientific method" as such, but the student uses the methods of science to become familiar with the processes of scientific investigation. He learns how to handle data, to utilize literature, and to develop experiments and ideas in investigating biological phenomena. The program concludes with a sequence of provocative essays dealing with science and society that provide valuable springboards for future study and discussion of biological relevance. Both the content and approach are quite different from those of other biological programs. The laboratory emphasis on the work of the scientist and the nonrepetitive nature of the content allows the student not only to build upon what he already knows but also to develop new insights concerning biology and the work of biologists.

The present volume began in 1962 with a three-year program that tested experimental editions in classrooms throughout the United States. On the basis of these trials, the first commercial edition appeared in 1965. This second edition is the result of modifications and modernizations made necessary by the intervening five years. It takes cognizance of the constructive comments of teachers, students, and others who used the volume during this period.

The BSCS, as an amalgamation of professional biologists, science teachers, and others interested in the improvement of biological education, extends its heartfelt thanks to those who have contributed to this volume and particularly to the current authorship under the direction of Mr. Robert G. Schrot. As a cooperative effort involving both the biological and educational communities, the BSCS has produced various sequences of innovative materials, of which *Biological Science: Interaction of Experiments and Ideas* is an important component.

As with previous works, you are invited to send your recommendations for improvement to the Director, Biological Sciences Curriculum Study, Post Office Box 930, Boulder, Colorado 80302. Many of those who commented on the previous work will be pleased to see their suggestions incorporated in this volume. Those who respond to the present volume may be similarly rewarded.

ADDISON E. LEE

Chairman of the Steering Committee
Biological Sciences Curriculum Study
University of Texas
Austin, Texas

WILLIAM V. MAYER, Director

Biological Sciences Curriculum Study
Post Office Box 930
Boulder, Colorado 80302

ACKNOWLEDGMENTS

The preparation of this book has involved the efforts of many individuals other than those listed as authors. Many of the experiments were adapted from the BSCS Laboratory Blocks developed under the direction of Addison E. Lee, University of Texas. The textual material on modern genetics was adapted from the BSCS Blue Version, *Biological Science: Molecules to Man.* We would also like to express our thanks to Merle Mizell for permitting adaptation of his unpublished experiments on tadpole-tail regeneration and frog-sperm irradiation conducted at Tulane University.

Through their contributions to the BSCS, Joseph J. Schwab, University of Chicago, and Evelyn Klinckmann, San Francisco College for Women, have provided much of the inspiration for presenting science as inquiry in this and other BSCS publications.

Leonard Reynolds, Chairman, Mathematics Department, Yuba City High School, gave valuable assistance in editing the sections involving mathematics.

We are grateful to the Literary Executor of the late Sir Ronald A. Fisher, F.R.S., Cambridge, to Frank Yates, F.R.S., Rothamsted, and to Messrs. Oliver and Boyd, Ltd., Edinburgh (Scotland) for permission to reprint Tables II and IV from their book *Statistical Tables for Biological, Agricultural and Medical Research.*

Our thanks to Nancy E. Stees and the BioSciences Information Service of Biological Abstracts for valuable assistance in the preparation of the "Literature of Biology" section.

These acknowledgments would not be complete without mention of the hundreds of BSCS writers, committee members, test teachers, students, and reviewers who have individually and collectively influenced the preparation of this book through their contributions to the BSCS program. Although it would be impossible to name each individually, we are grateful for their help.

ROBERT G. SCHROT, Supervisor

BIOLOGICAL SCIENCE: Interaction of Experiments and Ideas, Second Edition

The authors do not believe that any single textbook can provide, by itself, the entire substance for a meaningful course in biology. *Biological Science: Interaction of Experiments and Ideas* should, therefore, be regarded as a *guide to learning*. It is assumed that you will add to and expand on the knowledge you will gain in this course, through outside reading in a variety of journals and reference texts. To this end, a list of references has been provided in Appendix G, but it should not be considered all-inclusive: thousands of new articles and books are published each year. While the laboratory is the workshop of the biologist, it cannot be a productive laboratory without parallel readings in the literature. You will be expected to reflect on the laboratory Investigations, to seek a better understanding of experimental results, to study related subject-matter information, and to use your own intellectual ability purposefully in order to expand your knowledge of biology. To this extent, you are your own teacher

Today many problems in biology are studied by teams of scientists rather than by a single worker—and for good reason. Seldom can one person provide as much insight, imagination, creativity, and productive labor as can a team whose members work toward a common goal. Good teamwork demands that each member of the team strive to excel as an individual. An understanding of the value and problems of group effort should guide each team activity throughout the course; each team member, however, should accept opportunities to work alone occasionally. The excitement of a successful individual assault on a problem in science is something that the authors hope each of you will experience.

Keeping accurate records of laboratory results is essential, also. No effort should be spared to record accurate daily observations, to plan experiments carefully, and to discuss the significance of experimental results with others on the team. To enrich your understanding further, the experimental results and interpretations of different teams should be compared and discussed.

A primary goal of the course is to provide experiences that simulate biological research so that you will gain an understanding of science from direct experience rather than from a superficial dissertation on "the scientific method." Yeast metabolism; population dynamics; microbial genetics; plant and animal growth, regulation, and development; and animal behavior provide the major subject areas for laboratory study.

There are thirty-eight scheduled laboratory Investigations and several peripheral "Investigations for Further Study." Interspersed with the Investigations are supplemental materials. These include excerpts from the literature of biology, both contemporary and historic; instruction in the application of elementary statistics to an evaluation of data; subject-matter explanations that parallel the Investigations; and other pertinent materials. In the culminating activity of this course, the role of constructive controversy as an important mechanism of science is introduced.

This course and its predecessor were prepared by *teams* of biologists and biology teachers for those who find a challenge in learning more about life. The authors assume that you are ready to explore, in depth, many questions about life that invite the intellectually curious.

NORMAN ABRAHAM
Supervisor, First Edition

ROBERT G. SCHROT
Supervisor, Second Edition

CONTENTS

phase four **GROWTH, DEVELOPMENT, AND BEHAVIOR OF INDIVIDUALS, 164**

SECTION 10 GROWTH AND DEVELOPMENT IN PLANTS, 165

Meiosis, 165

Life History of a Flowering Plant, 167

PART 3   CONCLUSIONS AND BEGINNINGS

phase five   SCIENCE AND SOCIETY, 324

APPENDICES

IN BROKEN IMAGES

by ROBERT GRAVES

He is quick, thinking in clear images;
I am slow, thinking in broken images.

He becomes dull, trusting to his clear images;
I become sharp, mistrusting my broken images.

Trusting his images, he assumes their relevance;
Mistrusting my images, I question their relevance.

Assuming their relevance, he assumes the fact;
Questioning their relevance, I question the fact.

When the fact fails him, he questions his senses;
When the fact fails me, I approve my senses.

He continues quick and dull in his clear images;
I continue slow and sharp in my broken images.

He in a new confusion of his understanding;
I in a new understanding of my confusion.*

THE NATURE
OF BIOLOGICAL
SCIENCE

PART 1

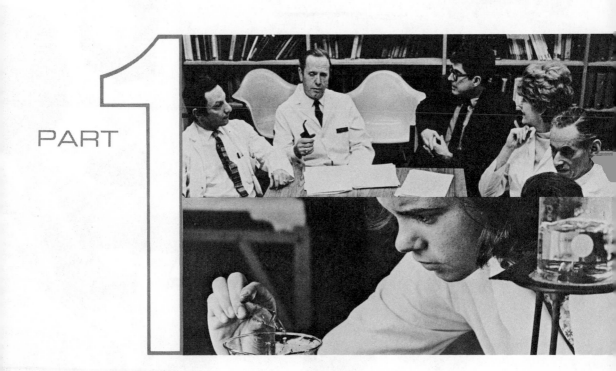

ORIENTATION

SECTION 1 THE MEANING OF SCIENCE

One measure of civilization is man's understanding of the natural world and of his place in nature. In his search for this understanding, man has called upon magic, demons, spirits, gods, and essences to help him explain some of the perplexing and fearful events of nature. The bolt of lightning in a severe thunderstorm, the plague that leaves death and despair strewn across an entire continent, the death of plants and animals as a result of drought or as victims of relentless insects, the blotting out of the sun by the moon, the sudden occurrences of violent earthquakes—all these and more struck fear in the heart of man and left him with a sense of helplessness. But man continued to struggle and to learn, and eventually began to understand more about the mysteries of nature.

The growth of scientific knowledge has been uneven. Early civilized man, especially the Greeks and Romans, made great strides in beginning to understand the relationships of similar kinds of living beings and the structure and function of the human body. Some of this knowledge, although imperfect, was applied to the practical problems of living. However, there was no steady progress of scientific advance until the 17th and 18th Centuries brought new insights and hope for dispelling the shadows of ignorance.

3

It was during this period that biology began to emerge as a science. Careful observation, classification, and experimentation as the primary methods of studying life, gradually replaced superstition, speculation, and a reliance upon "authority." The subsequent rapid growth of biology as a science has led many scientists to predict that we are now entering the age of a biological revolution at least as comparable in its effect upon the future of mankind as was the Industrial Revolution.

Science is, in effect, an approach to understanding nature. It is based upon the notion that there is order in the universe and that the mind of man can discern and understand this order. Further, it includes the concept that man is capable of testing which (if any) of his notions about the world are correct and which are not correct, and can thus, by a process of testing, arrive at truth.

Understanding what science means is a goal that you should achieve through participation in the laboratory investigations and supplemental activities associated with this course. The opinions of three scientists on the meaning of science are given to help set the stage for what the authors hope will be a continuous effort on your part to read widely on the meaning of science in a variety of books and journals.

Dr. James B. Conant, (pp. 24–26)[1]:

... limiting one's attention merely to the experimental sciences by no means provides a satisfactory answer to the question "What is science?" For, immediately, diversity of opinion appears as to the objectives and methods of even this restricted area of human activity. The diversity stems in part from real differences in judgment as to the nature of scientific work but more often from the desire of the writer or author to emphasize one or another aspect of the development of the physical and biological sciences. There is the static view of science and the dynamic. The static places in the center of the stage the present interconnected set of principles, laws, and theories, together with the vast array of systematized information: in other words, science is a way of explaining the universe in which we live. The proponent of this view exclaims "How marvelous it is that our knowledge is so great!" If we consider science solely as a fabric of knowledge, the world would still have all the cultural and practical benefits of modern science, even if all the laboratories were closed tomorrow. This fabric would be incomplete, of course, but for those who are impressed with the significance of science as "explanations" it would be remarkably satisfactory. How long it would remain so, however, is the question. . . .

The dynamic view in contrast to the static regards science as an activity; thus, the present state of knowledge is of importance chiefly as a basis for further operations. From this point of view science would disappear completely if all of the laboratories were closed; the theories, principles, and laws embalmed in the texts would be dogmas; for if all the laboratories were closed, all further investigation stopped, there could be no re-examination of any proposition. I have purposely overdrawn the picture. No one except in a highly argumentative mood would defend either the extreme static or the extreme dynamic interpretation of the natural sciences.

. . . . My definition of science is, therefore, somewhat as follows: Science is an interconnected series of concepts and conceptual schemes that have developed as a result of experimentation and observation and are fruitful of further experimentation and observations. In this definition the emphasis is on the word "fruitful." Science is a speculative enterprise. The validity of a new idea and the significance of a new experimental finding are to be measured by the consequences—consequences in terms of other ideas and other experiments. Thus conceived, science is not a quest for certainty; it is rather a quest which is successful only to the degree that it is continuous.

Gerald Holton and Duane H. D. Roller, (pp. 231–232)[2]:

. . . Since the methods and relationships of one field frequently suggest analogous procedures in another, the working scientist is ever alert for the slightest hints of new difficulties and of their resolutions. He proceeds through his problem like an explorer through a jungle, sensitive to every sign with every faculty of his being. Indeed, some of the most creative of theoretical scientists have stated that during the early stages of their work they do not even *think* in terms of conventional communicable symbols and words.

Only when this "private" stage is over and the individual contribution is formalized and prepared for absorption into "public" science does it begin to be really important that each step and every concept be made meaningful and clear. These two stages of science, which we shall have occasion to call science-in-the-making and

[2] Holton, Gerald and Duane H. D. Roller. *Foundations of Modern Physical Science.* 1958. Reprinted by permission of Addison-Wesley Publishing Co., Inc., Reading, Mass.

science-as-an-institution, must be clearly differentiated. Once an adequate distinction has been made between these levels of meaning in the term "science," one of the central sources of confusion concerning the nature and growth of science has been removed.

The American nuclear physicist H. D. Smyth has thus characterized the distinction: "We have a paradox in the method of science. The research man may often think and work like an artist, but he has to talk like a bookkeeper, in terms of facts, figures, and logical sequence of thought." For this reason we must not take at their face value either the chronology or the methods set forth in scientific papers and treatises, including those of Galileo and Newton. It is part of the game of science, simply because it promotes economy of thought and communication, to make the results in retrospect appear neatly derived from clear fundamentals, until, in John Milton's phrase,

> . . . so easy it seemed
> Once found, which yet unfound most
>   would have thought
> Impossible!

A research worker may hide months of torturous and often wasteful effort behind a few elegant paragraphs, just as a sculptor puts away his tools, preliminary studies, and clumsy scaffolds before unveiling his work.

A famous example shows how dangerous it is to refuse to accept provisional concepts into scientific work simply because they are not yet amenable to rigorous tests. Despite the large amount of indirect evidence for the hypothesis that matter is atomic in structure, a few prominent scientists, around the turn of this century, still rejected the atomic view stubbornly and vehemently as lacking "direct" confirmation. They eventually had to yield when all around them the atomic hypothesis led to an avalanche of testable conclusions and even to a revolution in classical science itself. Although perhaps an initial attitude of fundamental skepticism was justified, these men deprived themselves unduly long of a useful conceptual scheme (and the rest of science of their possible additional contributions) by waiting until the atomic picture had been fully fortified operationally. Today most scientists tacitly agree that their private creative activity must be unfettered by such preconceptions. "To set limits to speculation is treason to the future."

. . . It begins to appear that there are no simple rules to lead us to the discovery of new phenomena or to the invention of new concepts, and none by which to foretell whether our contributions will turn out to be useful and durable. But science does exist and is a vigorous and successful enterprise. The lesson to be drawn from history is that *science as a structure grows by a struggle for survival among ideas*—that there are marvelous processes at work which in time purify the meanings even of initially confused concepts. These processes eventually permit the absorption into science-as-an-institution (public science) of anything important that may have been developed, no matter by what means or methods, in science-in-the-making (private science).

Although each branch of science may have its own distinctive flavor, the philosophy presented in these quotations applies to all of science.

We will define the science of biology as *that human activity which is directed toward seeking knowledge about living matter.* Learning *how* knowledge is acquired will assume greater importance in this course than memorizing the details of what others have learned.

A serious study of biology requires a detailed study in books, reading articles in journals, identifying problems, asking questions, performing experiments, and making decisions. It requires asking questions about living things, questions they cannot answer directly, yet questions that must be asked if the riddles of life are to be investigated. The work is often dirty, sometimes tedious, and occasionally frustrating. However, with the frustration and the work *can* come some of the most rewarding experiences of life—those of discovery, of seeing for the first time a relationship between observations, or of sudden insights into previously obscure problems. This is the role of a biologist, a role that must be experienced to be appreciated.

PROBLEMS

1. Ask several students who are not enrolled in this course to give you their definition of science. Compare these opinions with those expressed in Section One.
2. In your own words, write a definition of science that takes into consideration all of the aspects of science mentioned in Section One.

QUESTIONS FOR DISCUSSION

1. Compare and discuss "Dynamic Science" and "Static Science."
2. Compare and discuss "Private Science" and "Public Science."

3. Describe the relationship between the terms used in questions 1 and 2. That is, relate dynamic to either private or public as these terms are used in the text.

4. A recent newspaper editorial ridiculed the U.S. Congress for approving funds to be used in various biological research programs. The projects criticized included an investigation of bee sounds, a study of the porpoise, and the maintenance of a colony of apes for behavioral studies. The editorial pointed out that this was a "foolish waste of tax money since studies of this kind are of no practical value."

   Comment on the rationale of this editorial in view of the meaning of science expressed in Section One.

SECTION **2**   THE INCREASE OF BIOLOGICAL KNOWLEDGE

For centuries man has been groping to achieve an understanding of the physical world and of the living creatures in it—and, in so doing, he has built up an enormous body of knowledge, observation, and surmise. About a quarter century ago, this reached the "critical mass"; that is, in a relatively few years, the development of new ideas, new explanations, and new ways of experimentation began to change our conceptions of how living beings were constructed, how they operated —in short, how they "lived, and moved, and had their being" (The Bible: Acts 17:28). These new approaches permitted a startling insight into the nature of living things, and today's student can understand much that was obscure to his teacher only a few years ago. Today's biology is quite clearly a modern science; a science textbook two decades old is almost useless today.

To visualize the staggering growth of *biological* knowledge alone is difficult. Professor Bentley Glass[1] of The Johns Hopkins University commented on this problem:

> The textbook of 1900 would have nothing about genetics, for Mendelian heredity was unrecognized, and the science of genetics did not really start until 1900. Biochemistry had begun in its modern sense only three years before. Edward Buchner's classic studies on the nature of the enzymes had just begun. Immunology did not exist. . . . Animal viruses had not been discovered at all. No one

[1] Glass, Bentley. "Revolution in Biology" (an address to participants at the BSCS 1960 Briefing Session for teachers in the BSCS 1960–61 Testing Program) 1961. *BSCS Newsletter No. 9,* Setember, pp. 2–5.

knew anything about specific vitamins in 1900. The science of experimental embryology was in its cradle. Pavlov was still to do his classic studies in experimental psychology showing conditioned reflex behavior. . . .

What will biology be in the year 2000? The biologists of 1930 would not have dreamed of what we know today, and I do not think I can dream of what biologists will know 30 years from now. But I can foresee, perhaps, a few directions in which our control over the forces of nature and the nature of life will extend.

We will probably learn not only how to increase the human life span, but also how to maintain the vigor of mature life into advanced years. So far, we have not increased the human life span at all. The *average* length of life has gone up, but the *maximum* does not seem to have changed at all. I would suspect that by 1990 biologists will have learned how to create some simple forms of living organisms, something at about the order of complexity of a virus, and that geneticists will have learned how to replace defective genes with sound ones. This will depend, of course, on advances made by embryologists, who will—I suspect before very long—show us how to maintain in artificial culture outside the body the reproductive organs of animals to the extent that spermatozoa and ova can be produced *in vitro,* that is, in a glass dish.

I certainly expect that before the next 30 years are finished man will have learned how to conduct artificial photosynthesis and so will have finally assured himself of an inexhaustible—and I hope palatable—food supply. Man will certainly have learned to accelerate his own evolution in a desired direction, though I wonder what direction he will desire. And he will probably have eliminated infectious disease completely.

These conjectures may be wrong, but this would seem to be the general direction in which history is moving: a logarithmic increase of human power in the biological as well as physical realm.

PROBLEM

1. Prepare a list of new and outstanding discoveries in biology which you think have been made since 1950.

QUESTIONS FOR DISCUSSION

1. What is meant by "a logarithmic increase of human power in the biological as well as physical realm"?

2. Discuss the statement "Man will certainly have learned to accelerate his own evolution in a desired direction, though I wonder what direction he will desire."

SECTION **3** **THE ROLE OF INQUIRY IN BIOLOGY**

It must be obvious that "coverage" of a body of knowledge is impossible and that the possession of a body of facts is not enough to qualify one as a person who understands science.

Rather than treat science as a series of absolute certainties, we should study science as a *process of inquiry*. We should see how an attitude of inquiry provides the mechanism for probing into the complexities of nature and how it enables the scientist to challenge his own conclusions. In this way the "conditional truth" of knowledge will be revealed and the full meaning of science understood.

Inquiry is broadly defined as a search for truth, information, or knowledge. It pertains to research and investigation and to seeking information by asking questions. Science is concerned with asking the *right kind of questions* so that the answers can be properly evaluated.

Patterns of Inquiry

A number of carefully planned teacher-student discussions entitled "Patterns of Inquiry" are provided throughout this book as an aid to understanding the nature of inquiry. Such understanding is necessary if proper use is to be made of the biology laboratory. The Patterns provide discussions that involve both teacher and student. These discussions will simulate, as closely as possible, the kind of thinking a scientist does in conducting research.

The Patterns of Inquiry involve the identification of problems, the formation of fruitful hypotheses, the design of experiments to test these hypotheses, and finally the analysis and interpretation of data resulting from the experimental work. The Patterns of Inquiry and the laboratory investigations which you will carry out throughout the course should complement each other.

Each pattern is a small sample of the operation of inquiry. It poses a problem and provides you with information which you are asked to use in the same way as if you were conducting the investigation. Additional problems may be introduced by the instructor. If you actively participate in these discussions, you will come to understand more fully that science involves much more than the learning of things which others already know. You will better understand Conant's characterization of science as an activity and a speculative enterprise. Also, you will better understand why Holton and Roller suggest that "the working scientist is ever alert" and "proceeds through his problem like an explorer through a jungle, sensitive to every sign with every faculty of his being."

As an illustration, the model Pattern will ask you to draw conclusions from very simple data. Later Patterns will involve other and more difficult aspects of the process of science.

PATTERN OF INQUIRY (MODEL)
Germination of Seeds

A student wished to learn what conditions were most favorable for the germination of seeds. He placed several bean seeds on moist filter paper in each of two glass dishes. One dish was placed in the light, and the other was placed in a dark box; both were kept at normal room temperature. He examined the seeds after a few days and found that all the seeds in each dish had germinated.

What inferences could you draw from the results of this experiment? Base your inferences solely on the data presented in this experiment and do not use facts which you may have obtained from other sources.

PROCESSES OF BIOLOGICAL INVESTIGATION

Introduction

The approach to an understanding of living creatures has roots deep in philosophy. For example, Aristotle (about 350 B.C.) said:

> While all knowledge is valuable, one kind may be more valuable than another, and surely we would place biology well in the front rank. Its aim is to grasp and understand the essential character of life.
>
> However, to attain any assured understanding of life is one of the most difficult things in the world. One might suppose that there was some single method of study which would be adequate. If so, we should naturally be well trained in this method.
>
> But suppose that there is no single or general method applicable to all the problems of biology. In this case, with what facts shall we begin our inquiry? For the facts which form the starting points in different subjects must be different. And the kind of training we undertake must likewise be different.

Over two thousand years later (in 1781) Kant, the German philosopher, pointed out that

> Reason should approach Nature with the view, indeed, of obtaining information, but not in the character of a pupil who listens to all

13

his master chooses to tell him, but rather in the character of a judge who compels the witness to reply to those questions he himself thinks fit to propose.

And Kant remarks that it was the understanding of this principle that "at last put man's wandering feet upon the path of certain progress." All science begins with observation of a phenomenon. It is true that not all persons are able to observe objectively (meaning that they must observe what happens, independently of their emotional state or previous belief); without observation, however, no science can progress. Paracelsus (about 1500) said, "Knowledge of nature as she is and not as we imagine her to be, constitutes science."

The next phase is the organization of observations into a "body of knowledge." For example, we observe that peas are green and grass is green, and (with a few exceptions) further observation shows that all creatures capable of using the energy of light to photosynthesize are green. We can thus organize this information into the generalization that light-using organisms are green. When we then suppose that being green is somehow related to the ability to use light, we are employing (perhaps subconsciously) another principle; that is, "that the sole invariable antecedent of a phenomenon, is *probably* its cause." If every time some event occurs or every time we observe a phenomenon, there is always something else happening, we may conclude that there is a relation between them. If every time A occurs (photosynthesis, for example) there is always B (green color), then there is likely to be a relationship between them. The closer this relationship, the more probable that one is the cause of the other.

The next stage, as is evident in the previous paragraph, is to assemble information in some sort of order for the purpose of recognizing the elements of sameness (that is, green color) amidst apparent diversity and to recognize the differences within apparent identity. After the collection of related facts, we search for a cause. In this search, one frequently builds an imaginary model showing how a given set of conditions is *always* followed by a particular result. These mental models are called hypotheses, and they serve to bridge, temporarily, the many gaps in our knowledge. But there are two aspects to this hypothesis making: (1) one should not make them too soon, and one should collect at least a few facts before making a model—for without observations, hypotheses are idle; (2) the hypothesis and its logical consequences must be tested by comparison with observation—not only with the observation previously made upon which the hypothesis was based, but also with new ones designed specifically to test the hypothesis; that is, the "if . . . then" situation which you will encounter in many experiments in this book.

A theory is a hypothesis which has stood several critical tests successfully, and it becomes a "law" when it and *it alone* is in harmony with all known facts. All hypotheses, theories, and laws are on probation: when facts which are not in harmony with them are discovered, revision is necessary; but, until such facts are discovered, we rely on them as representing the truth as we know it. A theory is normally an explanation of the cause of something. A law simply states a relationship without necessarily suggesting the cause.

SECTION 4 HYPOTHESES AND THE DESIGN OF EXPERIMENTS

When we attempt to interpret an observation or to understand the meaning of a series of events, we are likely to formulate some form of working hypothesis or explanation. This working hypothesis may be followed by an experiment designed to yield data that will either support or refute the hypothesis.

A hypothesis has been described as a logical linkage between *if* and *then*. Consider the hypothesis that nerves are necessary for the action of some organ. In this case we are proposing that, *if* nerves are necessary for the action, *then* cutting the nerves leading to the organ should result in failure of action. This assumption leads us to design an experiment in which we sever the nerves, and the results produced will be our data. If these data indicate that cutting the nerves resulted in a loss of action in the organ, we may say that our hypothesis has been supported, and as far as our information goes, it is a good one. However, if the organ continues to act normally after the nerves have been cut, then the data place our hypothesis in doubt, and we will probably reject it as being invalid on the basis of our new information.

When an experiment is performed to test a hypothesis, new information usually will be gained. The new information may or may not be useful in evaluating the hypothesis. If the investigator is alert, however, it can lead to new questions, new problems, and new experiments. In this way science progresses.

As you work with the Investigations in this book, practice stating the hypotheses in the "if . . . then" form. This will guide you in the design of experiments and in the evaluation of the data which they may yield.

Common baker's yeast has been selected as an organism with which to begin our laboratory studies because it is readily available, because it is relatively easy to work with, and because a study of the growth and metabolism of yeast will offer an opportunity to experience many aspects of biological research.

The yeast you are to investigate derives its name from an early observation that it is a sugar-consuming fungus, hence the name *Saccharomyces* (*saccharum* = sugar, and *myces* = fungus).

Studies of brewer's or baker's yeast, both of which are varieties of the species *Saccharomyces cerevisiae*, have provided much of our present knowledge of carbohydrate metabolism.

As early as 1838, Charles Cagniard-Latour and later, in 1860, Louis Pasteur attempted to explain how yeasts were able to convert sugar to ethyl alcohol (ethanol) and carbon dioxide. When maintained under anaerobic conditions (without oxygen), *S. cerevisiae* forms ethanol and carbon dioxide from sugar. This process is an ethanolic (alcoholic) fermentation. The following equation summarizes what happens in this kind of fermentation:

$$H_2O + C_{12}H_{22}O_{11} \xrightarrow{\text{yeast}} 4\,CH_3CH_2OH + 4\,CO_2 + \text{energy}$$
$$\qquad\quad (\textit{Sucrose}) \qquad\qquad\qquad (\textit{Ethanol})$$

INVESTIGATION 1
The Problem

Based upon what we have observed or what we may think to be probable, there arises in our minds a supposition or a question. We wish to ask a question of Nature. Our attempts to answer these questions are our experiments. We start with a certain amount of information which prompted the question in the first place. We try to obtain whatever other information may be available to us, and we then plan an experiment which might provide the answer to the question we wish to ask.

The previous paragraphs told us something about yeasts, and this has given us the notion that since yeasts change sugar to alcohol and CO_2 and grow, using this process, there must be a relationship between the sugar available and the amount of yeast growth. What is this relationship? Instead of measuring yeast growth, which might

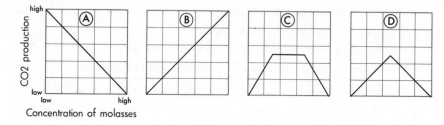

FIGURE 1. Concentration of molasses and CO_2 production.

be beyond our facilities or skills at the moment, we could measure the CO_2 produced. We think that the relationship could be that of A, B, C, or D of Figure 1.

Which would you guess you will find? The question we are asking Nature is: When we provide more and more sugar, do we get CO_2 production equivalent to the sugar provided? Since molasses contains a high proportion of sugar and since we know that yeast grows in molasses, we will use it as a starting material.

MATERIALS (per team)

1. One package of dry yeast or a yeast suspension
2. One graduated cylinder, 100-ml
3. Ten test tubes, 22 mm × 175 mm
4. Ten test tubes, 13 mm × 100 mm
5. Two Erlenmeyer flasks, 125- or 250-ml
6. Test tube rack

MATERIALS (per class)

1. One pint of commercial molasses
2. A supply of distilled water

Note: A well-organized and safe laboratory operation is possible only when certain rules of procedure are carefully followed by the investigator.

1. *Keep all glassware and other laboratory equipment clean and in the proper place. The use of chemicals and microorganisms introduces potential hazards; cleanliness should always be stressed.*

2. *Carefully handle all laboratory equipment such as microscopes and balances according to instructions.*

3. *Prepare for each investigation in a professional way. Label all necessary containers and arrange your equipment in an orderly manner.*

4. *Discard living materials and other wastes in the place specified by your instructor.*

PROCEDURE DAY I

1. Prepare by serial dilution, in the ten large test tubes, a series of molasses concentrations ranging from 100% to 0.19% molasses in water. This is easily done as follows:

Tube 1 = 100% molasses
Using the 100-ml graduated cylinder, measure 25 ml of pure molasses and add to tube 1.

Tube 2 = 50% molasses
Measure 25 ml of pure molasses in the graduated cylinder and add 25 ml of distilled water. Insure a uniform solution by pouring the molasses-water mixture back and forth between the graduated cylinder and a clean 125-ml (or larger) Erlenmeyer flask. Carefully pour one half (25 ml) of this solution into tube 2. Save the remaining half for tube 3.

Tube 3 = 25% molasses
Add 25 ml of distilled water to the 25 ml of molasses solution left over from the previous dilution. Again pour this molasses-water mixture back and forth between the graduated cylinder and the flask. Add one half (25 ml) of this dilution to tube 3.
Repeat this operation until a final dilution of 0.19% (approximately) molasses in water is obtained. Discard 25 ml of the 0.19% dilution; each of the ten tubes should contain 25 ml of solution.

Tube 4 = 12.5% molasses

Tube 5 = 6.2% molasses

Tube 6 = 3.1% molasses

Tube 7 = 1.6% molasses

Tube 8 = 0.78% molasses

Tube 9 = 0.39% molasses

Tube 10 = 0.19% molasses

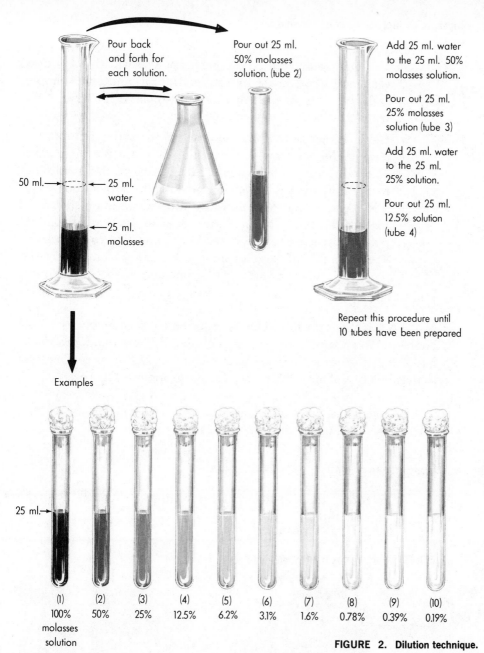

Pour back and forth for each solution.

50 ml. ← 25 ml. water

← 25 ml. molasses

Pour out 25 ml. 50% molasses solution. (tube 2)

Add 25 ml. water to the 25 ml. 50% molasses solution.

Pour out 25 ml. 25% molasses solution (tube 3)

Add 25 ml. water to the 25 ml. 25% solution.

Pour out 25 ml. 12.5% solution (tube 4)

Repeat this procedure until 10 tubes have been prepared

Examples

25 ml. →

| (1) | (2) | (3) | (4) | (5) | (6) | (7) | (8) | (9) | (10) |
| 100% molasses solution | 50% | 25% | 12.5% | 6.2% | 3.1% | 1.6% | 0.78% | 0.39% | 0.19% |

FIGURE 2. Dilution technique.

Note: The details of serial dilution will be omitted in future experiments. It should be seen that we have serially cut the concentration of each solution in half until the final desired concentration is reached. Each successive solution is $\frac{1}{2}$ the concentration of the preceding solution. Future experiments will require the same basic technique, although the dilution factor may vary.

2. Thoroughly shake the yeast suspension provided (0.2 g dried yeast or 1 g compressed yeast per 100 ml) and add 1 drop to each tube. Shake each tube of yeast-molasses mixture to distribute the yeast throughout the tube.

3. Invert one small test tube into each of the 10 larger tubes containing the yeast-molasses mixture. Stopper the large tubes with tight-fitting cotton plugs. Allow each of the small tubes to fill with the suspension by holding the larger tube on its side. When completed, each test-tube preparation should resemble the last tube in Figure 3. The amount of solution placed in each large tube will vary according to the relative sizes of small and large tubes. The suggested 25 ml works well when the tubes are 22 × 175 mm and 13 × 100 mm, respectively.

Note: A little care in technique is required. It is common for an air bubble to be trapped within the small tube. By *careful* and *gentle* rocking of the large tube, it is possible to completely fill the smaller tube so that it is free of air bubbles. See the illustration in Figure 3.

4. Label the tubes and place them in a convenient area where they will not be disturbed for 24 hours.

PROCEDURE DAY II

1. Examine the test tubes. Observe the gas collected in the smaller tube.

FIGURE 3. Preparing the fermentation tubes.

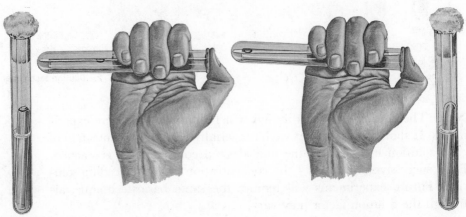

2. Record your observations and those of other members of your team in preparation for a class discussion on the Investigation.

OBSERVATIONS

1. Measure the comparative quantities of gas collected in the top (actually the bottom) of each small test tube by measuring the height of the gas column with a millimeter ruler.
2. The different concentrations of molasses should be plotted on the horizontal axis of linear graph paper, and the quantity of CO_2 produced in 24 hours should be plotted on the vertical axis.
3. Plot the average readings reported by each team. Does this curve approximate the one for your team?

INTERPRETATIONS

1. Compare the graph prepared from your experimental results with the graph selected prior to the experiment. Be prepared to explain any differences or to substantiate any similarities.
2. Based only on the results observed in this Investigation, describe the relationship that seems to exist between the concentration of available food and the production of CO_2 by yeast cells.

INVESTIGATION 2
A Study of Variables

The results of Investigation 1 provided some information, but undoubtedly they also raised some questions. These questions, which may occur to you, will not necessarily be the same for every Investigation, but the next step in scientific investigation is an attempt to answer them experimentally. This requires that first an attempt be made to find out the variables which influence the phenomena concerned. For example, we can reasonably suppose that it makes no difference in the results if the same experiment is done at one location in the laboratory compared to another location; that is, location is a variable, but it is not likely to influence the result. What variables, however, might affect the response?

One ought to consider that the yeast is alive and grows. Hence it needs carbon, hydrogen, oxygen, nitrogen, and mineral matters (principally, phosphorus, boron, copper, and zinc which are usually present as trace contaminants). The molasses provided all of these. Further, a source of energy was required. This was provided by the sugar in the molasses, which is sucrose (common table sugar). Any living organism has a pH range in which it can grow. In the case of yeast, this pH range is from 2 to 6. The molasses, as it is produced, is usually at a pH of about 4 to 5, but, if we wish to prepare a medium of pure sucrose rather than molasses, we shall need to add a buffer. While you already know what is meant by pH and buffers, we have outlined information in the following section for your review.

The first Investigation started with the question: When we provide more and more sugar, do we get CO_2 production equivalent to the sugar provided? From the results, we now wish to ask a slightly different question. For example, what can we use to grow yeast so that we will know the sugar content and so that we can measure growth easily? These inquiries reduce to this simpler question: What do we need to provide in order to permit yeast growth?

MATERIALS (per team)

   10 test tubes capable of holding at least 10 ml

   10 ml buffer (0.9 ml glacial acetic acid plus 0.54 g sodium acetate —($3H_2O$)—to 100 ml; 0.2 molar; pH 4)

   10 ml 10% sucrose (table sugar is chemically pure sucrose)

   1 ml yeast extract (1% of the powdered material or other extract)

   5 ml 1% $(NH_4)_2SO_4$ (ammonium sulfate)

   5 ml 1% KH_2PO_4 (monobasic potassium phosphate)

   3 ml vitamin solution

   3 ml soil extract (water solution from 10 g of soil in 20 ml water) yeast suspension

   10-ml pipette graduated in 1-ml units

Later you will design such an experiment by yourself, but, in order to save time and to show you how such experiments are designed, we will provide a suggested procedure which you can modify as you wish. To grow, any organism needs a suitable environment (pH, temperature, oxygen supply). For this reason we provide a buffer at pH 5 (which is good for yeast but stops the growth of most bacteria; that is, at this pH only yeast will grow). Room temperature is satisfactory, and yeast will grow with or without oxygen. An organism needs a source of energy. We will use sucrose and somewhat arbitrarily use 1% final concentration. You will note from Investigation 1 that tubes 7 and 8 showed activity. They contained 1.6% and 0.78% molasses. If the molasses was 60% sucrose, then these tubes had 0.96% and 0.48%, respectively. (How did we get these numbers?) If the molasses was about 30% sucrose, these tubes would contain 0.48% and 0.24%. Evidently about 0.5% to 1% sucrose is enough to get growth, and so we will use this amount of sucrose.

But energy and buffer are not all that are required. There must be a source of nitrogen, phosphorus, and sometimes trace elements. Sometimes living cells are unable to make particular substances, and these must be supplied in the medium. These required materials are sometimes vitamins, amino acids, or nucleic-acid components. One way in which we can obtain all these components is to extract them from the yeast itself. And we can also supply vitamins, ammonium sulfate $(NH_4)_2SO_4$ as a source of nitrogen, and KH_2PO_4 as a source of phosphorus.

PROCEDURE

1. To 10 tubes add 1 ml buffer and 1 ml sucrose to each. Then treat as follows:

   Tube #1: add 8 ml water
   2.: add 1 ml yeast extract and 7 ml water
   3: add 1 ml $(NH_4)_2SO_4$ and 7 ml water
   4: add 1 ml KH_2PO_4 and 7 ml water
   5: add 1 ml $(NH_4)_2SO_4$, 1 ml KH_2PO_4, and 6 ml water
   6: add 1 ml soil extract and 7 ml water
   7: add 1 ml vitamin extract and 7 ml water
   8: add 1 ml each of $(NH_4)_2SO_4$, KH_2PO_4, soil extract, and 4 ml water
   9: add 1 ml each of $(NH_4)_2SO_4$, KH_2PO_4, vitamins, and 4 ml water
   10: add 1 ml each of $(NH_4)_2SO_4$, KH_2PO_4, soil extract, vitamins, and 3 ml water

Tube 1 contains just sugar and buffer; tube 2 has, presumably, all the materials yeast needs to grow since it came from yeast; tube 3 has a nitrogen (and sulfur) source: tube 4 has a source of phosphate; tube 5 has both a nitrogen and a phosphate source; tube 6 has a soil extract that would provide trace amounts of minerals which might be required. Now, all of these may not be sufficient unless they are all there together. For example, if yeast needs vitamins and minerals, it will not grow when only one is present— both must be there. Tubes 7 through 10 provide these combinations. Some vitamin preparations contain minerals, but note that in tube 8 there are minerals and no vitamins, and in tube 10 there are minerals plus vitamins. If the vitamins are providing only minerals, there should be growth in tubes 8, 9, and 10. If the vitamin preparation supplies vitamins only (no minerals) and minerals are required, the growth should appear only in tube 10. If the vitamin preparation supplies both, and both are required, then growth should occur in tubes 9 and 10.

2. Inoculate each with 1 drop of yeast suspension made as in Investigation 1. (It may be the same suspension as that in Investigation 1 providing it has been held in the refrigerator not more than 4 days.) Mix well. The tubes do not need to be stoppered. Leave tubes in a convenient area and observe after 24 hours and at some later time.

3. Growth may be estimated by:
   a) looking at the amount of sediment (usually yeast cells) at bottom of the tube and grading it from − to + + + +; i.e., − represents none or very little sediment and + + + +, the most sediment present.
   b) shaking up the tubes and comparing their turbidity according to a similar −, +, + +, + + +, and + + + + scale.

INTERPRETATIONS

1. Which medium gave the best growth?
2. Which medium, giving good growth, had the least ingredients—that is, which is the simplest medium upon which yeast would grow well?
3. Which medium would you choose to study the effect of sucrose concentration—that is, one which supplies all the necessary materials for growth except sucrose?

PARALLEL READING: pH and Buffers

As you know from your previous studies, a water solution having the same number of H^+ and OH^- ions is neutral. If it has more H^+ ions, it is acid; if more OH^-, it is alkaline. Further, the product of H^+ and OH^- must equal 10^{-14}. A neutral solution thus has 10^{-7} $H^+ \times 10^{-7}$ $OH^- = 10^{-14}$. A solution having 10^{-6} H^+ has 10^{-8} OH^- (NOTE: 10^{-6} H^+ is *more* H^+ ions than 10^{-7}; 10-fold more). Instead of saying 10^{-6} H^+, we can use the "−6" as an index of acidity and call such a solution *p*H 6. By definition, therefore, *p*H is the negative logarithm of the H+ ion concentration. Neutrality is pH 7. pH 6 has 10^{-6} mols H+ ion per liter. pH 5 has ten times (pH 4, 100 times) as much acid as pH 6.

For example, a solution made with 1 mol of acetic acid per liter and 1 mol of sodium acetate per liter has a pH of 4. About 1 ml of concentrated acetic acid can be added to each 100 ml, and the pH will change only slightly (to pH 3.8–3.9). Such a material is called a buffer, and it prevents (within limits) a change in pH upon the addition of acid or alkali. A mol is the molecular weight of the substance in grams. In the example above, 1 mol of acetic acid per liter means 60 g of acetic acid per liter since the molecular weight of acetic acid is 60.

PARALLEL READING: Fermentation and Energy

Although some of the intermediate products may be of use in the synthesis of cell materials, the greatest value of fermentation to the yeast cell is probably the release of energy. All living things need the energy which is made available by exergonic (energy-yielding) reactions if they are to perform work. The work of the cell is done by endergonic (energy-using) reactions such as the various synthetic processes of cells. If the energy of exergonic reactions is evolved as heat, as when an organic substance undergoes combustion (burning of gasoline, and so forth), it is generally of little use to the cell. Instead of heat energy, living cells require chemical energy for most of their activities.

How do cells transfer the energy from exergonic to endergonic reactions? There may be several ways, but the most efficient is the use of a compound which will be active in both kinds of reactions. Such a compound may, by a change of its structure, pick up a major part of the energy released in an exergonic reaction. Then, by changing back to its original structure, it can give up the energy needed to drive an endergonic reaction.

The most common compound of this sort found in cells is adenosine triphosphate, commonly called ATP. It is produced from a similar compound, adenosine diphosphate (ADP). ATP is produced when the energy from some exergonic reaction is used to cause the reaction of ADP with H_3PO_4 (phosphoric acid). We can write this as follows:

ADP	+	H_3PO_4	+	energy	$\longrightarrow$	ATP
(*Adenosine Diphosphate*)		(*Phosphoric Acid*)		(*From an Exergonic Reaction*)		(*Adenosine Triphosphate*)

Conversely, an example of the utilization of ATP can be shown:

$$\text{glucose} + \text{ATP} \longrightarrow \text{glucose-phosphate} + \text{ADP}$$

The glucose-phosphate is much more reactive than was the glucose, partly because it contains most of the energy which was released when ATP gave up one phosphate group and became ADP. Again, look at the equation for alcoholic fermentation:

$$C_6H_{12}O_6 \xrightarrow{\text{yeast}} 2\ C_2H_5OH + 2\ CO_2 + \text{energy}$$

$\quad(Glucose)\qquad\qquad\qquad(Ethanol)\qquad(Carbon$
$\qquad\qquad\qquad\qquad\qquad\qquad\qquad\qquad Dioxide)$

One might think from this equation that the conversion of glucose to ethanol and carbon dioxide is a single step fermentation reaction. This is not so. The process involves a series of chemical reactions, each of which is controlled by a specific enzyme. Only two reactions, however, actually produce ATP, and only one step in fermentation actually uses ATP. Figure 4, shown on page 27, shows some of these steps.

Note that pyruvic acid is produced in the second to the last step; we will have more to say about the role of pyruvic acid in a later discussion of metabolism. The process by which glucose is converted to pyruvic acid is often called *glycolysis,* and the series of steps involved has been named (for three of the many men who worked out the reactions) the Embden-Meyerhof-Parnas (EMP) system.

We can summarize the steps shown on the chart:

$$C_6H_{12}O_6 + 2\ ADP + 2\ H_3PO_4 \longrightarrow$$
$$2\ CH_3CH_2OH + 2\ CO_2 + 2\ ATP + \text{heat}$$

The heat loss indicated in the equation is the energy which was in glucose but is not used in the formation of ATP or ethanol. Most of this heat will be of little use to the yeast cells and will be lost to the environment.

Principles of Experimental Design

An investigation begins with an awareness of a problem and a survey of related literature. It proceeds to the expression of a hypothetical answer and the formulation of experimental procedure. It concludes with data collection, analysis, and interpretation. The design of an experiment is a step in problem-solving consisting of a detailed plan

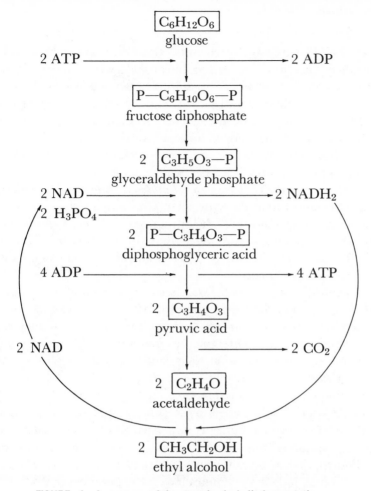

FIGURE 4. A summary of the steps in alcoholic fermentation.

to be followed in obtaining the needed data. Such a plan specifies what organisms, materials, and equipment will be used and details the step-by-step procedures for carrying out the experiment.

Ideally, the design of the experiment is one that will give the greatest amount of reliable information with the least expense and effort. Often such a design is difficult to achieve. The complexity of living systems and the variability inherent in populations of living organisms contribute to the difficulty of research in the biological sciences. Furthermore, there is no common blueprint that will serve as a guide to the design of experimental procedure. Each problem attacked by the investigator may require its own design.

Five important principles may be kept in mind concerning experimental design in biology. First, the investigator must choose a *suitable organism* for the experiment. He wants to use organisms which are readily handled in the experimental procedure and which will give the needed data as rapidly as possible at the lowest cost commensurate with reliability. The choice of organisms to be used is often determined by their availability. If satisfactory experimental organisms can be found locally, it is more economical to use them than to choose others which may have to be shipped long distances.

Second, the investigator must try to assure *representative selection* of experimental organisms from the whole population.

Third, he must be aware of *experimental variables.* An investigator hopes that he can limit the experimental treatment given to the living plant or animal specifically to the introduced variable. This is not easily done. Merely moving a plant or an animal from its native habitat to a laboratory may so affect the organism that the accuracy of the investigation can be questioned.

Regardless of the care exercised, experimental variables may unobtrusively enter the design and ultimately affect the data obtained. For example, suppose you are investigating the germination rate of specific seeds. You have attempted to maintain perfectly uniform conditions for all the germination trays, but can you put all the trays in your laboratory in identically the same place? This, of course, is a physical impossibility. Therefore, do all the trays receive the same light intensity? Are they all exposed to the same temperature and humidity? Do identical convection currents flow past all of them? Each of these numerous variables must be recognized and taken into consideration.

Fourth, the investigator should strive for *simplicity.* The casual observer often judges the importance of experimental work by the amount of elaborate instrumentation involved: an array of tubes, wires, pumps, stirrers, and dials appears to indicate that a significant experiment is under way. Expensive equipment may sometimes be essential to obtain required data, but it does not measure the true importance of the question being asked. In contrast, significant advances in biology have frequently resulted from apparently "simple" questions answered with a minimum amount of apparatus.

Finally, a *reasonable attitude* toward experimental organisms is essential in biological research. In all investigations, laboratory organisms must be conscientiously cared for. Animals must be fed and watered regularly, and kept clean. These practices, besides being dictated by humane considerations, are a practical necessity. Good experimental results cannot be expected from animals that have been subjected to the stress of hunger or to bad housing conditions. Plants must also be kept in good condition. A plant that becomes wilted or diseased through *neglect* is useless for experimentation.

The design of an experiment is dictated by the question it is to answer. The investigator has the responsibility for choosing organisms and procedures that will yield the maximum of reliable data.

QUESTIONS FOR DISCUSSION

1. Suppose that in Investigation 1 or 2 you took your yeast inoculum from the top of the flask while another student took his from the bottom. How might this have affected your results?

2. Compare the care needed for the experimental organisms when you use yeast to the care needed if you were experimenting with dogs or bean plants.

INVESTIGATION 3
Further Variables and Controls

We started, in Investigation 1, with the question "What is the relationship between the amount of sugar and growth?" From Investigation 2, we found out something about what was required for yeast growth when we gave 1% sucrose. This information will now permit you to design an experiment which ought to tell you the relationship between sugar concentration and growth. You have seen the necessity for vitamins, phosphate, and so forth. And you should be able to design a "base" medium—that is, one in which the organism will not grow if no sugar is provided but will grow abundantly when sugar is present.

PROCEDURE

1. From the information obtained in Investigation 2, devise a "base medium." This might consist of a buffer and such supplements (ammonium salts, phosphate, vitamins, and so forth) as required. It is better to have a medium composed of known constituents (as ammonium salts plus vitamins) rather than natural extracts (as yeast extract or soil extract), since these are more controllable. But, if you must use extracts to obtain growth, do so. We assume that the buffer, plus the chosen supplements, will not exceed 5 ml.

2. Set up 10 tubes of this base medium.

3. Having the same base medium in all tubes, vary the sugar concentration. For example, in tube 1, add no sugar. To tube 2 add 1 drop of (0.05 ml) 50% sucrose solution and enough water so that the final volume is 10 ml. You will have a final concentration of sucrose of 0.25%. If you add 5 ml of 50% sucrose

(to total volume of 10 ml), you will have 25% sucrose. If you add 5 grams of sucrose to a tube containing base medium, dissolve it, and make the final volume to 10 ml, you will have 50% sucrose. Therefore, in your 10 tubes, set up varied concentrations from 0 to 50% sucrose.

4. Inoculate each tube with one drop of a yeast suspension.
5. After growth, measure the turbidity quantitatively.

OBSERVATIONS

1. In a set of test tubes similar to the experimental tubes, you have been given a set of standards containing known quantities of yeast. These were made up by suspending 1 g of yeast in 10 ml of buffer + 90 ml of water and then diluting to the turbidities comparable to your growth tubes. Compare your tubes with these standards and estimate the amount of growth in each tube.
2. Plot the growth against the sugar concentration.

INTERPRETATIONS

1. Does more sugar mean more growth?
2. How do you explain what you have found?
3. What variables are still uncontrolled and unexplored in these experiments?
4. Suppose you increased the supplements in tube 7 and obtained more growth. What would you conclude? What new experiment would you do?

PATTERN OF INQUIRY 1
Interaction of Variables

A few decades ago, the average yield of corn grown in the South was much below that grown in the Corn Belt states of Illinois and Iowa. Why do you suppose this was true?

INVESTIGATION 4
A New Question

The results of Investigation 3 should give you an estimate of the relationship between sugar concentration and growth. However, they have also revealed something else—that is, that yeast does not grow on sugar alone; there seem to be some other factors involved. Some of these seem to be "organic materials" rather than just nitrogen or phosphate sources. Like many investigations, one may set out with a specific question to be answered, but in answering it, further new questions are raised. You will sometimes hear it said that a study uncovers more questions than answers. This does not imply that the study has not found answers, but it means that, with the increased knowledge derived, further understanding permits further questions, which thus lead to further knowledge.

The results of Investigations 1 through 3 raise some of these new questions. One, for example, is: What factors, in addition to sugar, nitrogen supply, and phosphate are necessary for yeast growth? Possibly you have other questions. You now have sufficient knowledge to design an experiment yourself. First pose your question; then decide what you would do to solve it.

PROCEDURE

You were given detailed directions for carrying out each of the previous Investigations. In contrast, no procedural detail will be given for Investigation 4. The problem has been outlined; the rest is up to you. Consider the problem carefully, reexamine the data gathered from previous experiments, consult the literature if necessary, and formulate a hypothesis that could lead to a solution to the problem. After formulating a hypothesis, make an "if . . . then" statement which can be tested experimentally. Design the experiment in such a way that the results obtained might either verify or negate your hypothesis.

It is quite likely that several directions for experimentation will emerge. This is also typical of research in science. Perhaps each team will perform a different experiment; and perhaps the results of each different experiment, when examined together, will provide the necessary evidence for a final solution to the problem.

PATTERN OF INQUIRY 2
Refining Hypotheses

It is difficult to say when science began. Certain aspects of science appear to be about as old as history. Modern science, however, is quite recent, and its rate of development has been rapid. The formulation of hypotheses and their gradual revision, because of more and better evidence, have been typical of attempts to understand biological processes.

The microscope was probably invented independently by Jansen in Holland and by Galileo in Italy about 1609. Even though inferior to present-day equipment, the microscope opened the door to observations of microorganisms. The first description of bacteria was probably rendered by Antony van Leeuwenhoek in 1676. This Dutch microscopist described a number of microorganisms in letters which were published in the *Philosophical Transactions of the Royal Society of London* during the period 1677–1684.

The process of alcoholic fermentation was known to the ancients, but the understanding of the nature of the process was not developed until the latter two-thirds of the 19th century.

In 1837, Theodore Schwann (the German zoologist who is given much of the credit for the development of the cell theory) published the results of a series of experiments on fermentation. He summarized these as follows[1]:

(1) A boiled organic substance or a boiled fermentable liquid does not putrefy or ferment, respectively, even when air is admitted, so long as the air has been heated.

(2) For putrefaction or fermentation or other processes in which new animals or plants appear, either unboiled organic substance or unheated air must be present.

(3) In grape juice the development of gas is a sign of fermentation, and shortly thereafter appears a characteristic filamentous fungus, which can be called a sugar fungus. Throughout the duration of the fermentation, these plants grow and increase in number.

(4) If ferments which already contain plants are placed in a sugar solution, the fermentation begins very quickly, much quicker than when these plants must first develop.

(5) Poisons which only affect infusoria and do not affect lower plants . . . [an alcoholic extract of *Nux vomica*] prevent the manifestations of putrefaction which are characteristic of infusoria, but do not affect alcoholic fermentation or putrefaction with molds. Poisons which affect both animals and plants (arsenic) prevent putrefaction as well as alcoholic fermentation.

The connection between the alcoholic fermentation and the development of the sugar fungus should not be misunderstood. It is highly probable that the development of the fungus causes the fermentation. Because a nitrogen-containing substance is also necessary for the fermentation, it appears that nitrogen is necessary for the life of this plant, as it is probable that every fungus contains nitrogen. The alcoholic fermentation must be considered to

---

[1] From the translation given by Brock, Thomas H. 1961. *Milestones in Microbiology*. Prentice-Hall, Inc., Englewood Cliffs, N. J. pp. 18–19.

be that decomposition which occurs when the sugar fungus utilizes sugar and nitrogen-containing substances for its growth, in the process of which the elements of these substances, which do not go into the plant, are preferentially converted into alcohol. Most of the observations on alcoholic fermentation fit quite nicely with this explanation.

What hypotheses did Schwann develop?

SECTION 5   PROBLEMS IN THE CONTROL OF VARIABLES: THE SIMPLIFICATION STEP

Uncertainty in Science

The problem of obtaining data needed for the evaluation of some hypotheses may lead to uncertainties concerning the validity of these data. In our studies with yeast, we have seen some of the difficulties which may arise when we try to control several variables individually. Perhaps a more common difficulty arises from attempts to study individuals or populations through time. What is the effect of our collection of data on the subsequent behavior of our experimental material? An animal which has been used in one experiment is no longer the same as it was before the experiment. A similar situation exists when we try to study changes in a small population. If our sampling removes individuals from the population or disturbs their breeding behavior in any way, we may never know what the population might have been like if it had not been disturbed. We can do little in the study of individuals or populations without some danger of disturbing the organisms we are studying.

Data (selected facts) have been called the raw materials of science, but data are seldom complete. We can define an experiment as a situation planned to provide the data needed for evaluating a hypothesis. The logical inferences which can be drawn from data are important. Data almost always involve variability which the scientist must interpret before he can properly draw inferences. Perhaps the major difficulty in carrying out an experiment is providing adequate controls for all important variables.

In Investigations 1, 2, 3, and 4, you saw that there were several variables. The differences in the concentration of molasses resulted in several of these—different concentrations of sugars, other organic compounds, and minerals. Attempts to repeat these experiments with different brands of molasses might be complicated by differences in the composition of the molasses. Also, you might have a problem of duplicating the amount of inoculum added, since different packages of yeast may contain different numbers of living cells. Length of time and other conditions of storage may influence the activity as well as the survival of yeast cells. You might find that changes in the weather will cause changes in the temperature of your laboratory, and this, in turn, may affect the experiment you are conducting.

T. H. Huxley (1860) pointed out that "all science begins with empirical knowledge, but Nature presents to our senses a panorama of phenomena, commingled with endless variety so that we are sometimes overwhelmed by the apparent complexity and contradiction of empirical knowledge." To find the cause of phenomena observed, one attempts to simplify the situation so that the "complexity and contradiction" are less. This approach is experimentation. An experiment is an attempt to observe phenomena under *simple* conditions. While it is an important tool, it is useful only when conditions obscure direct observation. A successful experiment does no more than to make a previously obscure fact as evident as one that was open to direct observation from the first. An experiment should include simplified conditions so that only one variable—or a few, at most—is altered. Leonardo Da Vinci wrote: "Experiment is the interpreter of Nature. Experiments never deceive. It is our judgement which sometimes deceives itself because it expects results which experiment refuses. We must consult experiment, varying the circumstances, until we have deduced genuine rules, for experiment alone can furnish reliable rules." When the investigators do the "same" experiment and obtain different results, it does not mean that one is right and the other wrong. It means that they did not do the same experiment—that somewhere an important controlling condition was different. The experiment is *always* right. Even when experimenters obtain the same results, they may disagree as to the meaning.

INVESTIGATION 5
Resting Cells

In earlier experiments we were studying some of the factors involved in the growth of yeasts. We used a small number of yeast cells and allowed them to grow. But growth is a very complex matter, and it would be helpful if we could simplify the situation. Suppose, for example, that we use cells already grown and let them ferment sugar. Since they don't need to grow, their requirements are probably simpler than those of growing cells. Do they need nitrogen, for example, or phosphate? It is a helpful step in the development of knowledge, when there are many variables involved, to attempt to simplify the system rather than to attempt to control all the variables. Knowledge gained from simpler systems can then be applied to more complex ones.

MATERIALS (per team)

Molasses
60% sucrose
Supplements from Investigation 2
Buffer—pH 4.0 (14 ml) (acetate buffer used in Investigation 2)
Yeast (suspension: 4 g to 100 ml buffer)
14 large and small test tubes for CO_2 trap
Graduated cylinder

PROCEDURE

1. Since we are going to set up a system in which the cells are already grown, we cannot use measurements of turbidity of growth as an index of activity. We shall therefore return to the measurement of CO_2 as used in Investigation 1.

2. Since we know that pH 4.0 is a good pH for yeast, permitting fermentation and growth while inhibiting bacteria, we will use this pH.

3. Prepare two solutions:
   A. Molasses: 25 ml molasses + 10 ml of water.
      This will dilute molasses sufficiently to allow pipetting. To make 25 ml of 50% molasses, we must add 12.5 ml of the original molasses (or $12.5 \times 35/25 = 17.5$ ml of the diluted molasses) to a 25-ml graduated cylinder, and bring up to a volume of 25 ml with water.
   B. 60% sucrose: 60 g of sucrose dissolved in water and diluted to 100 ml. To make 25 ml of a 50% solution of sucrose, one would add $\frac{5}{6} \times 25 = 20.8$ ml (we will settle for 20 ml as being close enough for our purpose) of the 60% solution to a 25-ml graduated cylinder, and bring up to a volume of 25 ml with water.

4. Set up 14 large test tubes as follows:
   1 ml acetate buffer pH 3.5 in all (% is in parentheses)
   A. 19 ml water (0)
   B. 17.5 ml diluted molasses + 1.5 ml water (50% molasses)
   C. 1.7 ml diluted molasses + 17.5 ml water (5%)
   D. 0.2 ml (4 drops) diluted molasses + 19 ml water (0.5%)
   E. 20 ml 60% sucrose (close to 50% sucrose)
   F. 2 ml 60% sucrose + 17 ml water (5%)
   G. 4 drops (0.2 ml) 60% sucrose + 19 ml water (0.5%)
   All the following contain 2 ml 60% sucrose ($= 5\%$):
   H. + 1 ml $(NH_4)_2SO_4$ + 16 ml water
   I. + 1 ml yeast extract + 16 ml water
   J. + 1 ml KH_2PO_4 + 16 ml water
   K. + 1 ml vitamins + 16 ml water
   L. + 1 ml $(NH_4)_2SO_4$ + 1 ml yeast extract + 15 ml water
   M. + 1 ml KH_2PO_4 = 1 ml vitamins + 15 ml water
   N. + 1 ml of all 4 additions (I–K) + 13 ml water.

5. To all tubes (A through N), add 5 ml of yeast suspension made by suspending 4 grams of yeast in 100 ml of formate buffer pH 3.5. Each tube thus has 0.2 g of yeast. Mix well.

6. Insert and fill the small tubes to serve as a CO_2 trap as described on page 20. You may omit the cotton stopper and use your fingers or a cork to invert the tubes. The same cork can be used for all tubes providing it is rinsed after each use.

7. Observe gas formation at the end of:

   30 minutes
   45 minutes or an hour
   2–3 hours, if possible.

   Measure (with a mm rule) the amount of gas in each tube at a suitable interval when some of the tubes are showing gas.

INTERPRETATIONS

1. What factors are required for the most rapid production of CO_2 by grown cells? Are they the same as those required for growth? Why?

2. This kind of system—that is, using cells already grown—is called a "resting cell" system. In one sense these cells are "resting"; in another sense they are not. What is the difference between these two senses of the term "resting cell"?

INVESTIGATIONS FOR FURTHER STUDY

1. Spoilage of certain materials containing high concentrations of sugar occurs most readily in hot weather. In ordinary laboratory media, however, yeasts will often grow best at the more moderate temperatures. Perhaps the optimum temperature for CO_2 production from sucrose solutions can also be changed with changes in sugar concentration. Design and perform an experiment to test this hypothesis. If significant support is found for this hypothesis, how might you expect varied concentrations of a single sugar such as glucose to compare with sucrose in changing the optimum temperature of fermentation?

2. Suggest an experiment for obtaining strains of yeast that will grow at a lower or at a higher temperature than does the parent strain.

PARALLEL READING: Enzymes and Fermentation

All chemical reactions are affected by the concentrations of the reactants. In our first investigations, the reactants considered were chiefly the carbohydrates. Another group of substances plays an important part in fermentation; these are the enzymes from the yeast. We might investigate the effects of varying yeast (enzyme) concentrations on the rate of fermentation.

The rates of chemical reactions are affected by catalysts. *Catalysts* are substances which change the rates of chemical reactions, but they themselves can be recovered from the reaction in an unchanged form. The catalysts of biochemical reactions are called enzymes.

An *enzyme* is commonly defined as a protein which acts as a catalyst. Enzymes are specific in their action; that is, ordinarily a given enzyme will act on only one kind of substrate and catalyze only one kind of reaction. Enzymes are said to be heat-labile, meaning they are destroyed by heat. Most proteins are coagulated by heat. (Recall what happens when the protein of an egg is heated.)

Some enzymes are made up only of amino acids. Others contain various chemical groups in addition to the protein. It is usually the nonprotein part of the enzyme which reacts with the substrate. The presence of the protein greatly increases the rate of the reaction and often narrows the specificity to a single reaction. If the nonprotein part of the enzyme is easily separated from the protein, it is called a *coenzyme.*

Some coenzymes are inorganic ions as Mg^{++}; others are organic compounds. Many of the B-vitamins serve as parts of coenzymes. In some cases, inorganic ions serve as activators; that is, they enhance the activity of the enzyme but do not become a part of it. In these cases, the ions probably do not enter into the primary reaction in the same way coenzymes enter.

The catabolism of glucose is a process which yields energy to the cell. In alcoholic fermentation, this energy comes from a series of reactions in which some chemical bonds are broken and others are formed. These reactions, although energy-releasing, do not occur in cells in an uncontrolled manner. All matter would exist as simple substances of minimum energy levels if complex molecules broke down spontaneously; no complex molecules would exist for long. Many molecules, however, such as those of glucose, are stable even though they possess large amounts of energy. Such stable molecules can react rapidly if they acquire sufficient energy. The energy necessary to cause the molecules to react is called the *energy of activation*.

At any temperature above absolute zero, the molecules of a substance will be in motion and will have kinetic energy, but not all the molecules in a given system have the same kinetic energy. Some are moving very rapidly; others are moving slowly. As the result of this motion and the corresponding collisions, some molecules will acquire enough energy to react; those which have little energy will not react. We find that the rates of chemical reactions will increase with increasing temperatures because the average kinetic energy of the molecules increases with increasing temperature. The increased average kinetic energy will cause an increasing number of the molecules to acquire the necessary energy of activation. The energy of activation may also be supplied by electricity or light. Photosynthesis uses light energy, but most other biochemical reactions depend on heat as the source of their energy of activation. How then, can an enzyme speed up a reaction when no additional heat is supplied?

Consider a simple reaction which is catalyzed by an enzyme. First the substrate may combine with the enzyme to yield a substrate-enzyme complex. This complex then breaks down to release the enzyme and form a product. We can write this as follows:

$$S \;+\; E \;\rightleftarrows\; ES \;\rightarrow\; E \;+\; P$$

(*substrate*) (*enzyme*) (*enzyme-substrate complex*) (*enzyme*) (*product*)

The substrate molecule must acquire a certain amount of activation energy in order to be changed into some other kind of molecule, such as P, even if P has less energy than S. It may have to form a complex with some other compound, or it may have its structure changed somewhat and become less stable. This relationship is shown in Figure 5, where *e* is the energy of activation.

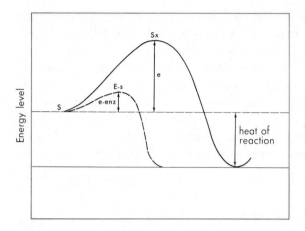

FIGURE 5. Energy of activation of a reaction with and without an enzyme.

In this figure are shown the relative energy levels of the substrate S, the substrate complex without an enzyme S_x, the enzyme-substrate complex E-S, and the product P. The energy of activation when no enzyme is present is represented by e. When the enzyme is present, the energy of activation is represented by e_{enz}. The net amount of energy produced is the heat of the reaction and is equal to the difference in the energy levels of S and P.

When an enzyme is present, a much lower energy of activation is required. This is explained by the assumption that less energy is required to form the substrate-enzyme complex E-S than to form a substrate complex S_x. Thus e is much greater than e_{enz}.

The heat of the reaction is the same regardless of whether or not the enzyme is present. In the cell, the temperature of the environment is sufficient to activate a reasonable percentage of the substrate molecules to react with the enzyme, but it is not sufficient to provide the energy of activation needed if no enzyme is present. The heat of the reaction will then provide enough energy to activate other molecules. This is probably the way most enzymes act in biological systems. The hydrolysis of sucrose to form glucose and fructose provides a suitable example.

In the hydrolysis of sucrose by HCl, the activation energy required is 26,000 calories, but the activation energy for the hydrolysis of sucrose by the enzyme invertase is only 13,000 calories. Thus the amount of energy which must be acquired for a sucrose molecule to be hydrolyzed by the enzyme is only half that required for hydrolysis by the acid. In some cases, the rate of a reaction in the absence of the enzyme at physiological temperatures is too slow to be measurable. In the presence of the enzyme, however, it may be quite rapid.

The names of most enzymes end in *ase,* and the prefix indicates something about the nature of the enzyme. For example, a dehydrogenase is an enzyme which catalyzes the transfer of hydrogen from a substrate to a hydrogen acceptor. In this case, the enzyme is named for the kind of reaction which it catalyzes. In other cases, enzymes are named on the basis of the substrate with which they react. *S. cerevisiae* produces an enzyme called maltase. Maltase splits the disaccharide, maltose, into two molecules of the monosaccharide, glucose. This reaction may be written to correspond with the generalized enzyme substrate reaction:

maltose $+$ H$_2$O $+$ maltase $\rightleftarrows$

(*substrates*)   (*enzyme*)

complex of maltose, water, and maltase $\rightarrow$

(*enzyme-substrate complex*)

glucose $+$ maltase

(*final product*)   (*recovered enzyme*)

INVESTIGATION 6
Making an Enzyme Preparation

"Resting cells," such as we have used in Investigation 5, are simpler than growing cells, but they still represent a complex situation. For example, the membranes of the cells are normally undamaged, and perhaps certain materials cannot enter. The fermentation of glucose is carried out by enzymes; perhaps we could both simplify our problem and learn something about fermentation by obtaining the enzymes which carry it out. For this purpose we could prepare a "yeast juice"— that is, the cell contents removed from the cell. This preparation is not alive, but it contains the enzymes involved in fermentation. On page 27, there is a diagram outlining some of the steps in the conversion of glucose to ethyl alcohol and CO$_2$. Twelve steps are involved, each requiring a separate enzyme. It may be rather difficult to prepare a juice which has all these enzymes active in it. However, in order to ferment sucrose, such as that in molasses or cane sugar, the yeast must split it into glucose and fructose. It has an enzyme for this purpose called "sucrase." An earlier name (still widely used) for this same enzyme was "invertase," and the product of its action on sucrose (a mixture of glucose and fructose) is called "invert sugar" because while sucrose bends a ray of polarized light to the right, the mixture of fructose and glucose bends it to the left and thus "inverts" the action of sucrose on polarized light. We could base our measurement of the enzyme on this inverting effect, but, since few laboratories have the necessary optical equipment, we will use a different method for measuring enzyme activity. Sucrose does not reduce Fehling's or Benedict's solutions— that is, it is not a "reducing sugar"—but glucose and fructose are reducing sugars. Therefore, we can base our measurements on this fact.

MATERIALS (per team)

Yeast acetone powder

Paper toweling

2 ml 0.5 M phosphate buffer, pH 4.5 (6.8 g KH_2PO_4 to 100 ml water)

Spot plate or 2 small dishes

Tes-tape

0.5 ml 1% sucrose

Toothpicks

Scissors

PROCEDURE

1. One g of dried yeast or 5 g of compressed yeast are suspended in 20 ml of cold acetone. Inasmuch as acetone is highly flammable, this step has been done for you by the teacher. This process kills the cells but does not destroy the enzyme. The appearance of the individual cells is essentially unchanged. The "cells," now an enzyme preparation, are removed from the acetone and dried. The resulting powder contains the enzyme activity.

2. From this acetone powder it is possible to extract a water-clear, cell-free enzyme solution, but this requires a good centrifuge. If you lack a good centrifuge, the enzyme may be measured without removing the cell debris. Suspend 0.1 g of the acetone powder in 2 ml of 0.5 M phosphate buffer, pH 4.5 (6.9 g, KH_2PO_4 to 100 ml). Stir vigorously for a few minutes.

If you have a good centrifuge, centrifuge down and take off the supernate to use as an enzyme. The suspension will be quite different from a yeast suspension. Usually it will be gummy and hard to clarify. If you do not have a centrifuge, use the suspension as it is. The cells are dead; only the enzyme remains.

3. "Tes-tape" is a material purchasable at most drug stores. It is used (primarily by diabetics) for detecting and estimating the presence of reducing sugar. The tape is yellow but when moistened turns varying colors, depending on the amount of reducing sugar present.

Glucose is a reducing sugar, but sucrose is not. Cut 16 pieces of Tes-tape each about ½ inch long and arrange them in two rows of 8 pieces each. Be sure to keep the strips far enough apart so that one does not moisten another. Do not handle them with the fingers. Label one row A; the other, B. Using a spot plate or small dish (plastic caps for screw-cap bottles are good), set up two reactions as follows:

| Dish A | 0.5 ml water |
| Dish B | 0.5 ml 1% sucrose |

Looking at your watch *at zero time,* add 0.5 ml of your enzyme suspension to each dish and mix. With a toothpick, immediately remove 1 drop of the reaction mixture to the first strip of Tes-tape: that from Dish A going onto the first piece of Tes-tape in the A row; that from B onto the first piece in the B row. After 2 minutes remove another drop from each dish to the second section of Tes-tape. If only a little green color shows up in B, you may want to extend the time until you take your third sample, but usually the third sample can be taken at 4 minutes, the fourth sample at 10 minutes, and the 5th, 6th, 7th and 8th samples at perhaps 10-minute intervals thereafter. It is important that the sample put on the tape is always the same size. Sometimes capillary tubes or "Pasteur" pipettes are useful. It is also possible to dip the Tes-tape into the reaction dishes, providing it is dipped to the same depth and for a very short interval.

4. Compare the color on the Tes-tape with the standards provided and estimate the amount of reducing sugar formed.

INTERPRETATIONS

1. Sometimes a reducing sugar is found in Dish A. What is its origin?

2. Sometimes if you watch the zero-time Tes-tape of Dish B, you will see a green color gradually develop while the paper is still moist. This does not usually develop in the comparable spot in Dish A. What could be its cause?

3. How do you know that the sucrose does not decompose to reducing sugar by itself?

INVESTIGATIONS FOR FURTHER STUDY

1. If you have facilities, you might try to get a water-clear (that is, it may have a yellow-to-brownish color) solution of the enzyme.

2. The acetone powder you have made contains other enzymes of the fermentation (or Glycolysis Pathway). You could set up methods for measuring one or more of these and determine whether they are present.

3. Since enzymes are proteins, their activity varies with pH. Also, they can be denatured (inactivated) by heat and poisoned by chemicals. You might study the properties of invertase.

4. Add different concentrations of known poisons to fermenting yeast cultures and determine which are the most effective inhibitors of the fermentation process. Determine if all inhibitors have the same degree of effectiveness at similar concentrations. CAUTION: *Remember, these are also poisonous to humans.*

5. To determine if all yeast preparations have the same enzyme activity, obtain suspensions of other yeasts and test each as was done in this experiment. You may use your own culture of yeast, other dried yeast, or compressed yeasts for this purpose. Be sure to determine the water content of non-dried yeast preparations and base all calculations on CO_2 produced per unit weight of dry cells.

6. Do you think it is necessary to have an intact yeast cell before its enzymes will operate? Explain. Suggest an experiment to test your hypothesis.

QUESTIONS FOR DISCUSSION

1. Define or describe:
   a. coenzyme
   b. enzyme-substrate complex
   c. activation energy

2. If a large number of yeast cells autolyzed in a solution of sugar in water, what effect do you think this might have on the survivors? Explain.

SECTION **6** **PROBLEMS IN MEASUREMENT**

The questions, "How many?" "How long?" and "How much?" are essential to the study of modern science.

Progress toward a theoretical explanation of hereditary events was very slow until Gregor Mendel counted the various kinds of offspring he obtained from crossing certain parent plants. Once Mendel knew how many of each kind of offspring resulted from a particular cross, he was able to explain his experimental results mathematically. Modern genetics owes much to Mendel's pioneer work in applying mathematics to the study of heredity.

The general theory that hormones regulate the bending of plants in response to light or to gravity was fairly well developed before a method was found to *measure* the amount of the hormone. Once such methods were developed, the theory was substantiated and practical application followed rapidly.

The accuracy with which a measurement is made depends on the measuring device used and the observer. For ease in converting one unit to another, most laboratories use the metric system. It is clearly much easier to convert 1788 centimeters into 17.88 meters than to convert 1788 inches into 149 feet.

The following rules help to make measurements easy and accurate:

1. Choose units of measurement that are convenient and meaningful. One would neither express the length of a table in miles nor the weight of a man in milligrams. The most common units of measurements used in the biology laboratory are centimeters or millimeters (for length), grams or milligrams (for weight), and liters or milliliters (for volume).

2. Choose a scale that offers the number of subdivisions necessary to permit the accuracy you need. A 1-ml pipette may be subdivided into either tenths of milliliters or hundredths of milliliters.

3. Read the scale to the nearest subdivision. When an object is measured with a meter stick subdivided into centimeters, it is difficult to express the length to the nearest millimeter. It is often possible to estimate points between the subdivisions of the scale, but estimates to more than one decimal place are likely to be of no significance.

4. When you add, subtract, multiply, or divide, remember that the answer is no more accurate than the least accurate measurement. For example, suppose you measured the heights of 100 bean plants with a meter stick, the smallest divisions of which were centimeters. As you made the readings, you estimated the heights to the nearest millimeter. You then summed the heights and found the total was 1953.4 centimeters. What is the average height of the bean plants? Of course, the arithmetical mean is 19.534 centimeters, but you probably would be stretching your limits of accuracy to say that you could tell the difference between 19.4 and 19.6 on your measuring scale. Therefore, it is misleading to imply that you have any faith in the last two figures. When you record the average as 19.5, there may be some doubt as to the accuracy of the .5, but very little about the 19. The figure in question should be rounded off to the nearest whole number. In rounding off numbers whose last digit is 5, it is customary to add the half when the number is odd and to drop the half when the number is even. Thus, 67.5 would be rounded off to 68, while 66.5 would be rounded off to 66.

QUESTIONS FOR DISCUSSION

1. Students with different meter sticks report the length of the same table as 179.73 cm, 180.00 cm, 180 cm, and 180.003 cm. What is the average of these measurements?

2. What is the difference between a measurement of 180 cm and 180.00 cm?

3. Four students determined independently the weights of each rat in the same experimental group of five white rats. The determinations were done on a balance with a scale indicating weights to the nearest tenths of a gram. The reported results follow:

TABLE 1. WEIGHTS OF EXPERIMENTAL RATS IN GRAMS

Student No.	1	2	3	4
Rat No.				
1	80.1	80	80.16	80.0
2	83.2	83	83.16	83.1
3	77.0	77	77.12	77.0
4	79.6	79	79.56	79.5
5	80.7	81	80.65	80.6

From these data, what would you consider a convenient and perhaps accurate-enough estimate to describe the weight of the group of rats? of the mean of the individual rats?

Compare the methods of reporting of each of the four students. Is the "accuracy" of student 3 useful? Is there any evidence of systematic error? What is your opinion of the techniques used by the four students?

PATTERN OF INQUIRY 3
Evaluation of Data

Each day for five days, four students were asked to conduct laboratory tests in order to determine the normality of an acid solution which was supplied to them. Although all samples were taken from the same bottle, the students were lead to believe that each might be different. They determined the normality of the acid by titrating it with a solution of a base whose concentration they had found to be 0.200 N.

In neutralizing an acid with a base (or vice versa), the volume of the acid times its normality is equal to the volume of the base times its normality. That is,

$$\text{ml} \times \text{Normality} = \text{ml} \times \text{Normality}$$
$$(acid) \qquad (acid) \qquad (base) \qquad (base)$$

The record of their measurements of the amount of base (in milliliters) required to neutralize the acid is as follows:

TABLE 2. TITRATION RESULTS

Day of Measurement	Student 1	Student 2	Student 3	Student 4
1	24.8	25.0	24.90	25.18
2	24.8	25.0	24.95	25.24
3	24.9	25.0	25.05	25.28
4	25.0	25.0	25.15	25.34
5	25.1	25.0	25.20	25.42

Suppose that you needed to use the acid in a subsequent experiment, and a reasonable estimate of its concentration (normality) was required. If this experiment were to involve many titrations, and thus many calculations, what estimate of the normality would you find most convenient and perhaps accurate enough?

If the experiment required the greatest accuracy of which you were capable, what value would you choose as the most accurate estimate of the concentration of the acid?

PARALLEL READING: Fermentation and Respiration

Like many other terms, fermentation and respiration are defined differently by different people. It would not be possible to repeat all these definitions here, nor would it be very useful. Most biologists probably think of fermentation as a process whereby food materials are only partially oxidized by microorganisms; that is, *some* of the products still contain energy which can be released by further oxidation.

To many, respiration means the process of breathing. (The word respiration is derived from the Latin word *respirare,* meaning to blow back or to breathe.) Ordinarily when a physician speaks of a patient's rate of respiration, he means how many times the patient inhales (or exhales) in a minute. Many biologists define respiration as a process in which food material is broken down and most of its energy released in the cell. Those who use this definition regard alcoholic fermentation as an example of *anaerobic respiration* because free oxygen is not utilized. If molecular oxygen *is* used, the process is called *aerobic respiration.*

Other biologists define respiration as a process in which energy is liberated from food materials and in which the final oxidizing agent is molecular oxygen. If we use this definition, respiration is always an aerobic process, and since alcoholic fermentation is anaerobic, it would not be called respiration. Acetic acid fermentation (the process in which bacteria of the genus *Acetobacter* convert ethanol to acetic acid and water) is an example of fermentation which involves respiration since molecular oxygen is used. Most researchers in the field of respiration consider incomplete oxidations, such as those in acetic acid fermentation, to be respiration if they involve the oxidation of hydrogen to water.

The conversion of sugar to carbon dioxide and water by complete oxidation provides more energy than the conversion of sugar to alcohol and carbon dioxide by fermentation. The summary equations for these two processes are,

1. alcoholic fermentation of glucose:

$$C_6H_{12}O_6 \rightarrow 2\ CH_3CH_2OH + 2\ CO_2 + 54\ kg\ calories$$

2. respiration of glucose:

$$C_6H_{12}O_6 + 6\ O_2 \rightarrow 6\ CO_2 + 6\ H_2O + 686\ kg\ calories$$

Those organisms which can ferment sugar may have an advantage over those which cannot when free oxygen *is not* available, but they are at a disadvantage if they cannot carry out respiration when oxygen *is* available.

If molecular oxygen is available, most cells, including yeast, can oxidize pyruvic acid to carbon dioxide and water. This is accomplished by a series of enzymatic reactions which have been called the Krebs cycle, the citric acid cycle, or the tricarboxylic acid cycle.

The net result of this complex series of simple chemical reactions is the production of 38 molecules of ATP from the respiration of one molecule of glucose. By comparison, recall that a net gain of only 2 molecules of ATP results from the alcoholic fermentation of glucose. See Figure 4 — "A summary of the steps in alcoholic fermentation," page 27.

Another respiratory mechanism called the pentose-phosphate pathway has been recognized in recent years. This mechanism is not quite as efficient as a combination of glycolysis and the Krebs cycle since only 36 molecules of ATP may be formed from a molecule of glucose. The results of a number of experiments indicated that the pentose-phosphate cycle is a common oxidative pathway in many microbes and in most plant tissue. A discussion of the pentose-phosphate cycle is beyond the scope of this book, but it can be found in several recent biochemistry texts.

It may be useful to examine some of the characteristics of respiratory processes. We can measure the rate of respiration by measuring the rate of consumption of either oxygen or food, or the rate of production of carbon dioxide, water, or heat.

While respiration occurs both in the light and in the dark, the release of oxygen during photosynthesis may mask the utilization of oxygen involved in green-plant respiration. Here we see the importance of the proper choice of experimental organisms. It would be extremely difficult to measure respiration in a photosynthesizing green plant. For this reason, germinating seeds, which have not yet begun photosynthesis, are often used in studying respiration.

INVESTIGATION 7
Measuring Rates of Respiration

Precise measurements of the rate of respiration require elaborate equipment. We can, however, obtain reasonably accurate measurements using simpler methods. This is often done by placing the living materials in a closed system and measuring the amount of oxygen which goes into the system or the amount of carbon dioxide which comes out. By using suitable techniques, we can measure the amounts of one or both of these gases over a given period of time and determine the respiration rate. A simple volumeter can be set up as illustrated in Figure 6.

The volumeter should be arranged as follows. The material for which respiration measurements are desired is placed in one or more

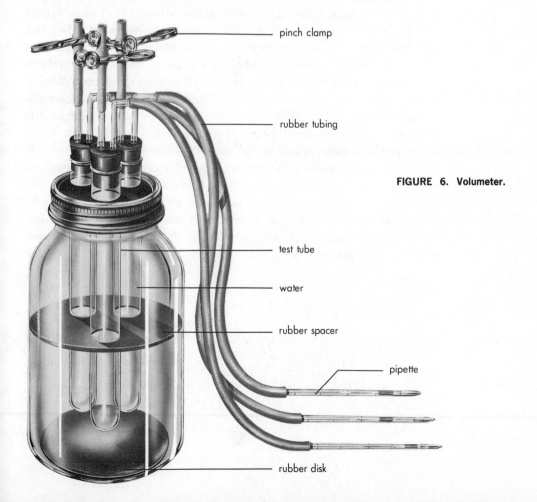

pinch clamp

rubber tubing

FIGURE 6. Volumeter.

test tube

water

rubber spacer

pipette

rubber disk

test tubes of uniform size. Each tube contains a stopper and pipette as shown in the illustration. One of the test tubes contains an inert material such as glass beads or washed gravel and is used to correct changes in temperature and pressure which cannot be completely controlled in the system. This tube is called a *thermobarometer*. Equal volumes of both test and inert materials must be placed in all the tubes. This precaution is necessary to assure that an equal volume of air is present in each tube. A very small drop of colored liquid is inserted into each pipette at its outer end. This closes the tube so that, if there is any change in the volume of gas left in the tube, the drop of colored liquid will move. (The direction of movement depends on whether the volume of gas in the system increases or decreases.) The distance of movement over a given period of time can be read from a ruler placed on the side of the pipette. The *volume* of gas added or removed from the system can be read directly from the calibrated pipette.

In attempting to measure respiration with the equipment just described, we must take into consideration not only that oxygen goes into the living material (and thus out of our volumeter test tube), but also that carbon dioxide comes out of the living material (and thus enters into the volumeter test tube). If we are to measure the oxygen uptake in our respiring material, we must first trap the carbon dioxide as it evolves. This can be done by adding any substance (ascarite is commonly used) which will absorb the carbon dioxide as fast as it is evolved. Efficient removal prevents the carbon dioxide from being added to the volume of gas in the tube.

Each team should set up one volumeter and compare the respiration of dry seeds with those which have been soaked for 24 hours. The work involved in setting up the volumeter and in obtaining measurements is difficult to complete in one laboratory period. It is very important that certain preparations be made in advance, and that each member of the team understands clearly what is to be done.

MATERIALS (per team)

1. One volumeter (complete)
2. One thermometer
3. One hundred Alaska pea seeds
4. Germination tray
5. 100-ml graduated cylinder
6. Glass beads
7. Three beakers, 150 ml
8. Solution of dye
9. Cotton
10. Ascarite
11. Eyedropper

PROCEDURE DAY I

Each team should place 40 pea seeds in a germination tray between layers of wet paper towels and allow them to soak for 24 hours. (Label the trays as to team, class, experiment, and date.)

PROCEDURE DAY II

1. Determine the volume of the 40 soaked seeds. This volume will be used as a standard for preparing materials for the other two test tubes in the volumeter. (Volumes of solid objects, including seeds, can be determined readily by adding them to a measured volume of water in a graduated cylinder and reading the volume of displaced water.)

2. Determine how many glass beads must be put in the tube with the dry seeds so that the volume of air in the tubes with soaked and with dry seeds will be the same. To do this, place 25 ml of water in a 100-ml graduated cylinder. Add the dry seeds. Then add enough beads so that the *increase* of the water level in the cylinder containing both seeds and beads is equal to the volume of the seeds soaked for 24 hours. Dry the 40 seeds and the glass beads by blotting them with paper toweling or cleansing tissue. Place the dried seeds and beads together in a beaker. Label the beaker and store it in the laboratory until you are ready to use the volumeter the following day.

3. Obtain the same volume of glass beads as that determined for the soaked pea seeds. Place these in a beaker, label the beaker, and store it in the laboratory until you are ready to set up the volumeter on the third day.

4. Mix about 25 ml of a dilute solution of vegetable dye (food coloring) in water and add a drop of detergent.

5. Set up the volumeter as illustrated in Figure 7. Add water to the jar in which the test tubes are immersed, but do not add anything to the test tubes.

PROCEDURE DAY III

1. Remove the stoppers from each of the three test tubes. Add the 40 soaked pea seeds to one tube; add the dry pea seeds and glass beads which you measured out in Step 2 to the second tube; and add the glass beads measured out in Step 3 to the

third tube. Loosely pack cotton over the material in each tube to a depth of $\frac{1}{2}$ inch. Add $\frac{1}{4}$ teaspoon of ascarite or sodium hydroxide to the top of the cotton in each tube. CAUTION: *Ascarite is caustic. Be very careful not to get it on your hands, your body, or on your clothes. If some is spilled, clean it up with a dry paper towel or paper cleansing tissue. Do not use damp cloth or paper as ascarite reacts strongly with water.* The tube should now be packed as illustrated in the illustration below.

2. Replace the stoppers and arrange the pipettes so that they are level on the table.

FIGURE 7. **Volumeter tubes after preparation.**

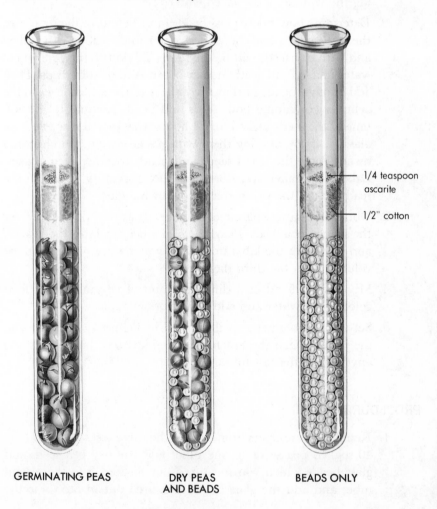

1/4 teaspoon ascarite

1/2" cotton

GERMINATING PEAS

DRY PEAS AND BEADS

BEADS ONLY

3. With a dropper, add a small drop of colored water to each of the three pipettes. (See Step 4 of Procedure, Day II.)

The diagram shows setup of stopper and pipettes attached to each tube in volumeter. After colored water indicator has been introduced at outer end of pipette, it can be adjusted by opening pinch clamp and drawing air from system or pushing it into system with eyedropper inserted into rubber tube at top of apparatus.

Adjust the marker drops so that the drop in the thermobarometer is centered in the pipette and the other drops are placed near the outer ends of the pipettes.

4. Allow the apparatus to sit for about 5 minutes before making measurements.

5. For 20 minutes, at 2-minute intervals, record the distance the drop moves from its starting point. (If respiration is rapid, it may be necessary to readjust the drop with the medicine dropper as described in Step 3. If readjustment is necessary, add the new readings to the old readings so that the total change during the time of the experiment will be recorded.) Record your results as illustrated below.

GAS VOLUME CHANGES IN A CLOSED SYSTEM CONTAINING GERMINATING PEAS AND DRY PEAS

Time	Thermo-barometer readings ml	Readings for germinating peas ml	Readings for dry peas ml

Note: If the drop in the thermobarometer pipette moves toward the test tube, subtract the distance it moves from the distance the drop moves in each of the other pipettes. If the drop in the thermobarometer pipette moves away from the test tube, add the distance it moves to the distance the drop moves in each of the other pipettes.

The readings in each case should be recorded as the change in volume from the original reading. If the observed volumes are corrected to volumes at standard temperature and pressure, the equivalent weights of glucose used may be calculated with greater accuracy.

QUESTIONS FOR DISCUSSION

1. What is the effect of moisture on the germination of pea seeds?
2. Would adding more water to the soaked seeds result in an increased rate of respiration?
3. What is the significance of the difference in the respiration rate of dry seeds compared with that of germinating seeds as far as the ability of the seed to survive in nature is concerned?

INVESTIGATIONS FOR FURTHER STUDY

1. Design a modification of this experiment which will allow you to measure the amount of *carbon dioxide* given off by seeds during respiration.
2. Measure the effects of temperature on the respiratory rates of two different insects.
3. Compare the rates of respiration of different kinds of plant tissues. You might use tissues such as carrot root, potato tuber, or leaves. If green tissues are used, keep them dark by use of black paper or cloth.

PATTERN OF INQUIRY 4
The Respiratory Ratio

After completing Investigation 7, a student wished to study other aspects of respiration in seeds. He decided to see if the *respiratory quotient* or *ratio* is different in different kinds of seeds. The respiratory quotient is defined as the ratio between the volume of CO_2 produced and the volume of O_2 used ($RQ = CO_2/O_2$). He experimented with seeds of wheat and castor bean, and obtained the results shown in Table 3 on page 55.

Plot the data for each species on graph paper with milliliters of carbon dioxide produced as the ordinate and milliliters of oxygen used as the abscissa. Can you connect all the points for either species with a straight line? Why?

TABLE 3. PRODUCTION OF CARBON DIOXIDE AND UTILIZATION OF OXYGEN BY GERMINATING SEEDS OF WHEAT AND CASTOR BEAN

Milliliters of carbon dioxide produced		Milliliters of oxygen used	
Wheat	Castor bean	Wheat	Castor bean
11.5	7.0	11.3	9.0
13.7	4.5	13.9	7.0
5.5	20.0	5.2	28.5
20.0	14.5	19.4	19.5
17.6	3.1	17.9	4.2
6.2	8.0	6.4	10.5
7.8	10.0	8.0	15.0
15.7	12.5	15.8	18.3

SECTION 7 **STATISTICAL EVALUATION OF DATA**

Assume that your class used the volumeters described in Investigation 7 to study the difference in respiration rates between germinating seeds of pea and corn plants. Further assume that a number of readings were made and recorded as shown in Table 4.

What is the difference between the rates of respiration of germinating seeds of pea and corn at 25°C? Is the difference between corn and pea seedlings a real difference? Could you expect similar differences between the two sets of pea seedlings and the two sets of corn seedlings if you were to repeat the experiments?

In Pattern of Inquiry 3, you became familiar with two kinds of error: systematic error and random error. Can you detect evidence of systematic error in the data above? How might you check to determine if this source of systematic error is present?

If we assume that the differences are not the result of differences in the apparatus, what are the chances that you might get differences of this magnitude simply as a result of random error? Notice that, in reading numbers 2 and 5, more oxygen was used by corn than by pea seedlings.

TABLE 4. AMOUNT OF O_2 USED BY GERMINATING SEEDS OF CORN AND PEA PLANTS

Reading number	Milliliters of O_2 Used per Hour at 25°C	
	Corn	Pea
1	0.20	0.25
2	0.24	0.23
3	0.22	0.31
4	0.21	0.27
5	0.25	0.23
6	0.24	0.33
7	0.23	0.25
8	0.20	0.28
9	0.21	0.25
10	0.20	0.30
Total	2.20	2.70
Mean (Average)	0.22	0.27

Fortunately, mathematicians have developed techniques which are useful in determining what the probability is that differences such as those suggested here may be due to chance. These techniques are included in the branch of mathematics called *statistics*. Many decisions you make in regard to experimental data may be stated in terms of probabilities, and some understanding of probability and statistics is fundamental to an understanding of research in science. Statistical applications are based on probability statements, despite the fact that quite often you hear that anything can be proved with statistics. The reverse is true—nothing can be proved definitely with statistics. All you can do is report the probability that similiar results would occur if you were to repeat the experiment. This probability is based upon the data that you have collected. It must be emphasized that unless proper care is taken in planning investigations, the use of statistical procedures may not lead to any valid conclusions.

Statistics deal with numbers, and in order to decide what type of statistic to use, a biologist must be familiar with the nature of the numbers obtained in collecting his data. These numbers may be referred to as *variables*, and we classify them as either *discrete* variables or *continuous* variables.

Discrete Variables. Numbers of this type are often referred to as counting or categorical data. Numbers of boys or girls, number of students preferring biology to engineering, numbers of green and yellow corn plants, number of students ranked according to grades, number of seeds germinating—all of these are examples of discrete data. Families can only be made up of a discrete number of children. They are not made up of 1.2 children. In other words, this kind of variable can take on only a limited number of values.

Continuous Variables. Numbers of this type are associated with measuring and weighing. The data may take any value in a continuous interval of measurement. Hence, the weight of students, the height of pea plants, and the time it takes for plants to flower are all examples of continuous variables. Although 1.2 inches is a very acceptable measurement, you cannot have 1.2 children in a family. (Of course, you might find that a given group of families have an average of 1.2 children each.)

PROBLEMS

What kind of data are the following?

1. the numbers of persons preferring Brand X in 5 different towns
2. the weights of high school seniors
3. the lengths of oak leaves
4. the number of seeds germinating
5. 35 tall and 12 dwarf pea plants

Populations and Samples. A population includes all members of any specified group. For example, all the students in the United States constitute a population. Even though this is a rather large group of individuals, populations are not always large. The number of students in a given school can also constitute a population. Thus, no absolute number of individuals is required to make up a population; it is the researcher who sets up the limits that define the population with which he is concerned.

Nevertheless, the researcher often wishes to make inferences about large populations. If he defines his population as a single biology class, then conclusions from his data can only apply to this one class. However, if the researcher defines his population as all the biology classes in the United States, he can make inferences about this population from data obtained on small groups (that is, samples) within the population.

Populations do not always consist of intact organisms. An investigator may deal with populations of parts of either organisms or objects of various kinds. For example, one might be interested in the heights, weights, or metabolic rates of individuals, the numbers of red blood cells in individuals of a certain population, or in the types of textbooks used in various schools.

Samples are parts of populations, and statistics are the values used to describe samples. You are a member of a population consisting of all the biology students in the United States. Assume that we are interested in the average reading level of students in these biology classes. Gathering these data could go on for a long time if we attempted to compute the average reading level of students from the total of *all* the reading level scores of *every* student in *all* biology classes in this country. However, we can readily give a reading test to selected students and compute an average reading level score for this group. This group of selected students would constitute a *sample*. Statistics describing samples are used as estimators of the corresponding population.

Naturally, there will be a great deal of variation in the type of students represented by the sample. When samples are used to make inferences about a population, it is important that the individuals selected be a fair representation of the population. An experimenter attempts to insure representative samples by taking *random samples*. These are samples drawn with as little bias as possible from a population. In other words, extreme care is taken to make sure that all individuals or elements within the population have an equal chance of being selected.

For example, assume that there are 100,000 students in the United States who are studying advanced biology. We want to estimate the reading level of these 100,000 students by using data from a sample of only 100 students. How may we obtain this random sample? We could assign a different number to each of the students; the numbers could be written on separate slips of paper and placed in a container. A number could be drawn, recorded, and returned to the container. After thorough mixing of the numbers, this procedure could be repeated until 100 numbers were drawn. These 100 numbers representing 100 students would be our random sample representing the population of 100,000 students.

In theory, each number drawn should be returned to the container. However, in actual practice, neither is this always possible nor is it always necessary with large samples.

Suppose that instead of being selected at random, the 100 students represented groups with above-average reading abilities. The sample average would be too high to use as an estimate of the reading ability of the 100,000 students. The estimate is *biased* upward. Random sampling is intended to produce unbiased estimates of the population.

INVESTIGATION FOR FURTHER STUDY

1. In scientific research one should obtain *as many samples as practical* to insure that even random samples may not be giving an unrealistic estimate of the population. To demonstrate, place 10 pennies in a closed container. Shake well and allow them to drop on a table top. Record the numbers of heads and tails for this toss. Repeat this operation 25 times. Total your results from the 25 tosses and plot the total number of heads and tails on a bar graph as follows:

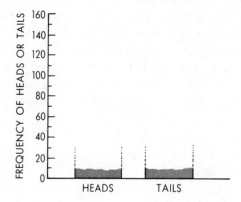

Now select the tosses which turned up the most heads and tails, respectively. Plot these two tosses. How do the profiles of your 3 graphs compare? Which profile do you think is most representative of the population? Why?

PROBLEM

If you had 1,000 rats in a cage and you wanted to select a sample of 20, what might be the bias if you selected the first 20 that you could catch?

There are many examples which would illustrate the danger in forming judgments based on biased samples. The field of politics has provided many such examples. One which has become classic was the public opinion poll conducted by a national magazine during the

1936 presidential campaign. Several million postcard ballots were circulated to obtain a large sample of the voting population. On the basis of the returns of these ballots, the magazine predicted that the winning candidate would be Alfred Landon and the loser would be Franklin D. Roosevelt. The magazine selected its sample from listings in telephone directories and automobile registration records. Apparently those who were neither telephone subscribers nor car owners (and who were more numerous in 1936 than they are now) voted in a manner that completely reversed the prediction. The magazine made the error of making inferences about this presidential election from samples which were not random and which gave a biased estimate of the proportion of votes for Landon. Many advertising claims seen in the newspapers and on TV are based on biased samples.

As an exercise, you should attempt to give other examples of inferences based on biased samples.

As pointed out earlier, populations are usually so large that they cannot be dealt with in their entirety. Also, a study of entire populations is too expensive and requires too much time. In fact, many are considered to be infinite in number. Consequently, the scientist usually works with samples that represent a population, and on the basis of the data from his samples, he makes inferences about this population. Of course, the accuracy of the inferences depends upon the degree to which the sample was representative.

Statistical Computations. In the handling of data concerning populations and samples of populations, certain statistical tools are of particular value. These tools represent simple devices for using familiar number relationships in order to help describe data.

Mean. In some of your earlier work with the Investigations and the Patterns of Inquiry, you used the "average" or arithmetic mean to help describe your data. Of course there are mathematical means other than arithmetic means, but they do not apply to this course. Throughout this text we shall refer to the arithmetic mean simply as the mean. The mean is used in computing average grades in school, average weight of football players, average temperature for the month of June, and so on. The sample mean or average, symbolized by $\bar{x}$ (x-bar), is a basic computation in statistics. A mean is the summation (symbolized by the Greek letter sigma, Σ) of the individual observations ($x_1, x_2, x_3, \ldots$ and so forth) divided by the number of observations (n). Thus we may write:

$$\bar{x} = \frac{(x_1 + x_2 + x_3 \ldots, \text{etc.})}{n}$$

or, in mathematical shorthand,

$$\bar{x} = \frac{\sum_{i}^{n} x_i}{n} \quad \text{(Formula 1)}$$

The symbol $\sum_{i}^{n} x_i$ represents the sum of the individual measurements; that is, $x_1 + x_2 + \cdots + x_n$ where i takes on all integral (whole number) values from 1, 2, . . . , n. For example, $\sum_{i}^{4} x_i$ would represent the sum of 4 measurements, $x_1 + x_2 + x_3 + x_4$.

For example, five groups of students conducted an experiment regarding the effect of temperature on the growth of peas. In one part of the experiment, each group grew a plant in the refrigerator at 10°C and measured the length of the shoot at the end of a given time. The results were as follows:

TABLE 5. SAMPLE DATA FOR CALCULATION OF THE MEAN

Group	Length of shoots in inches
A	10
B	7
C	6
D	8
E	9

$$\sum_{i}^{5} x_i = 40$$

$$\bar{x} = \frac{10 + 7 + 6 + 8 + 9}{5} = \frac{40}{5}$$

$$\bar{x} = 8$$

The values for the individual members of a "normal" population or of a sample tend to fall on either side of a particular value. The mean is an estimate of this value.

It is often useful to construct a graph of laboratory data in order to obtain a profile of the distribution of the values. To do this, first construct a distribution chart.

The distribution chart shown below was constructed from the height of 100 control plants in a particular experiment.

DISTRIBUTION CHART OF HEIGHTS OF 100 CONTROL PLANTS

Class (Height of Plants in cm)	Number of Plants in each Class
0.0–0.9	3
1.0–1.9	10
2.0–2.9	21
3.0–3.9	30
4.0–4.9	20
5.0–5.9	14
6.0–6.9	2

Note that each of these 100 measurements was assigned to classes of equal size. This frequency distribution chart can now be used to construct a graph known as a histogram (see figure). If one then draws a line connecting the top centers of each of the bars in a histogram, a frequency polygon results. In this case, it can be seen quickly that the individual members of the sample tend to cluster about the mean in a nearly symmetrical fashion. This is a random sample typical of what one might expect from a "normal" population.

Refer to Figure 8 which shows the frequency distribution of the heights of 100 control plants from Experiment X on day Y.

INVESTIGATION FOR FURTHER STUDY

1. Determine the heights in inches of fifty randomly selected boys and fifty randomly selected girls from a class in your school, being careful to note your method of sampling. Plot the heights of each sample as a histogram and a frequency polygon on the same graph. Are two "normal" populations indicated?

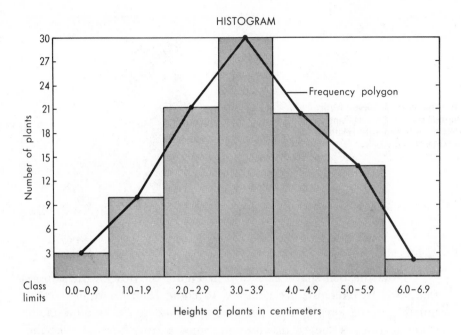

FIGURE 8. Histogram and polygon of the heights of 100 control plants from experiment X on day Y.

PROBLEMS

Find the means for the following sets of data:

1. 7, 7, 5, 3, 3
2. 10, 9, 8, 7, 6, 5, 4, 3, 2
3. 20, 16, 10, 6, 2

Normal Distribution. The relation of individual values to the mean is often important in statistics. Graphs which represent the frequency of occurrence of different values in a population are called frequency distributions, or simply, distributions. To graph a distribution, we can place the frequencies of our values either on the vertical or on the horizontal axis. As a matter of convenience, the frequencies are generally placed on the vertical axis, and the classes of the measured values for which the frequencies are determined are placed on the horizontal axis. A population with a bell-shaped distribution of values about its mean is said to have a "normal" distribution. We place the word "normal" in quotes because most populations appear to vary somewhat from the predictable normal curve.

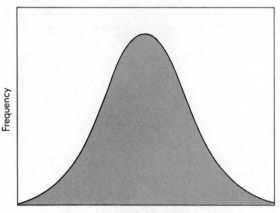

FIGURE 9. The normal curve. Notice that in a normal curve the measurements cluster around the mean with relatively few measurements in the tails.

Note the bell shape as shown in Figure 9. Most of the samples examined by biologists are assumed to come from "normally distributed" populations. It is important, therefore, to know some of the characteristics of such a distribution. Since a true "normal" distribution is an ideal theoretical distribution based on an infinite number of measurements, it is unlikely that the curve plotted from most data will look exactly like the curve of normal distribution.

Variance (s^2). We can see from the data on the amount of O_2 used by the corn and pea seedlings (Table 4, page 56) that the results vary for the different readings. It is often convenient to have a way of expressing this variation; one such expression is called variance. Variance is a measure of the degree of variation of scores from the mean. A large variance indicates that the individual scores in the sample deviate considerably from the mean, whereas a small variance indicates that the scores deviate little from the mean.

The method of computing variance requires some explanation. Individual deviations from the mean can be described in terms of the number of units between a measurement and the mean. If the mean is 8, then a measurement of 10 is 10 minus 8, or positive 2 units from the mean. Similarly, a measurement of 6 is 6 minus 8, or negative 2 units from the mean. To compute the variance of a set of deviations, we are tempted to sum all the deviations and find an average. However, the sum of the positive and negative deviations would always be zero for all samples. This would be of no use to us. To avoid this useless average of zero, each deviation from the mean is squared, and these squares are then added. Squaring the deviations gives us a series of positive numbers, because when a negative number is squared the result is positive.

The variance is then calculated by dividing the sum of the squared deviations by $n - 1$, the total number of measurements minus 1. We use $n - 1$ instead of n because it has been shown that when deviations are taken from a sample mean instead of from a population mean, division by $n - 1$ gives an unbiased estimate of the population variance; division by n does not.

The formula for variance is therefore:

$$s^2 = \frac{\sum\limits_{}^{n} (x_i - x)^2}{n - 1} \quad \text{(Formula 2)}$$

As we have mentioned earlier, the variance helps to characterize the data concerning a sample by indicating the degree to which individual members within that sample vary from the mean.

The data in Table 6 can be used to demonstrate both the calculation of the mean and the variance.

$$\bar{x} = \frac{\sum\limits_{}^{5} x_i}{5} = \frac{40}{5} = 8$$

$$s^2 = \frac{\sum\limits_{}^{5} (x_i - \bar{x})^2}{n - 1} = \frac{10}{5 - 1} = \frac{10}{4} = 2.5$$

TABLE 6. HEIGHTS IN CENTIMETERS OF FIVE RANDOMLY SELECTED PEA PLANTS GROWN AT 8—10°C

Plant	Heights of pea plants	Deviations from mean	Squares of deviations from mean
	(x_i)	$(x_i - \bar{x})$	$(x_i - x)^2$
A	10	2	4
B	7	−1	1
C	6	−2	4
D	8	0	0
E	9	1	1
	$\sum\limits_{}^{5} x_i = 40$	$\sum\limits_{}^{5} (x_i - x) = 0$	$\sum\limits_{}^{5} (x_i - x)^2 = 10$

PROBLEMS

1. Compute the mean and variance for an experiment similar to the above but in which the heights of pea plants were 9, 9, 8, 7, and 7 centimeters. Is the variance greater or less than in the previous example? Do your results bear out the conclusions that can be made by merely examining the data?

2. Compute the variance for the data on the amount of oxygen used by corn and pea plants as shown in Table 4, page 56.

3. Compute the variance for scores of 10, 9, 8, 7, 6, 5, 4, 3, and 2.

4. Compute the variance for scores of 20, 15, 10, 5, and 2.

INVESTIGATION FOR FURTHER STUDY

1. Compute the mean and variance for the heights of the fifty boys and girls. Randomly sample the heights from each group and compare the mean and variance of these subsamples with the original sample. Account for any differences noted.

Standard Deviation. This is an important statistic that is used in many statistical operations and is associated with the interpretation of the normal curve. It is another measure of variation that is used to describe samples. We will use the symbol (s) as a designation for standard deviation. The standard deviation is calculated by taking the square root of the variance.

$$s = \sqrt{\frac{\sum_{i}^{n}(x_i - \bar{x})^2}{n - 1}} \qquad \text{(Formula 3)}$$

The standard deviation of the heights of our sample of pea plants (Table 6) is as follows:

$$s = \sqrt{\frac{10}{4}}$$

$$= \sqrt{2.5}$$

$$= 1.6$$

While the standard deviation is shown to be 1.6, it refers to a measurement $\pm$ 1.6 centimeters.

Examine the relationship of the standard deviation of a sample to the "normal" distribution curve. Consider a curve in which each

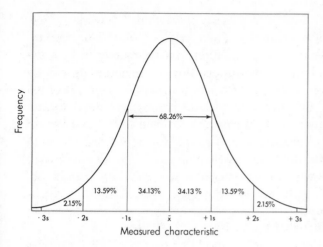

FIGURE 10. Percentage of individuals in a normal distribution falling within a plus or minus 1, 2, or 3 standard deviations from the mean.

interval on the *x*-axis is a unit of standard deviation (*s*). The numbers of individuals are plotted on the *y*-axis. In our example, $s = 1.6$. The concept of a normal distribution may be used to predict the percentage of individuals in a sample which should fall within a particular range of standard deviations. This relationship is shown graphically in Figure 10.

Figure 11 shows that about 68 percent of the measurements in a normal population have values which are within plus or minus 1 standard deviation from the mean. About 95% will fall within plus or minus 2 standard deviations, and nearly all members (over 99%) will fall within plus or minus 3 standard deviations from the mean.

One standard deviation in the example of pea plants was shown to equal 1.6. Since one standard deviation from the mean can be either

FIGURE 11. Percentage of classes in a normal distribution falling within a plus or minus 1, 2, or 3 standard deviations in a population with a mean of 8.

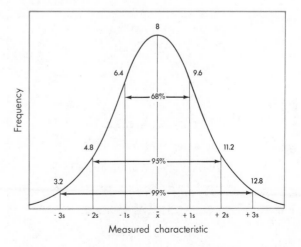

positive or negative, 1.6 *added* to a mean of 8 equals 9.6 while 1.6 *subtracted* from a mean of 8 equals 6.4. Therefore, in a sample representative of a normally distributed population, approximately 68% of the same species of pea plants, grown under the same conditions, can be expected to range from a height of 6.4 centimeters to a height of 9.6 centimeters. If 2 × 1.6 (two standard deviations) is added to and subtracted from the mean of 8, it can be predicted that about 95% of the same population of pea plants will range in height from 4.8 cm to 11.2 cm. Three standard deviations (in this case 3 × 1.6) will yield a range of 3.2 to 12.8 which should account for 99% of the population. From this it can be predicted that the probability of finding a pea plant that is grown under the same condition and is over 12.8 cm in height is remote—less than 1 out of 100.

For another example, consider the litter sizes of a certain species of dog. Assume the population has a mean of 10 pups per litter and a standard deviation of 2. We can then predict that slightly over 68% of the litters of this species will contain from 8 to 12 pups (±1 standard deviation from the mean); 95+% will have from 6 to 14 (±2 standard deviations from the mean); and 99+% of our sample will be in the range from 4 to 16 pups (±3 standard deviations from the mean). These relationships can be seen in Figure 12.

To illustrate further the relationship between the standard deviation and the mean, consider the plot of dog-litter sizes in Figure 12. If we write the litter sizes of all individual litters on slips of paper and place the slips in a hat, what is the probability of drawing a score of 16? It is rather unlikely; however, the chances of selecting a score of 11 are rather good. The normal distribution curve provides the basis for making these probability statements.

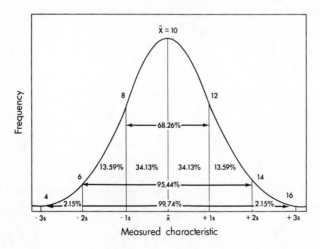

FIGURE 12. Samples (litters of pups) with a mean score of 10 pups and a standard deviation of 2.

Thus the standard deviation, s, can be a valuable tool in statistics because it reveals predicted limits within which one has a stated chance of being correct. It is also valuable in further statistical analysis.

PROBLEMS

1. Refer to the data for oxygen consumption by germinating corn and pea seeds given in Table 4, page 56. The standard deviation for corn was 0.02 and for peas was 0.035. If you made additional readings under similar conditions, between what values would you expect:
   a. 68% of the readings for corn to fall?
   b. 68% of the readings for peas to fall?
   c. 95% of the readings for corn to fall?
   d. 95% of the readings for peas to fall?

2. Compute the standard deviations for the data on oxygen consumption in germinating corn and peas (Table 4, page 56).

3. In any normal distribution, what percentages of the individuals will be found under that part of the curve which extends from the mean to plus 2 standard deviations from the mean?

4. With a specific disease, the white blood counts of randomly selected patients were (cells per mm^3) 11,200, 10,600, 10,600, 11,800, 11,000, 11,200, 10,400, 12,200, 10,800, and 11,600. We can expect 95% of all persons having this disease to fall in what white-blood-count range?

Standard Error of the Mean ($s_{\bar{x}}$). The mean, the variance, and the standard deviations are helpful in estimating characteristics of the population from a single sample. Statisticians have shown that if many random samples of a given size, n, are taken from the same population, the means ($\bar{x}$) of these samples would themselves form a normal distribution. This distribution of sample means would have a mean either equal or nearly equal to the population mean.

From this distribution a standard deviation of sample means can be estimated. The standard deviation of sample means is called the standard error of the mean ($s_{\bar{x}}$) or simply the standard error. Because there is less dispersion in the distribution of the sample means, the standard error is less than the standard deviation. We will use the idea of standard error later on in our work to test the reliability of our data. It is usually either impractical or impossible to take a large number of samples, compute the means of all the samples, and then determine the standard error of this distribution of means. It is important that we have a rather simple method for estimating the standard error. As was mentioned earlier, the dispersion in the distribution of means is very small, and, therefore, an estimation based on

a single sample will be adequate. It can be shown that our best estimate of the standard error is given by the following formula:

$$s_{\bar{x}} = \frac{s}{\sqrt{n}} \qquad \text{(Formula 4)}$$

where $\quad s_{\bar{x}} =$ standard error

$\quad\quad s =$ previously calculated standard
deviation of a sample of size (n)

Assume that a sample of ten corn plants revealed a standard
deviation of 0.02.

$$s_{\bar{x}} = \frac{0.02}{\sqrt{10}}$$

$$= \frac{0.02}{3.16} = 0.006$$

Thus, 0.006 represents one standard deviation in the distribution of the
means of samples of size 10.

Note that a larger sample size (n) lowers the standard error. In
the above example, an n of 100 with a standard deviation of 0.02
reduces the standard error from 0.006 to:

$$s_{\bar{x}} = \frac{0.02}{\sqrt{100}} = \frac{0.02}{10} = 0.002$$

This is logical, since one would expect a larger sample to represent
the population better.

In this section we have examined some of the basic ideas of the
normal distribution curve. The mean, variance, and standard error
of the mean will take on more significance as they are applied throughout this course.

PROBLEMS

1. What is the standard error of the mean for the data on oxygen consumption
by pea plants (Table 4, page 56)?

2. What is the standard error of the mean for the data on the height of pea plants
(Table 6, page 65)?

3. Five randomly selected boys received the following grades on a standardized examination: 76, 90, 85, 72, and 81. A second grouping of ten boys received the following scores on the same examination: 79, 84, 83, 86, 89, 76, 90, 85, 72, and 81. Calculate and compare the standard errors for these two groups.

INVESTIGATION FOR FURTHER STUDY

1. Continuing with your data concerning the heights of fifty girls and fifty boys from your class, calculate the standard deviation for each group. Determine the heights of two more randomly selected samples of twenty boys and girls each from your class. What percentage of these have heights within $\pm$ 1 standard deviation of the means of the original samples? within $\pm$ 2 standard deviations of the means of the original samples?

Probability and Tests of Significant Differences

In dealing with populations, the experimenter often compares two samples which have been treated differently. He wants to know if the two samples differ sufficiently to warrant a decision that they represent two different populations. This may be done by determining that the difference in the means of the two samples did not come about by "normal" sampling variations. To make such a decision, it is necessary to know how large the difference must be between the two samples (usually a treated sample and an untreated control sample) in order to state that the difference was caused by the treatment and not by "normal" sampling variations. To make such a decision requires a knowledge of the "laws" of probability.

The Laws of Probability

1. *The results of one trial of a chance event do not affect the results of later trials of the same event.* No matter how many times in a row a coin comes up heads (if it is an honest coin), the next tossing of the coin has the same probability value of being either heads or tails, namely, $p = 0.5$.

2. *The chance that two or more independent events will occur together is the product of their chances of occurring separately.* (Two or more events are said to be independent if the occurrence or nonoccurrence of one in no way affects the occurrence or nonoccurrence of any of the others.) If you roll a dice, what is the chance that a 3 will come up? Since a dice is six-sided and the 3 occupies only one of the sides, the probability of rolling a 3 is $\frac{1}{6}$. However, if there are two dice, the probability

of rolling a 3 on both dice is the product of the individual probabilities, or $\frac{1}{6} \times \frac{1}{6} = \frac{1}{36}$. Remember that all the probabilities must add up to 1. Thus there are 35 chances out of 36 that you may roll some combination other than two threes. For this, $p = \frac{35}{36}$. Assuming the sex of an individual is a chance event, what is the probability that both fraternal twins will be boys? Both will be girls?

3. *The probability that either of two or more mutually exclusive events will occur is the sum of their probabilities.* (Events are said to be mutually exclusive if when one of them happens on a particular occasion, the other cannot happen.) What is the chance of rolling a total of either two or twelve with a pair of dice? The probability of rolling a two (a one on each dice) is $\frac{1}{6} \times \frac{1}{6}$ or $\frac{1}{36}$. Similarly, the probability of rolling a twelve (a six on each dice) is $\frac{1}{36}$. Therefore, the p for rolling a two or a twelve is $\frac{1}{36} + \frac{1}{36} = \frac{2}{36}$ or $\frac{1}{18}$.

PROBLEMS

1. Assume that the probability that your car will have a tire blowout next month is 0.06 and that the probability that a second blowout will occur next month is 0.25. (This is predicted on the likelihood that a first blowout may indicate that all your tires are more worn than those found on the average car.) What is the probability that you will have two blowouts next month?

2. The following is called a two-way frequency table. The population which it represents is a hypothetical class of 1000 biology students.
    a. What is p for a particular student's being a male?
    b. What is p for a particular student's being a brunette?
    c. What is p for a particular student's being a male brunette?
    d. What is p for a particular student's being a female blond?
    e. What is p for a particular female's being a blond?
    f. What is p for a particular female's being either brunette or redhead?

TABLE 7.　TWO-WAY FREQUENCY TABLE

Sex	Hair Color			Totals
	Brunette	Blond	Redhead	
Male	300	150	25	475
Female	310	160	55	525
Totals	610	310	80	1000

3. In a single toss of two pennies, what is the probability both will come up heads? Both will come up tails? One will come up a head and the other a tail?

INVESTIGATION FOR FURTHER STUDY

1. Continuing your investigation for further study using the samples of the heights of fifty boys and fifty girls in your class, what is the probability the height of a given girl will exceed that of the mean height of the group by more than plus two standard deviations? Now record the hair colors of the girls in your sample as blond, brunette, or redhead. What is the probability that a given girl will be a tall blond? brunette? redhead? Repeat this operation for the sample of boys. Do you think your results are representative of the population of students in your school? in the schools of the United States? in all the schools of the world? Discuss the reasons for your conclusions.

Null Hypothesis. When we use probability as a guide in deciding whether two samples (such as one from a treated and one from a control group) are really different, we may make the assumption that any differences are due to chance and are not valid reflections of the populations. This assumption is called the *null hypothesis*. The null hypothesis assumes that there will be no difference as a result of the experimental treatment. If, after assuming that treatment A is no better than treatment B, we find evidence to the contrary, we reject the null hypothesis. That is, we say that the noted differences probably did not result from chance alone.

Statisticians use two common formulas to determine the probability that the null hypothesis is valid. These are known as the t test and the chi-square (χ^2) test. Each test has its own use.

The t Test. The t test is a valid technique for random samples of continuous variables (page 57) from normally distributed populations. When the conditions are met, the t test can determine the probability that the null hypothesis concerning the means of two small samples is correct; that is, the probability that the two samples are representative of a single population or of different populations. The t statistic was proposed in 1908 by a statistician, W. S. Gosset, who called himself "Student." It was developed as the relationship between series of sample means and the true mean of a population from which the samples were drawn. "Student" showed the areas under the distribution curve to be inconsistent with that of the normal curve when the sample size varies.

Usually we use the t test to determine the probability that two samples, A and B, came from the same or from different populations. We can then compute a standard error of the difference between the means based on the consideration that all the measurements in the two samples represent a single sample. If we divide this standard error of the difference in the means into the difference between the means of the two samples,

$$\frac{\bar{x}_1 - \bar{x}_2}{s_{\bar{x}_1} \, s_{\bar{x}_2}}$$

the result is the number of deviation units which separates the mean of Sample A from the mean of Sample B.

For example, if Sample A has a mean of 8, Sample B a mean of 12, and the standard error of the difference is 1, then if these two measurements are from the same population, they are separated by $\frac{12 - 8}{1}$ or 4 standard deviation units. To illustrate, assume that 8 is actually a true estimate of the mean of the population from which Sample A was taken. Suppose we plot a normal distribution about $\bar{x} = 8$ (see Figure 13) and visually estimate the probability that Sample B was derived from the same population as was Sample A. As can be seen, when the standard error of the difference is 1, the probability of drawing two samples with means of 8 and 12 from the same population is rather unlikely. A more probable conclusion would be the alternative possibility that Samples A and B represent samples from different populations.

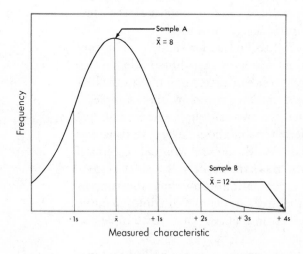

FIGURE 13. A comparison of two samples, assuming they are from the same population.

There are several formulas by which t may be determined, the choice depending on several factors. For data collected in this course, the following formula may be used:

$$t = \frac{\bar{x}_1 - \bar{x}_2}{\sqrt{\dfrac{(n_1 - 1)s_1^2 + (n_2 - 1)s_2^2}{n_1 + n_2 - 2} \cdot \left(\dfrac{1}{n_1} + \dfrac{1}{n_2}\right)}} \qquad \text{(Formula 5)}$$

where

$\bar{x}_1$ = mean of Sample 1
$\bar{x}_2$ = mean of Sample 2
n_1 = the number in Sample 1
n_2 = the number in Sample 2
s_1^2 = the variance of Sample 1
s_2^2 = the variance of Sample 2

If the samples are equal in size, then $n_1 = n_2 = n$, and the above formula can be simplified as follows:

$$t = \frac{\bar{x}_1 - \bar{x}_2}{\sqrt{\dfrac{s_1^2 + s_2^2}{n}}} \qquad \text{(Formula 6)}$$

It should be noted that the value of t depends on both the numerator and denominator of this equation. Assuming a constant value for the denominator, we find that the larger the numerator $(\bar{x}_1 - \bar{x}_2)$ is, the larger is the value of t.

The denominator is the standard error of the difference between the means. In Formula 5, the assumption is made that the variance of Sample 1 is similar to the variance of Sample 2. You may recall that we assumed both samples were from the same population. If this assumption is not valid and the two variances are in fact so different that they cannot be pooled, we would need to use a different formula for t. This should not be necessary in this course. If you are interested in further refinement of the t test, the several references on statistics in the back of this book may be consulted.

We will use the t test to determine if the data on the rate of oxygen consumption in corn seedlings is significantly different from that of pea seedlings. (See Table 4, page 56.) First we must state the hypothesis we wish to test. In this case, perhaps the most useful hypothesis to test is as follows: The rates of O_2 consumption in the two kinds of seeds are not different. (We did observe a difference, however; and if our null hypothesis is not rejected, we must be able to assume that there is a reasonable probability that such a difference might be obtained by chance sampling of the same population.)

The two sample sizes are equal—that is, $n_1 = n_2$. Therefore, we may use Formula 6.

$$t = \frac{\bar{x}_1 - \bar{x}_2}{\sqrt{\dfrac{s_1{}^2 + s_2{}^2}{n}}}$$

We have already found:

$$\bar{x}_1 = 0.22$$
$$\bar{x}_2 = 0.27$$
$$s_1{}^2 = 0.0004$$
$$s_2{}^2 = 0.0012$$
$$n = 10$$

Substituting these values in the formula, we have:

$$t = \frac{0.22 - 0.27}{\sqrt{\dfrac{0.0004 + 0.0012}{10}}}$$

$$= \frac{-0.05}{\sqrt{0.00016}}$$

$$= \frac{-0.05}{0.0126}$$

$$= 3.97$$

This indicates that the means of Sample 1 and Sample 2 are almost 4 standard deviations apart. From our earlier discussion we would not expect to draw two such samples from the same population by chance, and we can immediately guess that we may reject our null hypothesis. However, we can make more critical use of our value of t by examination of a table of the distribution of t (Table 8).

This table shows what value for t may be expected at the various levels of probability. It employs, along the top, a series of p values, and along the left side a listing of the degrees of freedom (d.f.). Degrees of freedom may be defined as the number of individuals or events, or sets of individuals or events, which are free to vary in a given sample. For example, if the total of five numbers is 20, the first four numbers can be a combination of quite a few numbers, but the fifth number will be determined by the first four numbers. If our total is 20 and

TABLE 8. DISTRIBUTION OF *T* PROBABILITY

		Probability			
		0.1	**0.05**	**0.01**	**0.001**
D	1	6.314	12.706	63.657	636.619
e	2	2.920	4.303	9.925	31.598
g	3	2.353	3.182	5.841	12.941
r	4	2.132	2.776	4.604	8.610
e	5	2.015	2.571	4.032	6.859
e	6	1.943	2.447	3.707	5.959
s	7	1.895	2.365	3.499	5.405
	8	1.860	2.306	3.355	5.041
o	9	1.833	2.262	3.250	4.781
f	10	1.812	2.228	3.169	4.587
	11	1.796	2.201	3.106	4.437
F	12	1.782	2.179	3.055	4.318
r	13	1.771	2.160	3.012	4.221
e	14	1.761	2.145	2.977	4.140
e	15	1.753	2.131	2.947	4.073
d	16	1.746	2.120	2.921	4.015
o	17	1.740	2.110	2.898	3.965
m	18	1.734	2.101	2.878	3.922
	19	1.729	2.093	2.861	3.883
	20	1.725	2.086	2.845	3.850
	21	1.721	2.080	2.831	3.819
	22	1.717	2.074	2.819	3.792
	23	1.714	2.069	2.807	3.767
	24	1.711	2.064	2.797	3.745
	25	1.708	2.060	2.787	3.725
	26	1.706	2.056	2.779	3.707
	27	1.703	2.052	2.771	3.690
	28	1.701	2.048	2.763	3.674
	29	1.699	2.045	2.756	3.659
	30	1.697	2.042	2.750	3.646
	40	1.684	2.025	2.704	3.551
	60	1.671	2.000	2.660	3.460
	120	1.658	1.980	2.617	3.373
	∞	1.645	1.960	2.576	3.291

Note: *The shading in this table emphasizes the areas of increasing probability that the null hypothesis should be rejected—that is, the areas of increased confidence that the two samples represent two different populations. In evaluating data from the Investigations in this course, p values of less than 0.05 will generally be considered as adequate for rejection.*
Source: Abridged from Table III of R. A. Fisher and F. Yates: *Statistical Tables for Biological, Agricultural, and Medical Research,* published by Oliver and Boyd Ltd., Edinburgh, by permission of the authors and publisher.

our first four numbers are 1-3-5-7, the fifth number must be 4. Of n numbers, with a fixed mean, only $n - 1$ are free to vary. In our data, we had 10 readings in Sample 1 and 10 readings in Sample 2. Each sample has $n - 1$ degrees of freedom and the total is $(n_1 - 1) + (n_2 - 1) = 18$. Therefore, we enter the table with 18 degrees of freedom. With 18 d.f., our t must be 2.101—if the difference between the means is to be significant at the 5% level of probability. At the 1% level it must be 2.878, and at the 0.1% level it must be 3.922. Our value of 3.97 is greater than any of these.

Conclusion: We reject the null hypothesis that the rates of oxygen consumption are the same. We reject it at the $p = 0.001$ level of significance. In so doing, we are running the risk of being in error, but only 1 time in 1000.

If such data were reported in the literature, the authors would usually report the data with the means and point out that "the difference between the means is highly significant." In most biological research, if $p < 0.05$, we say the results are *significant* (we reject the null hypothesis); if $p < 0.01$, we say the results are *highly significant*.

Our table of t includes only p values of 0.1, 0.05, 0.01, and 0.001. Any one of these levels of probability might be used for rejection of the null hypothesis in a given research problem. The nature of the problem will determine what level we will use. Other tables may include p values of 0.2, 0.3, 0.4, or even 0.9. These values are sometimes useful even if we do not use them for rejection of a null hypothesis.

In order to gain more experience in the use of the t test, consider the data in Table 9. We are hopeful, of course, that other conditions such as light intensity, soil fertility, available moisture, and others were the same in both samples. If any differences do occur, we are hopeful that they are due to the difference in temperature. What, then, should our null hypothesis be?

Having stated the hypothesis, compute the value of t. Because the sample sizes are equal, we can use Formula 6. Note, however, that you will need to know the values of the following: $\bar{x}_1$, $\bar{x}_2$, s_1^2, s_2^2, n_1 and n_2. Determine each of these values and then substitute them into the formulas. Remember that $n_1 = n_2 = n$ in the formula.

Now enter the table for the distribution of t. How many degrees of freedom are there? With your values for t and d.f., what is the probability that you might have obtained the observed differences in the means by chance alone? Can you reject the null hypothesis? If so, at what level of confidence? What conclusion might you draw from the data?

TABLE 9. HEIGHTS OF PEA PLANTS GROWN UNDER DIFFERENT TEMPERATURES

Plant	Room Temperature (22° C) (Sample 1)			Temperature in Refrigerated Growth Chamber (10° C) (Sample 2)		
	Heights of plants in inches x_i	Deviations from mean $(x_i - \bar{x})$	Squares of deviations from mean $(x_i - \bar{x})^2$	Heights of plants in inches x_i	Deviations from mean $(x_i - \bar{x})$	Squares of deviations from mean $(x_i - \bar{x})^2$
A	12	1	1	10	2	4
B	12	1	1	7	−1	1
C	10	−1	1	6	−2	4
D	11	0	0	8	0	0
E	10	−1	1	9	1	1
Totals	$\sum_{}^{5} x_i = 55$	$\sum_{}^{5}(x_i - \bar{x}) = 0$	$\sum_{}^{5}(x_i - \bar{x})^2 = 4$	$\sum_{}^{5} x_i = 40$	$\sum_{}^{5}(x_i - \bar{x}) = 0$	$\sum_{}^{5}(x_i - \bar{x})^2 = 10$

The following is a summary of the requirements for use of the t test in evaluation of a null hypothesis:

1. Samples must be randomly chosen.
2. Samples must have the characteristics of a "normal" distribution.
3. Measurements must be of continuous variables.

If these criteria can be met, proceed as follows:

1. State your null hypothesis.
2. Compute the means (Formula 1, page 61).
3. Compute the variances (Formula 2, page 65).
4. Determine which formula for t is applicable. If n_1 and n_2 are different, use Formula 5 (page 75); if $n_1 = n_2 = n$, use Formula 6 (page 75).
5. Substitute the proper values in the formula and calculate t.
6. Determine the number of degrees of freedom (page 76).
7. Refer to a table of t and determine the proper level of probability for your data.
8. State your conclusion.

PROBLEMS

On page 63 data were given for three samples with instructions to calculate the means $(\overline{x})$. These data are listed below as Samples A, B, and C. Compute the variances (s^2) for each of the samples.

Sample A: 7, 7, 5, 3, 3

Sample B: 10, 9, 8, 7, 6, 5, 4, 3, 2

Sample C: 20, 16, 10, 6, 2

Assume that each set of data meets the three requirements listed for application of the t test.

1. a. Hypothesize that the means of Samples A and B are not different.
   b. Which t formula should you use to test this null hypothesis?
   c. Find t for Samples A and B.
   d. How many degrees of freedom are there?
   e. Interpet the t value.

2. a. Hypothesize that the means of Samples A and C are not different.
   b. Which t formula should you use to test this null hypothesis?
   c. Find t for Samples A and C.
   d. How many degrees of freedom are there?
   e. Interpret the t value.

3. a. Hypothesize that the means of Samples B and C are not different.
   b. Which formula should you use to test this null hypothesis?
   c. Find t for Samples B and C.
   d. How many degrees of freedom are there?
   e. Interpret the t value.

INVESTIGATION FOR FURTHER STUDY

1. Continuing with your data concerning the heights of the students in a particular class in your school, do the following:
   a. State a null hypothesis concerning the heights of boys and girls in the class.
   b. Select the proper formula and calculate a t value.
   c. Determine the degrees of freedom.
   d. State your conclusions about the null hypothesis.
   e. Randomly select 10 heights to be dropped from the sample of boys, calculate the new variance, and repeat the above procedures. Account for the similarities (or differences) in the results, using the different size samples of boys' heights.

   Chi-Square. You will recall that one of the requirements for application of the t-test was that the data be made up of continuous variables. When we have data that are made up of discrete variables, we often use a *chi-square* $(\chi)^2$ *test* to determine if our null hypothesis is tenable. (The chi-square method was devised in 1900 by Karl Pearson of England.) It is most often used to evaluate differences between experimental data and expected or hypothetical data, and it can be used with two or more samples.

Suppose that we have collected a number of snails, and that after examining a few, we have the feeling that $\frac{3}{5}$ of all snails twist clockwise and $\frac{2}{5}$ of all snails twist counterclockwise. Suppose that we have caught and observed 1000 snails. According to our theory, 600 should twist clockwise and 400 counterclockwise. Now suppose that our actual study population of snails turns out to have 615 which turn clockwise and 385 which turn counterclockwise.

As we look at these numbers, it is very hard to know whether they are an accidental departure from our theorized 600 and 400 or whether they are a somewhat unusual departure from, say, 500 and 500. This is the question we want to test. For this kind of testing we first make a null hypothesis and then test it by the chi-square method. To use chi-squares, it is necessary only to identify, count, or classify samples. Once a chi-square value has been determined, it can be used for estimating the probability of the null hypothesis in much the same way as we used the t value.

Often when chi-square is applied, a modified form of the null hypothesis is used. Rather than compare two experimental samples, researchers may predict that certain results will occur and then note how closely their actual results approximate the predicted ones. When this technique of experimentation is used, the predicted results are analogous to what are often the control samples. In such cases the null hypothesis states, in effect, that "no difference exists between the hypothetical population (or predicted population) and the population from which the experimental sample was drawn." The chi-square value for the sets of data is then calculated. The probability that the hypothesis is tenable can then be determined by consulting a table of chi-square values. Small values of chi-square lend support to the null hypothesis, while large values indicate that it should be rejected.

Chi-square is easy to calculate. The following formula is used:

$$\chi_2 = \Sigma \frac{(\text{observed No.} - \text{expected No.})^2}{\text{expected No.}}$$

The formula states that chi-square is determined by squaring each difference between the number of a certain attribute expected (or hypothesized) and the number actually observed. The value squared is then divided by the expected number in each case. The quotients are then added together.

We will examine our problem concerning the direction in which snails twist. Our hypothesis is that $\frac{3}{5}$ of all snails twist clockwise and $\frac{2}{5}$ twist counterclockwise. Our null hypothesis states that there will be no difference between our predicted ratio of clockwise to counterclockwise twisting and the ratio actually found in the 1000 snails that make up our sample.

Of the 1000 snails, our hypothesis (not the null hypothesis) predicts that 600 will twist clockwise and 400 will twist counterclockwise. The null hypothesis states that there will be no difference between the hypothesized sample and the sample actually observed. In our actual sample we found 615 twisted clockwise and 385 counterclockwise. From the formula,

$$\chi^2 = \Sigma \frac{(\text{observed} - \text{hypothetical})^2}{\text{hypothetical}}$$

$$\chi^2 = \frac{(615 - 600)^2}{600} + \frac{(385 - 400)^2}{400}$$

$$\chi^2 = \frac{(15)^2}{600} + \frac{(15)^2}{400} = \frac{225}{600} + \frac{225}{400}$$

$$\chi^2 = 0.375 + 0.562 = 0.937$$

Table 10 is a modified table of chi-square values. In chi-square tests used in this text, the degrees of freedom (d.f.) are one less than the number of attributes being observed, or $n - 1$, where n equals the number of attributes. The attributes in our snail problem are two: one twisting clockwise and one twisting counterclockwise. Hence, in this case we have two attributes but only one degree of freedom. We enter Table 10 with one degree of freedom and find that our obtained chi-square of 0.937 for the sample problem falls between $p = 0.5$ (0.455) and $p = 0.2$ (1.642).

In conclusion, we can say that if we had selected $p = 0.05$ as our level of rejection, we would not reject the null hypothesis. (The null hypothesis states that our hypothetical value and that of the actual sample were from the same population.) Our observed data, we would say, are consistent with our hypothesis that the ratio of clockwise to counterclockwise twisting is $\frac{3}{2}$.

TABLE 10.   CRITICAL VALUES OF χ^2

Values of χ^2 equal to or greater than those tabulated occur by chance less frequently than the indicated level of p.

d.f.	p = 0.9	p = 0.5	p = 0.2	p = 0.05	p = 0.01	p = 0.001
1	.0158	.455	1.642	3.841	6.635	10.827
2	.211	1.386	3.219	5.991	9.210	13.815
3	.584	2.366	4.642	7.815	11.345	16.268
4	1.064	3.367	5.989	9.488	13.277	18.465
5	1.610	4.351	7.289	11.070	15.086	20.517
6	2.204	5.348	8.558	12.592	16.812	22.457
7	2.833	6.346	9.803	14.067	18.475	24.322
8	3.490	7.344	11.303	15.507	20.090	26.125
9	4.168	8.343	12.242	16.919	21.666	27.877
10	4.865	9.342	13.442	18.307	23.209	29.588

For another example, suppose that, in studying a genetic cross between two kinds of tomato plants, it is expected that half the off-spring will have green leaves and half will have yellow leaves. (This supposition is based on the hypothesis that a single pair of genes is responsible for the difference between green leaves and yellow leaves.) In one actual experiment, it happened that of 1240 seedlings there were 671 with green leaves and 569 with yellow leaves. This is clearly different from the 620 of each kind which our hypothesis has predicted would occur. The null hypothesis is that there will be no difference in the hypothesized 1 — 1 ratio and the actual results observed in the experiment. Is the observed difference large enough to cause a rejection of our null hypothesis? In other words, is it likely that the difference between the observed and expected values is due to sampling variation rather than to some real difference not accounted for by our hypothesis?

In this case, we would expect or predict that half of our seedlings would have green leaves and half would have yellow leaves. This expectation is based on our knowledge of what is inherited by the off-spring when a homozygous recessive parent is crossed with a heterozygous parent and one of the characters is dominant over the other.

Our null hypothesis states that there will be no difference between the expected number (based on the ratios from Mendel's principles) and the observed value in the actual sample.

Expected were 620 yellow plants and 620 green plants; observed were 569 yellow plants and 671 green plants.

We now apply Pearson's chi-square formula to these data.

$$\chi^2 = \frac{(671 - 620)^2}{620} + \frac{(569 - 620)^2}{620}$$

$$\chi^2 = \frac{(+51)^2}{620} + \frac{(-51)^2}{620}$$

$$\chi^2 = 8.4$$

Our attributes are two in number—green and yellow leaves—and 2 minus 1 gives us 1 degree of freedom. From Table 10, we see that a chi-square of 8.4 with 1 degree of freedom indicates that there is less than one chance in one hundred ($p < 0.01$) that these or more extreme values would occur in a sample if the expected values were true of the population. If we decided that 0.01 would be our level of rejection, then the null hypothesis should be rejected at this level. In this case, since the two hypotheses are similar, rejection of the null hypothesis means rejection of the original hypothesis. The original hypothesis states that the crossing of those two strains of tomato should yield progeny which are $\frac{1}{2}$ green and $\frac{1}{2}$ yellow. This does not exclude the possibility that we may be rejecting the null hypothesis when actually the difference between the observed and predicted results are due to chance. A difference of this magnitude can be expected to occur, however, less than one time in 100 ($p < 0.01$).

At this point, the investigator has another important question to ask. How consistently will repetitions of the same experiment produce differences of nearly the same magnitude as that between actual and expected numbers? If repeated crosses produce about the same results, it is time to search for a suitable explanation for this departure of the obtained ratios of green to yellow from those which are expected. (In the case of the tomato plants, further crosses did show about the same results. The investigator found that the difference was caused by a loss of yellow-leaved plants. They were constitutionally less sturdy than the others, and fewer of them germinated and lived.)

Up to this point in our discussion of χ^2, we have noted how an observed result can be compared with an expected one. Often, however, we have no way of knowing what the expected result should be. For example, if a new drug is being compared with a control in the treatment of a certain disease, the expected results from either treatment may be unknown. Data of this type may be treated by a technique known as the "2 × 2" contingency table. Use of this table will require that you combine your previous knowledge of χ^2 with that which you have learned about the rules of probability.

Suppose 100 persons had a disease. And suppose an investigator treated 55 of them with the control (placebo) and 45 with the new drug under test. Of the former, 10 were cured within five days while 15 of those treated with the drug were cured within the same length of time. The results could be recorded as follows on a "2 × 2" contingency table.

	Control (Placebo)		Drug		Total
	Observed	Theoretical	Observed	Theoretical	
Cured	10	(13.75)	15	(11.25)	25
Not cured	45	(41.25)	30	(33.75)	75
Total	55		45		100

The null hypothesis in this case is that the placebo and drug results were drawn from the same population. The expected numbers for use in our χ^2 formula can be estimated by determining the probability for each box in the table. Thus, the probability by chance alone that a case would be a placebo is 55/100 while the probability of a case being a cure would be 25/100. Now according to probability rule 2, page 71, the chance that two or more independent events will occur together is the product of their chances of occurring separately. Hence, the probability by chance alone of a case being a cured placebo would be:

$$\frac{55}{100} \times \frac{25}{100} = \frac{13.75}{100}$$

If we consider that there are 100 patients involved in the test, the expected number of placebo cures by chance alone would thus be:

$$\frac{13.75}{100} \times 100 = 13.75 \text{ patients}$$

Similarly for the other squares in the table we may calculate:

Placebo (no cures):

$$\frac{55}{100} \times \frac{75}{100} = \frac{41.25}{100} \text{ or } 41.25 \text{ patients}$$

Drug cures:

$$\frac{45}{100} \times \frac{25}{100} = \frac{11.25}{100} = 11.25 \text{ patients}$$

Drug (no cures):

$$\frac{45}{100} \times \frac{75}{100} = \frac{33.75}{100} = 33.75 \text{ patients}$$

Now we may calculate the contribution χ^2 of each box in the usual manner remembering that:

$$\chi^2 - \frac{(\text{observed-hypothetical})^2}{\text{hypothetical}}$$

Thus we obtain for the placebo cures:

$$\frac{(10 - 13.75)^2}{13.75} = 1.0$$

Similarly one may calculate the χ^2 contribution of the placebo (no cures) to be 0.3; the drug cures, 1.3; and the drug (no cures), 0.4, and:

$$\chi^2 = 1.0 + 0.3 + 1.3 + 0.4 = 3.0$$

Since the marginal totals are fixed for the table, it should be noted that, after one value is filled in, the other three can easily be filled in. Only one value, therefore, is free to vary and the "2×2" table has 1 degree of freedom.

In our case a χ^2 of 3.0 with one degree of freedom would yield a p value for our null hypothesis greater than .05 but less than 0.2. While insignificant by our arbitrary criteria, the investigator may want to pursue the effect of this drug and its deviations further because there are certainly indications of being on the right track.

PROBLEMS

1. The same investigator looked into the inheritance of red flesh and yellow flesh in tomatoes. He hypothesized that the F_2 generation should have a 3:1 ratio of red to yellow. (Once again, the expectations were based on classical genetics and the assumptions (1) that one pair of genes determines the difference between red-fruited and yellow-fruited tomatoes and (2) that one allele is

dominant over the other.) The red-fleshed class turned out to have 3629 fruits, and the yellow-fleshed class had 1176. From the expected 3:1 ratio of the theory, we might expect to have 3604 red-fleshed and 1201 yellow-fleshed. Is the difference between the expected and observed numbers due to the action of something we had not taken into account in our theory, or is it only a chance departure from what our theory says should occur? What is the value? What is your conclusion?

Using the example, what would have happened if our hypothesis predicted that the ratio would be 4 to 1? 2 to 1?

2. The following experiment with the vaccine against polio was performed by Dr. Salk and his colleagues. Equal numbers of students received the vaccine and the placebo (plain shot without the vaccine). In this group of students, 862 cases of polio occurred. Of these cases 112 received the vaccine and 750 received the placebo. To evaluate such data, we would use the chi-square statistic. Why? Interpret your results.

Chi-Square with More Than Two Attributes. When more than two attributes occur in the experimental design, it is necessary to use degrees of freedom other than 1 in determining the probability that chance alone could have been responsible. This new number of degrees of freedom is needed when we want to consider the significance of results for several classes in some series of ratios. For example, there are four attributes and three degrees of freedom $(n - 1)$ in a 9:3:3:1 ratio for the F_2 generation of a hypothetical, dihybrid cross.

PROBLEMS

1. A new surgical technique is tried on 30 victims of a disease which normally is 50% fatal. Twenty-two of the thirty patients survive the disease. Is the technique effective?

2. An investigator was determining the effect of a certain treatment on the germination of seeds after planting. His results were as follows:

	Treated Seeds	Untreated Seeds	Total
Germinated	30	20	50
Not germinated	20	30	50
Total	50	50	100

Did the treatment have a significant effect?

3. When pink four-o'clocks are crossed, it is expected that the offspring will turn out to be red-, pink-, and white-flowered in a 1:2:1 ratio. This is equivalent to

saying that $\frac{1}{4}$ of the flowers will be red, $\frac{2}{4}$ or $\frac{1}{2}$ pink, and $\frac{1}{4}$ white. An experimenter made the cross and found that he had 66 red-flowered plants, 115 pink-flowered plants, and 55 white-flowered plants. The total number of plants was $66 + 115 + 55 = 236$.

Use the chi-square test to determine if the observed data are consistent with the hypothesis that such a cross should produce a 1:2:1 ratio of phenotypes in the offspring.

a. State your null hypothesis.
b. What was the expected (or hypothetical) number of (a) red-flowered plants, (b) pink-flowered plants, and (c) white-flowered plants?
c. Substitute in the formula for χ^2 and compute its value.
d. Determine the number of degrees of freedom.
e. Consult the table for χ^2. What p value does this value of χ^2 with your d.f. represent?
f. Will you reject the null hypothesis?
g. What is your conclusion?

4. A geneticist wishes to see if a 9:3:3:1 ratio exists in frequency of red: orange: yellow: blue flowers on a certain plant. He counts 1600 flowers and finds 802 red, 194 orange, 399 yellow, and 205 blue ones. State a null hypothesis and determine the probability that the 9:3:3:1 ratio exists for the flowers of this plant.

Comparison of t and Chi-Square Tests. One might legitimately ask at this point: What is the consequence of making an error in the selection of a test of significance? The t is a very powerful statistic and will make maximum use of the data; you must be able, however, to meet the assumptions of random sampling and normal distribution, and your data should consist of continuous variables. If you use the chi-square test in its place, you run the risk of not using your data to its greatest advantage. On the other hand, to use chi-square you do not have to meet the assumption of normal distribution, and you can use discrete data. Chi-square does not have the power of t, but it serves a purpose in being usable for some types of data that cannot be tested by t. If you use t when you should use chi-square, you run the risk of overstating the confidence levels provided by your data.

Conclusion. In analyzing data with statistics, it should be emphasized that the end product of research experiments are considered "conditional truths." In research there is always the probability of error whether you reject or do not reject the null hypothesis. It is relatively easy to see the potential importance of research where the null hypothesis is rejected, but quite often the value of negative results are underestimated. Negative results are valuable because they can eliminate some hypotheses that, at first thought, might seem to be possible solutions to a problem. Also, negative results may serve as an impetus to refine a given bit of research. They may be the first clues, and they may reveal insights that eventually lead to possible answers.

For example, in comparing the use of a new method of skin transplanting with an old method, you might get a t value which indicates a probability level of 0.2 that the two methods are equally effective. This p value, while not small enough to reject the null hypothesis, might indicate that further experimentation with this method is justified. Perhaps a slight change in the technique or use of an additional drug might be all that is needed to make this method acceptable and beneficial.

Statistics are valuable, whether the results are significant or not, because they provide tools with which we can quantitatively analyze data.

SECTION 8 THE LITERATURE OF BIOLOGY

All knowledge about the natural world results from ideas and observations. Many of the observations are data recorded in experiments designed to test hypotheses or to better describe certain phenomena. From your own experimental observations, you have learned much about the metabolism of yeast cells. Scientific work, carried out in a similar manner, may lead to understandings of many other aspects of the natural world. The more careful the work, the more accurate is our knowledge of nature.

Any experiment, observation, or idea is interesting to the person who does the work, but it is not really a contribution to scientific knowledge until it is published for others to read and ponder. The work then becomes a part of the *scientific literature* to which other persons who are interested in the same or related problems can refer.

Although the great bulk of new scientific information and understanding results from the work of professional scientists, today—as in the past—amateur naturalists and experimenters also make important contributions to science. In North America, for example, hundreds of members of the National Audubon Society make a careful count of birds they observe around Christmas time. These data, when assembled, provide a "bird census," which is important information about the abundance of bird species, and which would be impossible to obtain without the help of thousands of amateur observers. Without their aid, we would know much less about our natural populations of birds.

Scientific Journals. The hundreds of thousands of scientists in the world work in many different fields of science. Scientists work on a problem until they think they have something of interest and importance to report to other scientists. Then they write an article, or "paper," which summarizes their work and thoughts. The paper is submitted to the editors of one of the thousands of scientific journals published throughout the world. It is reviewed by other scientists who are acquainted with the subject with which it deals, and who are able to judge whether the article makes a contribution to science. If it does, the editor arranges to have it published in a forthcoming issue of the journal.

As a rule, each scientific journal publishes papers that deal primarily with a single field of science; the name of the journal usually gives a clue to the subject matter involved. In the field of biology, for example, it is not difficult to tell what kinds of articles are probably found in *The American Journal of Anatomy, American Journal of Botany, British Journal of Nutrition, Canadian Journal of Zoology, Japanese Journal of Genetics,* or *Journal of Cell Biology.* Sometimes, just the name of the subject matter makes up the name of the journal, such as the publications entitled *Developmental Biology, Ecology, Evolution, Genetics, Growth, Heredity,* and *Human Biology.* Other journals carry the names of the societies that publish them, such as *Journal of the American Medical Association, Proceedings of the Indian Academy of Sciences,* and *Transactions of the American Microscopical Society.*

In addition to the journals which publish articles primarily on biology, chemistry, physics, or geology, there are some which publish articles of interest to scientists in all fields. The weekly journal *Science* is published in the United States and covers all areas of science; its British counterpart is called *Nature.* These two journals contain not only reports of original research but also general articles, announcements, and advertisements of interest to scientists. Other periodicals of general scientific interest are *Scientific American* and *American Scientist.*

Scientific journals are published in many countries. The articles may be written in the language of the country, but this is not always the case; many foreign journals publish articles in English. Some journals (such as *Experientia* and *Die Naturwissenschaften*) accept articles in more than one language, such as English, French, German, or Italian. Other journals publish papers in one language but provide summaries of each selection in one or two other languages. Journals that contain articles written in languages with distinctive alphabets or symbol systems—such as Chinese, Japanese, and Russian—may have summaries or aids in English or in other languages. A number of

FIGURE 14.   A few of the many scientific journals that are received by a university library.

Russian science journals are now regularly translated into English. Scientific names of organisms are given in Latin and thus appear the same around the world to people of all languages.

Consequently, the search for knowledge about natural events recognizes no national boundaries. Libraries subscribe to the journals of the world. One measure of the value of a library to scientists is the number of scientific journals to which it subscribes.

Scientific Papers. There is no single way in which a scientific paper must be written, but most papers are similar in structure. The title of the paper should tell, in a few words, as much as possible about the content of the paper. Some titles are very general—for instance, "Biology and ecology of tree species." Other titles may give experimental conclusions, such as "Classification of functional liver tests." Some papers primarily present data, for example, "Further data on the over-dominance of induced mutations." Papers presenting a theoretical point of view may have titles such as "Selective forces in the evolution of man" or "The bearing of philosophy on the history of science."

Examine a scientific paper on a topic of interest to you to see how it is organized. The paper on "*Zea* shoot development in response to red light interruption of the dark-growth period: 1. Inhibition of first internode elongation" appeared in the July 1969 issue of *Plant Physiology* and is reproduced on pages 93 through 96. The names and addresses of the authors, S. O. Duke and J. L. Wickliff, are listed below the title of the paper. Scientists who want to learn more about the material presented in a paper may then correspond with the authors.

An abstract should appear at the heading of the paper, before the text of the paper *per se*, so the reader can preview the article to see whether it contains information pertinent to his investigation. In fact, many abstracting services request that the abstract be prepared by the author for publication.

In most journals the body of the paper follows the abstract and begins with the introduction which states the problem and reviews briefly the work of others related to it.

The investigator then describes the experimental procedure including the organism(s) used as well as the media, methods, and various techniques employed. This detailed information is necessary if the reader is to evaluate the results or to repeat the experiment. In the next section of the paper, the author reports his results in some detail; data are listed in tabular form as seen in this paper. Sometimes graphs or other means of presenting the data are also used. After the results are presented, the author discusses the significance of the work in the section titled "Discussion."

Following this, there is a brief summary or conclusion of the work and acknowledgments to those who aided in the investigation. Bibliographic references should be listed at the end of the paper as "Literature Cited." A list of references enables the reader to refer to any paper or book mentioned in the article. Names of journals are usually abbreviated and standard abbreviations for journals can be found in the *Annual List of Serials* published by *BioSciences Information Service*, 2100 Arch Street, Philadelphia, Pennsylvania 19103.

Briefly, then, the paper should follow an outline consisting of the title, name and address of the author, an abstract, the introduction, experimental procedure, conclusion, acknowledgments, and literature cited.

Plant Physiol. (1969) *44*, 1027–1030

Zea Shoot Development in Response to Red Light Interruption of the Dark-Growth Period. I. Inhibition of First Internode Elongation

S. O. Duke and J. L. Wickliff

University of Arkansas, Fayetteville, Arkansas 72701

Received December 23, 1968.

Abstract. Brief, low energy (approximately 400 Kerg cm^{-2}) red light interruption of the early dark-growth period of *Zea mays* L. cv F-M Cross induced inhibition of first internode elongation which was maximal whenever light interruption occurred from 2.5 to 4 days of seedling age. Two-thirds of the maximal inhibition occurred in the tissues constituting the region 0 to 2 mm below the coleoptilar node at time of light treatment. The coleoptile tip showed the greatest sensitivity for red light reception in the internode response. Far-red light exposures following red light treatments reversed the red light effect. However, far-red light alone inhibited first internode elongation as effectively as red light of similar dosage.

Low intensity red irradiation (600–700 nm) has been routinely employed in investigations of the physiology of etiolated grass seedlings as a pretreatment for more uniform shoot development and frequently as a manipulation light source (15). A developmental effect of red light under such conditions is an inhibition of first internode elongation. Inhibition of *Avena* first internode elongation by low intensity red light exposures is nearly complete (2, 3), but the response in *Zea* and other grasses is generally only partial, even with continuous red light exposures (14). The growth responsive and photoreceptive regions of etiolated *Avena* shoots in the red-light-induced inhibition response have been investigated (2, 10, 12), but comparable studies of etiolated shoots of *Zea* have not been reported.

This report presents results of our studies of *Zea* first internode elongation in response to red light interruption of the dark-growth, including the response to red and far-red light. Photoreceptor and developmental response regions of the shoot have been identified.

Materials and Methods

Experimental Procedure. Seeds of *Zea mays* L. cv F-M Cross (Ferry-Morse Seed Company) were surface sterilized in 0.2 % sodium hypochlorite (w/w), washed in running deionized tap water, then sown embryo side up on wet paper towel pads in covered plastic containers. After 40 hr development in the dark, uniformly developed seedlings were selected and transferred to 0.67 % agar in 14 × 2 cm petri dishes under dim green illumination. Subsequent development of the seedlings, except during the brief red or far-red light exposures, took place in the dark at 25 ± 1° and 80 to 100 % relative humidity. Under these culture conditions the logarithmic phase of first internode elongation persisted from about

2 days until about 6 days of age (time after imbibition). Routinely, seedlings were harvested at 6 days of age and the excised internodes were measured to the nearest millimeter. For the controlled exposures to red or far-red light, samples of 10 to 20 seedlings each were randomly selected and irradiated outside the controlled environment of the dark germination cabinet, then returned for further dark development. Control samples were manipulated similarly, including brief green light exposures, except that they received no other light treatment.

Light Sources. Transmission characteristics for the 3 light systems employed are shown in Fig. 1. The dim green manipulation light was a 22 watt cool white fluorescent lamp covered with 2 sheets of green and 1 sheet of amber filter mediums (Cinemoid Nos. 39 and 33). (All Cinemoid filters were obtained from: Century Lighting, Incorporated, 3 Entin Road, Clifton, New Jersey 07014.) This light source, at the usual working distance of 100 cm, induced no measurable effects on internode elongation when exposures were less than 15 min. Light from a 500 watt incandescent reflector flood lamp

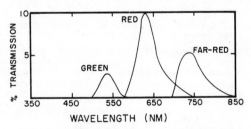

Fig. 1. Transmission spectra of the colored filter systems based on measurements with a Perkin-Elmer Model 202 Spectrophotometer (350–750 nm) and a Beckman Model DU Spectrophotometer (750–850 nm). Base line represents 0.2 % transmission.

filtered through 5 cm of water, then through appropriate combinations of liquid and solid filters was used for the red and far-red light sources. The red spectral region was isolated by 3 cm of 0.1 M $CuSO_4$ contained in a sealed plexiglass cell and a solid filter composed of 2 sheets of red and 1 of amber filter mediums (Cinemoid Nos. 6 and 33). Far-red light was obtained by using a 3 cm liquid filter consisting of 1 Kg $FeSO_4(NH_4)_2SO_4 \cdot 6H_2O$ dissolved in 3 liters of 0.5 N H_2SO_4 and a solid filter composed of 2 sheets of Cinemoid No. 6, one of Cinemoid No. 5A, and 9 sheets of blue cellophane.

Seedlings were irradiated from above, except for 1 set of experiments where irradiation was unilateral. The internode elongation response could be saturated with red light (100–1000 Kerg cm^{-2}) with exposures of 5 min or less. Intensities were measured with a Photovolt Electronic Photometer, Model 501-M, using phototube E calibrated to erg cm^{-2} sec^{-1} with a YSI Model 65 Radiometer. Calibration of the latter is ultimately traceable to a standard lamp.

Statistical Interpretation. Standard errors of the means (S.E.) and t-tests for significant differences between paired treatment and control means were calculated by methods presented in Snedecor (13).

Results

Red Light Response vs. *Age at Treatment.* Red light interruption of the dark-growth period of *Zea* seedlings induced the same degree of inhibition regardless of seedling age at time of interruption from 2.5 to 4 days (table I). This developmental period

Table I. *Elongation Response of* Zea *First Internodes to Red Light Interruption (400 Kerg cm⁻²) of the Dark-growth at Various Developmental Ages*

Age at interruption	Internode mean length
days	*mm ± S.E.*[1]
2.5	41.0 ± 2.6
3.0	41.4 ± 2.2
3.5	41.8 ± 2.1
4.0	40.6 ± 2.1
Dark control	48.7 ± 2.4

[1] All treatment means significantly different from control mean at P = 0.05.

corresponded to the early logarithmic elongation phase of the first internode.

Localization of Growth Response to Red Light. Preliminary experiments using India ink markings on the internode had indicated that most of the elongation of etiolated internodes, during seedling development from 2.5 to 6 days of age, occurred in the apical tissues of this organ. Results of a subsequent experiment are presented in table II. In this particular experiment duplicate samples of seedlings selected for uniform shoot development were irradiated with 400 Kerg cm^{-2} red light, then the internodes were marked, along with those of unexposed control seedlings, with India ink 1 mm and 2 mm below the coleoptilar node. The greatest increase in length of the control internodes during the subsequent dark-growth period occurred in the region composing the apical 0 to 2 mm of this organ at the time of marking, while 67 % of the total internode inhibition in the red light exposed seedlings occurred in this same region.

Table III. *Effects of Brief Irradiation of Different Regions of 66-Hr-old* Zea *Shoots With Red Light Sufficient to Saturate the Internode Elongation Response*

Shoot region irradiated	Internode mean length	Response as % of maximal inhibition
	mm ± S.E.	
None (dark control)	62.0 ± 3.5	
Entire shoot	40.1 ± 2.6[1]	100
Apical 0 - 3 mm	41.4 ± 2 6[1]	96
Nodal 2 - 4 mm	46.5 ± 3.8[1]	71
Basal 2 - 4 mm	55.2 ± 1.8	31

[1] Treatment means significantly different from control mean at P = 0.05.

Photoreceptor Region for the Red Light Response. Dark-grown seedlings with uniform shoot development were irradiated unilaterally with 400 Kerg cm^{-2} red light. Illumination of localized portions of the shoots was achieved by use of a slotted opaque screen placed in the unilateral light beam. Results of a representative experiment are presented in table III. Typically, the amount of internode response decreased as red light exposure was moved down the shoot; the

Table II. *Elongation of Different Regions of* Zea *First Internodes in Response to Brief Red-light Interruption of the Dark-growth Period at 66 Hrs of Age*

Internode region at 66 hr age	Control internode		Red treated internode	
	Length	% of Total elongation	Length	% of Total inhibition
	mm ± S.E.		*mm ± S.E.*	
Apical 0 - 1 mm	34.2 ± 2.9	66	25.1 ± 2.4[1]	52
Apical 1 - 2 mm	7.9 ± 1.0	14	5 2 ± 0.2[1]	15
Basal 2 - 10 mm	17.9 ± 1.2	20	12.1 ± 0.9[1]	33

[1] Mean length significantly different from that of corresponding control section at P = 0.05.

Table IV. *Effects of Far-red Light Exposures on Dark-grown and Red Light Exposed* Zea *Seedlings*

Light treatment	Internode mean length
	$mm \pm S.E.$
Dark control	57.7 ± 3.3
Red	45.3 ± 3.8[1]
Far-red	45.0 ± 2.5[1]
Red/Far-red	52.6 ± 3.4

[1] Treatment means significantly different from control mean at $P = 0.05$.

coleoptile tip was the most sensitive site of photoreception, while the tissues in which most of the growth response occurred were only partially responsive to the red irradiation.

Effects of Far-red Light Exposures. Involvement of a phytochrome system in the photoresponse was investigated by irradiation of seedling samples with red (400 Kerg cm^{-2}) and far-red (500 Kerg cm^{-2}) light. Each exposure was of 5 min duration and, in the case of red followed by far-red, the interval between exposures was always less than 5 min. Results of a typical experiment (table IV) show that the red response could be reversed by a subsequent exposure to far-red of similar dosage. However, the same far-red dosage which reversed the red effect, when applied alone, induced inhibition to the same extent as the red light exposure.

Discussion

Results of investigations reported here show that the red-light-induced inhibition of *Zea* first internode elongation is localized in the apical region of this organ (table II). This region responds maximally at an early age and maintains a constant responsiveness well into the logarithmic phase of internode elongation (table I). Since the *Zea* first internode has a persistent meristem located immediately below the coleoptilar node (1), our data suggest that cell enlargement and/or cell division were affected by red irradiation. However, the anatomical studies confirming this interpretation have not been done as reported for *Avena* (2, 10).

The photoreceptor region for red-light-induced *Zea* internode inhibition need not correspond to the tissue region in which the response occurs (table III). A similar phenomenon has been observed in *Avena*, but no relative comparisons of effectiveness between photoreceptor regions were made (10). Our data indicate that the coleoptile tip is the most effective receptor for the red-light-induced response of the internode. The limited internode inhibition in response to a red light interruption might be accounted for by a red-light-induced interruption of auxin flow, if there were direct dependence on auxin for internode growth as suggested for *Avena*

(10, 11). Such interpretation is further suggested by evidence that red light suppresses the diffusion of auxin from excised *Zea* coleoptile tips, a response which decays during the dark period following red irradiation (4).

Reversal of the red light effect by subsequent far-red irradiation indicates mediation of the internode elongation response by a phytochrome system (table IV). The involvement of a phytochrome photoreceptor in developmental responses can be inferred from reports of phytochrome localization in the *Zea* shoot (5, 6, 7). Briggs and Chon (5, tables I and II), in their study of the red light effect on *Zea* phototropic responses, have shown that the distribution of red light sensitivity in the *Zea* shoot correlates with relative phytochrome distribution. Our results (table III, column 3) show a similar distribution of red light response. However, far-red exposures of the same dosage that reversed the red light effects, when applied alone, induced an inhibition equal to the red light inhibition. Evidence that prolonged, low intensity far-red irradiation can cause the same effects as red light in *Zea* shoot development by maintaining a low, active, photostationary level of far-red absorbing phytochrome has been reported (8, 9). That such a photostationary level was established during our far-red irradiation experiments seems unlikely since the effects of the far-red light are complete, but different depending on the presence or absence of prior red light exposure. Neither can our data be explained by the presence of a low energy, far-red-irreversible red light response which has been described for *Avena* (3, 10). The complete reversal of the red effect by subsequent far-red exposures in our experiments preclude such an interpretation.

In conclusion, our results support the interpretation that the red-light-induced internode inhibition of dark-grown *Zea* shoots is mediated by a phytochrome system in the coleoptile tip. The physically separated sites of photoreception and growth response in the shoot suggests this organism is potentially useful for studies of phytochrome-mediated control of early growth of the grass shoot.

Literature Cited

1. AVERY, G. S., JR. 1930. Comparative anatomy and morphology of embryos and seedlings of maize, oats, and wheat. Botan. Gaz. 89: 1–39.
2. AVERY, G. S., JR., P. R. BURKHOLDER, AND H. B. CREIGHTON. 1937. Polarized growth and cell studies in the first internode and coleoptile of *Avena* in relation to light and darkness. Botan. Gaz. 99: 125–43.
3. BLAAUW, O. H., G. BLAAUW-JANSEN, AND W. J. VAN LEEUWEN. 1968. An irreversible red-light-induced growth response in *Avena*. Planta 82: 87–104.
4. BRIGGS, W. R. 1963. Red light, auxin relationships, and the phototropic responses of corn and oat coleoptiles. Am. J. Botany 50: 196–207.

1030 PLANT PHYSIOLOGY

5. BRIGGS, W. R. AND H. P. CHON. 1966. The physiological *versus* the spectrophotometric status of phytochrome in corn coleoptiles. Plant Physiol. 41: 1159–66.

6. BRIGGS, W. R. AND H. W. SIEGELMAN. 1965. Distribution of phytochrome in etiolated seedlings. Plant Physiol. 40: 934–41.

7. BUTLER, W. L. AND H. C. LANE. 1965. Dark transformations of phytochrome *in vivo*. II. Plant Physiol. 40: 13–17.

8. BUTLER, W. L., H. C. LANE, AND H. W. SIEGELMAN. 1963. Nonphotochemical transformations of phytochrome *in vivo*. Plant Physiol. 38: 514–19.

9. CHON, H. P. AND W. R. BRIGGS. 1966. Effect of red light on the phototropic sensitivity of corn coleoptiles.. Plant Physiol. 41: 1715–54.

10. GOODWIN, R. H. 1941. On the inhibition of the first internode of *Avena* by light. Am. J. Botany 28: 325–32.

11. JOHNSTON, E. S. 1937. Growth of *Avena* coleoptile and first internode in different wave-length bands of the visible spectrum. Smithsonian Miscell. Collections 96, No. 6, p 1–19.

12. SCHNEIDER, C. L. 1941. The effect of red light on growth of the *Avena* seedling with special reference to the first internode. Am. J. Botany 28: 878–86.

13. SNEDECOR, G. W. 1956. Statistical Methods, 5th ed. Iowa State University Press, Ames.

14. WEINTRAUB, R. L. AND L. PRICE. 1947. Developmental physiology of the grass seedling. II. Inhibition of mesocotyl elongation in various grasses by red and by violet light. Smithsonian Miscell. Collections 106, No. 21, p 1–15.

15. WENT, F. W. AND K. V. THIMANN. 1937. Phytohormones. The Macmillan Company, New York.

Review Articles. Not all papers are reports of original experimental work. Since the amount of scientific knowledge grows so rapidly, it is difficult for an individual scientist to keep up with all that is done in his own field, to say nothing of related fields. The problem is partially solved by special "review" articles, which attempt to summarize in brief form the state of knowledge in limited fields. Some review articles list and classify significant papers in a specific field which have appeared during the past year or other recent periods of time.

Some journals are devoted solely to review papers, many of which appear quarterly (every three months), while books of review articles may appear annually or even less frequently. Some examples of review journals are *Bacteriological Reviews, Physiological Reviews,* and the *Quarterly Review of Biology* (which also contains reviews of recent books in the biological sciences). Examples of review volumes are *Annual Review of Physiology, Advances in Genetics,* and *Advances in Enzymology.*

Searching the Literature

How does a scientist conduct a literature search to discover what has been done previously on a particular problem? This necessary task is becoming more difficult every year. It has been estimated that there are in the world more than 15,000 publications that carry articles relating to some phase of biology. How is it possible for a scientist to keep himself informed of all these articles?

The major job of organizing information about articles that appear in the field of the biological sciences is done by information services that abstract and index these articles. The information service employs a system by which they assign bio-scientists to read articles appearing in certain journals. The scientists, in turn, write short summaries, or abstracts, of the important points made in the article. The abstracts are sorted according to subject area and are given a number. The numbered abstracts are then published in an issue of the abstracting journal.

BioSciences Information Service of Biological Abstracts (BIOSIS). BioSciences Information Service of Biological Abstracts, which is located in Philadelphia, Pennsylvania, is the world's largest information service for the life sciences. During 1970 more than 230,000 research papers, representing 7,600 primary publications originating in 97 countries, will be covered in BIOSIS' publications. Abstracts of 140,000 of these will be published in *Biological Abstracts.* Citations to the remaining 90,000 will be published in *BioResearch Index.*

To illustrate how a particular article is handled by BIOSIS, the publishers of *Biological Abstracts*, let us return to the article, "*Zea* shoot development in response to red light interruption of the dark-growth period. 1. Inhibition of first internode elongation," by Duke and Wickliff. The abstract was prepared by the authors. After the abstract was edited and indexed by BIOSIS, it was assigned No. 16504. It was published on page 1590 of Volume 51 of *Biological Abstracts.* The abstract contains a description of the research conducted, the organism used, the type of study, the instrumentation, and the results—together with such pertinent information as the identity of the authors and their location. The abstract of this paper is shown on page 99.

Once abstracts are numbered and published, the indexes in *Biological Abstracts* help the scientists to find recent articles on a particular topic. A most important aid in this respect is the subject index B.A.S.I.C. (Biological Abstracts Subjects In Context). This index is compiled from key words derived from two sources: the author's title and supplemental terms. The supplemental terms are words taken from the body of the paper by a member of the editorial staff at BIOSIS and are added to the title to better describe the contents of the paper. The key words describing each paper that has been abstracted (the title along with the supplemental terms) are then placed on cards which can be read by a computer-printer. The computer identifies all the terms or key words that have been listed and arranges them in alphabetical order with modifying terms adjacent to each.

In looking at the actual pages of a subject index, we can learn how the system works. Two pages from the subject index are reproduced on pages 100 and 101. On page 100, which lists the titles of papers that deal with internode, we find listed the paper of Duke, S.O., and Wickliff, V. L., and its abstract number (16504) at the end of the line. You will note that the few words both before and after "internode" give the reader a good idea of what the article is about. If a scientist wanted to find more about this paper of Duke and Wickliff, he would look up the abstract. There he would find, as we have seen, the complete title, the author(s) and the address(es), and the journal issue in which the paper originally appeared, in addition to the abstract.

On page 101 we see another example of the subject index. In this case the paper is listed under the word "light." A given paper will be listed under each key word from its title and the supplemental terms added by the editors at BIOSIS.

BIOSIS provides not only the subject index just described, known as B.A.S.I.C., but also Author, Biosystematic (major taxonomic categories), and CROSS (Computer Rearrangement Of Subject Specialties) Indexes.

The *Author Index* is found in the yellow pages of *Biological Abstracts*. It is particularly useful for locating an abstract if the author is known, if you are following a particular author's work, or when compiling a bibliography. An example of the author index is found on page 102.

The *Biosystematic Index* is also found in the yellow pages of *Biological Abstracts*. Here abstracts are listed according to phylum, class and order (and in some cases families) of the organism(s) studied in the paper. When interested in searching for a given genus and species, the student must refer to B.A.S.I.C. (Example on page 103).

Page from BIOLOGICAL ABSTRACTS with the abstract of the article by Duke and Wickliff.

16504-16515   PLANT PHYSIOLOGY, BIOCHEMISTRY AND BIOPHYSICS   [Vol. 51] 1590

formation is controlled by a gene regulator of the corresponding operon; the substances R_1 and R_2, formed in the protoplasm as the output of a system of inter-related cycles, can play the role of inductors or repressors of the regulator: here R_1 is the inductor of short-day ⁚ies, while R_2 is the inductor of ⸱ay species; the active form of ᵁnd R_2, reverses the sign of their ⸱ellular periodic system (biological ⸱he endogenous rhythm and on the ⸱n of the substances X and Y, which ⸱cles and thus the formation of R_1 ⸱e photoperiodic reaction of the ⸱sing the proposed model.

† 16504. DUKE, S. O., and J. L. WICKLIFF. (Univ. Arkansas, Fayetteville, Arkansas, USA.) Zea shoot development in response to red light interruption of the dark-growth period: 1. Inhibition of first internode elongation. PLANT PHYSIOL 44(7): 1027-1030. Illus. 1969.--Brief, low energy (approximately 400 Kerg cm^{-2}) red light interruption of the early dark-growth period of Zea mays L. cv. 'F-M Cross' induced inhibition of 1st internode elongation which was maximal whenever light interruption occurred from 2.5 to 4 days of seedling age. Two-thirds of the maximal inhibition occurred in the tissues constituting the region 0 to 2 mm below the coleoptilar node at time of light treatment. The coleoptile tip showed the greatest sensitivity for red light reception in the internode response. Far-red light exposures following red light treatments reversed the red light effect. However, far-red light alone inhibited 1st internode elongation as effectively as red light of similar dosage.

† 16505. HALAI N. J., USA.) Eff movement of a sl Illus. 1969.--St and far-red (FR) Coleus blumei X illumination with the period of the continuous darkn was significantly FR, the period length was not significantly different from the dark control. It was observed that under continuous FR illumination, the leaves tended to oscillate in a more downward position. Eight-hr red light signals were effective in advancing the phase of the rhythm as compared to a control under continuous green light. Blue light signals were effective in delaying the phase of the rhythm. FR light signals were ineffective in producing either delay or advance phase shifts. Far-red light did not reverse the effects of either red or blue light signals. It is suggested that pigments which absorb blue or red light, rather than phytochrome, mediate the effect of light on the circadian rhythm of leaf movement.

† 16506. MacINNES, CLAIRE B., and LUKE S. ALBERT. (Dep. Bot., Univ. R. I., Kingston, R. I., USA.) Effect of light intensity and plant size on rate of development of early boron deficiency symptoms in tomato root tips. PLANT PHYSIOL 44(7): 965-967. 1969.-- Young tomato plants (Lycopersicon esculentum Miller, cv. 'Rutgers') grown in solution culture at 27° at 2 light intensities with adequate B (0.1 mg/l) and treated with these 2 intensities in the absence of adequate B developed root B deficiency symptoms. The typical deficiency symptoms of decreased root elongation, increased depth of brown color and decreased RNA content of tips developed more rapidly at high than at low light intensity, and plant size influenced results. Plants supplied with adequate B did not exhibit deficiency symptoms.

† 16507. BOISARD, JEAN. (Lab. Physiol. Veg., Mont-Saint-Aignan, Fr.) Role du phytochrome dans la photosensibilite des akenes de Laitue variete "Reine de Mai". [The role of phytochrome in the photosensitivity of the suds of the lettuce cultivar 'Reine de Mai'.] PHYSIOL VEG 7(2): 119-133. Illus. 1969. [Engl. sum.]--The presence of P_{730} molecules immediately after sowing shown by in vivo spectrophotometry explains the dark germination of lettuce seeds cv. 'Reine de Mai'. The photoinhibition of germination requires exposures to far-red light because of the conversion of inactive P_{660} to active P_{730} which occurs in darkness. De novo synthesis of phytochrome takes place at the end of the 'germinating phase'. The neo-phytochrome does not seem to have the same properties as preexisting phytochrome.

16508. MILAEV, Ya. I. Opticheskie svoistva list'ev u gibridov kukuruzy i ikh roditel'skikh form. [The optical properties of the leaves of hybrid corn and their parental varieties.] SELEK SEMENOVOD RESPUB MEZHVED TEMAT NAUCH SB 7. 88-95. 1967. From: REF ZH OTD VYP RASTENIEVOD, 1968, No. 10.55.269. (Translation)--Studies were made of the optical properties of the leaves (transmission coefficient, reflection, and absorption) and chlorophyll and water content of the hybrid corn 'Bukovinskii 3' and 'Iskra' and their parental forms. The optical properties of the leaves changed during the day and from one level to the next. In the transition from the lower levels to the higher, the transmission coefficient declined, while the coefficient of reflection, on the other hand, rose. The maximum absorption shifts from the leaves of the lower levels to the higher ones as the plant grows. According to all these parameters the hybrids do not differ materially from their parents.

16509. MOSHKOV, B. S., S. L. PUMPYANSKAYA, and L. Ya. FUKSHANSKII. Obshchaya model'naya skhema fotoperiodicheskoi reaktsii rastenii. [A general diagrammatic model of the photoperiodic reaction of plants.] SB TR AGRON FIZ 15. 5-42. 1968. Translated from REF ZH BIOL, 1969, No. 6G137.--A description is given of a diagrammatic model of the photoperiodic reaction of flowering in which the following assumptions are made: The impetus to pass through the photoperiodic reaction is the appearance of enzymes whose

action.] SB TR AGRON FIZ 15. 83-91. 1968. Translated from REF ZH BIOL, 1969, No. 6G178.--The motion of Perilla leaves during 24-hr photoperiods was recorded by slow-motion cinematography. When the night was 3-5 hr long, the leaves drooped 2 and sometimes 3 times. With any duration of day or night except a 4-hr day and a 2-hr night, the maximum droop of the leaves occurred on the average 3 hr before the middle of the night. In this case the middle of the period of rise of the leaves either coincided precisely with the middle of the night, or else it occurred 1.1 hr either before or after it. When there was alternation of long (15-hr) and short (9-hr) nights, the phases of movement of the leaves, also alternating, moved toward times that were about 3 hr earlier or later (it differed in 2 experiments), depending on whether the night always began at the same time or whether it was different. It is concluded that the location of the phases of motion of the leaves with respect to day and night is determined by the middle of the periods between turning the light on and turning it off.

16512. MOSHKOV, B. S., G. A. ODUMANOVA-DUNAEVA, and N. V. KHOVANSKAYA. (Agr. Phys. Res. Inst., Leningrad, USSR.) Rol' uglekisloty v protsesse viviparii u Kalanchoe daigremontiana (Hamm. et P. de la Bath.) Jacobson na raznykh svetovykh rezhimakh. [Role of carbon dioxide in the process of vivipariy in Kalanchoe daigremontiana (Hamm. et P. de la Bath.) Jacobson under various light regimes.] BOT ZH 54(8): 1186-1196. Illus. 1969. [Engl. sum.]--An important role of both the photosynthetic and the dark CO_2 fixation in the photoperiodic response of the vegetative reproduction of K. daigremontiana by epiphyllous buds was shown. In an atmosphere devoid of CO_2 the formation of leaf-buds is increasingly inhibited with the increasing exposure, both in the light and dark. The role of light in the process of vivipariy was established.--N. F. G.

16513. CARRICABURU, PIERRE, and PAULETTE CHARDENOT. (Lab. Biophys. et Lab. Biochim., Alger, Algeria.) Action de la lumiere polarisee sur le beta-carotene soumis a un champ magnetique. [The action of polarized light on beta-carotene subjected to a magnetic field.] C R HEBD SEANCES ACAD SCI SER D SCI NATUR PARIS 266(9): 937-939. Illus. 1968[recd. 1969].--Polarized light acting on a solution of beta-carotene, possessed different photochemical properties according to the angle formed by the polarization. An attempt was made to orient the molecules by means of a magnetic field.--A. C. S.

16514. DUBROV, A. P. Geneticheskie i fiziologicheskie effekty deistviya ul'trafioletovoi radiatsii na vysshie rasteniya. [The genetic and physiological effects of the action of ultraviolet radiation on higher plants.] 250p. Illus. Nauka: Moscow. 1968. Pr. 1 ruble 20 kopecks. Translated from REF ZH BIOL, 1969, No. 2G19 K.-- Consideration is given to the primary action of UV radiation, its effect on genetics, metabolism, growth and development, photosynthesis, plant respiration and transpiration, and questions of photoreactivation and use of UV radiation in agriculture.--S. T.

16515. SHYMACH, L. M. Do pytannya pro vyznachennya oriyentatsiyi lystya roslyn u prostori. [Determination of the orientation of plant leaves in space.] INTROD AKLIM ROSL UKR RESPUB MIZHVIDOM ZB 3. 271-275. 1968. Translated from REF ZH BIOL, 1969, No. 2G39.--A simple apparatus is proposed for measuring the angle between the source of light (the sun) and the leaf plane. It consists of an upper sighting disk 2 cm in diameter with a 0.5 mm hole in the center and parallel to it a lower disk 8 cm in diameter with a protractor scale applied to it. The disks are located at a distance of 1 cm and fastened with 3 fasteners. Examples are given

iziologiya rastenii. [Plant photo- meteoizdat: Leningrad, USSR. TR AGRON FIZ 15. 1-212. 1968. ᵁ969, No. 6G1 K.--This collection ᵁatory of photophysiology of ᵁd from 1962 to 1965. The principal ᵁg agricultural plants under various ᵁudying physiological processes.

ᵁizheniya list'ev perilly maslichnoi ᵁyakh s sutochnym periodom. ᵁ⸱ymoides under daily photoperiodic

100 PROCESSES OF BIOLOGICAL INVESTIGATION

Page from the Subject Index of BIOLOGICAL ABSTRACTS.

S110 INSUFFICIENCY INTESTINAL [Vol. 51(3)]

OTIC FLUID IN PLACENTAL	INSUFFICIENCY HUMAN CENTRIFUGATION A	15167
ASSOCIATED WITH AORTIC	INSUFFICIENCY HUMAN ROENTGENOGRAPHY	15230
SMIN DURING ACUTE RENAL	INSUFFICIENCY HUMAN/ ERYTHROPOIETIN	13330
OGRAPHY IN RENAL	INSUFFICIENCY HUMAN/ INTRA VENOUS UR	13296
OF CHRONIC RESPIRATORY	INSUFFICIENCY IN ASTHMA AND PULMONAR	14415
ON IN ACUTE RESPIRATORY	INSUFFICIENCY IN THE CASE OF CHRONIC	13418
URRENT CEREBRO VASCULAR	INSUFFICIENCY OF CARDIO VASCULAR ORI	13096
ETES/ EFFECT OF INSULIN	INSUFFICIENCY ON THE CARDIAC CONTRAC	13598
G/ TREATMENT OF CARDIAC	INSUFFICIENCY PATIENTS STILL REQUIRI	14335
Y AND PULMONARY CARDIAC	INSUFFICIENCY TREATED WITH EUROPEAN	14331
TS HAVING CHRONIC RENAL	INSUFFICIENCY WITH A LOW PROTEIN DIE	12706
AND URINE DURING KIDNEY	INSUFFICIENCY WITH EXTRARENAL KIDNEY	13504
H-D UNDER CONDITIONS OF	INSUFFICIENT LIGHT INTENSITY GREENHO	16802
CHILDREN UNDER GENERAL	INSUFFLATIVE ANESTHESIA HUMAN/ TONSI	14478
GRADE IN THE ABSENCE OF	INSULIN /STUDIES ON INSULIN ACTION P	13571
OF INSULIN/ STUDIES ON	INSULIN ACTION PART 3 LINEAR DISTRIB	13571
RTIES OF CYANOETHYLATED	INSULIN AN INSULIN DERIVATIVE WITH B	13572
ORMS/ TAENIA-CRASSICEPS	INSULIN AND CARBOHYDRATE METABOLISM	17333
ENIA BEING TREATED WITH	INSULIN AND CHLORPROMAZINE HUMAN CEN	14568
UNOASSAY/ SERUM GLUCOSE	INSULIN AND GROWTH HORMONE IN CHRONI	13502
IS BLOOD GLUCOSE PLASMA	INSULIN AND GROWTH HORMONE LEVEL DUR	15263
LEARANCE OF CRYSTALLINE	INSULIN AND OF VARIOUS RADIO IODINAT	13579
ANOETHYLATED INSULIN AN	INSULIN DERIVATIVE WITH BLOCKED AMIN	13572
DE NERVATED KIDNEYS DOG	INSULIN DIAGNOS-DRUG/ FUNCTION OF TR	13269
DIABETES MELLITUS HUMAN	INSULIN EPINEPHRINE GLUCAGON GROWTH	13596
TRANSFERASE IN THE RAT	INSULIN GLUCAGON HORMONE-DRUGS ALLOX	13588
ORE AND AFTER OPERATION	INSULIN HORMONE-DRUG/ DYNAMICS OF SO	13683
TRON MICROSCOPE ALLOXAN	INSULIN HORMONE-DRUG/ FATTY-ACID OXI	13589
ENSITIVITY TO ESTRADIOL	INSULIN HORMONE-DRUGS ALLOXAN/ IN-VI	13564
DIABETES MELLITUS HUMAN	INSULIN HORMONE-DRUGS ORAL HYPO GLYC	14406
OLESTEROL IN THE RABBIT	INSULIN HORMONE-DRUGS/ INHIBITOR EFF	14405
ART 2 EFFECT OF OBESITY	INSULIN HYDRO CORTISONE DEXAMETHASON	12629
ANCE OF IMMUNOASSAYABLE	INSULIN IN URINE HUMAN/ THE SIGNIFIC	13574
C DISORDERS AFTER RAPID	INSULIN INACTIVATION RAT GUINEA-PIG/	13591
ND PLASMA IN NORMAL AND	INSULIN INJECTED RATS HORMONE-DRUG/	13577
XAN DIABETES/ EFFECT OF	INSULIN INSUFFICIENCY ON THE CARDIAC	13598
THESIS ALLOXAN DIABETES	INSULIN METAB-DRUG/ THE EFFECTS OF N	12469
NE GLUCAGON EPINEPHRINE	INSULIN METAB-DRUGS ALLOXAN DIABETES	12569
METAB-DRUGS/ ACTION OF	INSULIN ON SODIUM EFFLUX FROM THE TO	13597
GLUCOSE INTOLERANCE AND	INSULIN RESISTANCE IN PATIENTS WITH	13582
/ GLUCOSE TOLERANCE AND	INSULIN RESPONSE DURING DAILY CONTIN	14397
TOLERANCE AND	INSULIN SENSITIVITY PIG DOG/ GLUCOSE	13566
Y AND DIABETES MELLITUS	INSULIN TESTOSTERONE METAB-DRUGS/ A	15264
OF GLUCAGON IN THE RAT	INSULIN THYROCALCITONIN HORMONE-DRUG	13567
TH HUMAN GROWTH HORMONE	INSULIN TOLERANCE TEST METAB-DRUG/ T	13632
EALAND OBESE MICE HYPER	INSULINISM LIPOGENESIS/ ROLE OF LIVE	12637
VARIOUS RADIO IODINATED	INSULINS DOG/ COMPARISON BETWEEN REN	13579
AL ACTIVITY OF MODIFIED	INSULINS MOUSE/ BIOLOGICAL AND IMMUN	13565
RETION OF BILE SALTS BY	INTACT AND ISOLATED RAT LIVERS GAS C	12835
E SYNCHRONIZED SLEEP IN	INTACT CATS AND IN CATS WITH SINO AO	13002
INGLE PULMONARY VEIN IN	INTACT DOGS CARDIAC CATHETERIZATION/	13057
ANTED NEPHRECTOMIZED OR	INTACT DOGS RAT RABBIT TAMARIN MAN/	13203
/ MEMBRANE EXPANSION OF	INTACT ERYTHROCYTES BY ANESTHETICS H	14441
G/ AN ATTEMPT TO INFECT	INTACT RABBIT KIDNEYS INTRA VENOUS A	15703
RAL	INTAKE AND BO	
EET-D PULP ON VOLUNTARY	INTAKE BY DAI	
Y WEIGHT AND DAILY FOOD	INTAKE IN CAPTIVE SHREWS SOREX-ARANE	17494
THE REGULATION OF FOOD	INTAKE IN MAN METRECAL/ PRE LOADING	12679
RE AS A CONTROL OF FOOD	INTAKE IN PUPPIES HUMAN PRESSURE REC	12620
OF TYBAMATE ON ETHANOL	INTAKE IN RATS DURING PSYCHOLOGICAL	14570
-M HAY/ GROWTH AND FOOD	INTAKE OF FRIESLAND CALVES WEANED AT	15293
HE RELATIONSHIP OF FOOD	INTAKE TO THE STOMACH CONTENTS IN TH	12833
N INSECTICIDE PARASITES	INTEGRATED CONTROL/ CURRENT STATUS O	17027
E/ A SIMPLE	INTEGRATING THERMOMETER FOR FIELD US	11772
IN THE MECHANISM OF DNA	INTEGRATION BY LYMPHOMA CELLS/ THE I	12018
RAPY AND PSYCHO THERAPY	INTEGRATION OR DISINTEGRATION HUMAN/	14197
RIPS SERUM PROTEIN DISK	INTEGRATOR COMPUTER/ AUTOMATIC CALCU	12341
THE PUPARIUM AND ADULT	INTEGUMENT OF THE FLESH FLY SARCOPHA	17411
-LINEATUM PHOSPHATE-32/	INTEGUMENTAL ABSORPTION VERSUS ORAL	17412
ENT OF SANCTIONS HUMAN/	INTELLECTUAL COMPETANCE AS A VARIABL	14147
OMAZINE ON BEHAVIOR AND	INTELLECTUAL FUNCTIONING HUMAN CENT-	14572
ND	INTELLIGENCE CHILDREN/ BIRTH ORDER A	14126
ONMENTAL INTERACTION IN	INTELLIGENCE MODELS OF BEHAVIOR COMP	12081
HE RELATION OF MEASURED	INTELLIGENCE TO BIRTH ORDER AND MATE	14133
N/ PHENYL KETONURIA AND	INTELLIGENCE TRIMODAL RESPONSE TO DI	14244
OL/ PERITONEAL DIALYSIS	INTENSIFIED BY ULTRA FILTRATION AS I	13229
HNIQUE/ EFFECT OF LIGHT	INTENSITY AND CARBON DI OXIDE ON APP	16398
NTUM-D/ EFFECT OF LIGHT	INTENSITY AND PLANT SIZE ON RATE OF	16506
E TO CANOPY COVER LIGHT	INTENSITY AND THROUGHFALL PRECIPITAT	12155
LOSS HUMAN/ RESPONSE TO	INTENSITY CHANGE AT THRESHOLD IN SEN	13883
NG INTER AURAL TIME AND	INTENSITY CHLORALOSE NEMBUTAL CENT-D	13960
S OF INSUFFICIENT LIGHT	INTENSITY GREENHOUSE YIELD/ EFFECT O	16802
A CORN-M CANOPY AND THE	INTENSITY OF DIRECT RADIATION ON THE	12180
MERIZATION AND STAINING	INTENSITY OF POLY SACCHARIDES BACTER	11767
E HELICAL CONFORMATION/	INTENSITY OF THE CHARACTERISTIC VIBR	12394
IZING ORGANISMS AND THE	INTENSITY OF THE STIMULATING LIGHT G	16415
ESIS/ DEPENDENCE OF THE	INTENSITY OF ULTRA WEAK RADIATION OF	16416
S TUMOR/ CHANGES IN THE	INTENSITY OF UV FLUORESCENCE OF IRRA	15005
EA GREECE DIATOMS LIGHT	INTENSITY PHOSPHATE DEPLETION/ ANNUA	12238
HALENEACETIC-ACID LIGHT	INTENSITY TEMPERATURE/ METHODS OF IN	16455
FT HUMAN/ COCKPIT NOISE	INTENSITY 15 SINGLE ENGINE LIGHT AIR	16061
ON THE IMPLANTATION OF	INTENSIVELY FED BEEF CATTLE WITH HEX	15286
ERSICUS/ STUDIES ON THE	INTER ACTION OF BLOOD OF HENS RECOVE	15754
LAR FAILURE BY MEANS OF	INTER ATRIAL AND OTHER SHUNTS RIGHT	13051
TION EFFECTS OF VARYING	INTER AURAL TIME AND INTENSITY CHLOR	13960
G IMPORTANCE OF ROLE OF	INTER BACTERIAL INHIBITION IN MAINTA	15400
LACRIMAL GLANDS/ ON THE	INTER CELLULAR INTERSTITIUM OF THE S	14904
CE TO THE NATURE OF THE	INTER CELLULAR MATERIAL/ THE CYTO TR	15126
TES OF SECRETION AND OF	INTER CONVERSION OF STEROID HORMONES	13496
HUMAN/ EFFECTIVENESS OF	INTER EPIDEMIC VACCINATION AGAINST I	15897
GRADES DETERMINANTS OF	INTER PERSONAL ATTITUDES/ PERSONALIT	14129
AND ITS IMPLICATIONS IN	INTER PLANT NUTRIENT CYCLING PINUS-T	12150
STAGES OF THE DISEASE/	INTER RELATIONSHIP BETWEEN THE ALDO	12873
GENERA AFFECTING FEED/	INTER RELATIONSHIPS BETWEEN FUSARIUM	16911
965 BOOK SHELTER BELTS/	INTER REPUBLIC ZONAL CONFERENCE ON P	16841
EY DOG OX RABBIT HUMAN/	INTER SEGMENTAL ANASTOMOSES BETWEEN	13902
ITIONING EFFECTS OF AGE	INTER STIMULUS INTERVAL AND INTER TR	12131
R STIMULUS INTERVAL AND	INTER TRIAL INTERVAL YOUNG/ RABBIT E	12131
AL CREATION OF MUSCULAR	INTER VENTRICULAR SEPTAL DEFECT DOG/	15283
MAN/ PERFORATION OF THE	INTER VENTRICULAR SEPTUM BY TRANS VE	13045
LECITHIN ELECTROSTATIC	INTERACTION /NMR SPECTROSCOPIC STUDI	14437
IS GTP CODON ANTI CODON	INTERACTION /PARTIAL CHARACTERIZATIO	12444
E CLINICIAN IN CLINICAL	INTERACTION A TAXONOMY AND A REVIEW	14174
L 7 BOOK PATHOGENS HOST	INTERACTION ANATOMY MORPHOLOGY HISTO	11710
SUGAR BEET-D BULGARIA/	INTERACTION BETWEEN CLIMATE AND FERT	16757
ESS-DRUG/ NATURE OF THE	INTERACTION BETWEEN CORTICAL PRIMARY	13981
PARASITE/ ANTAGONISTIC	INTERACTION BETWEEN STRIGEID AND SCH	17330
GTP CHROMATOGRAPHY/ AN	INTERACTION BETWEEN THE TRANSFER FAC	15494
IDOL AND ELECTRIC SHOCK	INTERACTION EFFECTS ON TREMORS AND O	14514

E BY HERD ENVIRONMENTAL	INTERACTION FOR MILK PRODUCTION GRAI	15320
SM OF GIBBERELLIN AUXIN	INTERACTION III THE EFFECT OF GIBBER	16474
OF THE CONFORMATION AND	INTERACTION IN DI NUCLEOSIDE MONO PH	12379
/ GENETIC ENVIRONMENTAL	INTERACTION IN INTELLIGENCE MODELS O	12081
A SYNTHESIS/ HOST VIRUS	INTERACTION IN RNA BACTERIO PHAGE IN	15551
AND HERRING SPERM DNA/	INTERACTION OF BASIC OLIGO L AMINO-A	12369
TEMS/ MODALITIES OF THE	INTERACTION OF BERYLLIUM WITH DNA AN	14661
OUTH WALES BRITAIN/ THE	INTERACTION OF BIOTIC AND CLIMATIC F	16738
T-2 PHAGE CALF THYMUS/	INTERACTION OF LYSINE RICH HISTONES	12389
4 6 TRI CHLORO PHENATE/	INTERACTION OF SPORE CONCENTRATION A	16948
RY COMMUNICATION HUMAN/	INTERACTION OF THYROXINE WITH BLOOD	13661
LLELOPATHY/ BIOCHEMICAL	INTERACTION OF WOODY PLANTS BOOK PIN	12166
OCHRONE C PHOSPHO LIPID	INTERACTION STRUCTURAL TRANSITIONS A	12392
HROME P-450/ FATTY-ACID	INTERACTION WITH THE HYDROXYLATING E	12543
PHOSPHATE AND THE FINAL	INTERACTION WITH THYROCALCITONIN/ ST	13669
RPRETATION CAT NEURONAL	INTERACTIONS /SIMULTANEOUSLY RECORDE	13947
POLY NUCLEOTIDES PART 2	INTERACTIONS BETWEEN NITROGENOUS BAS	12355
TURE OF INTRA MOLECULAR	INTERACTIONS DETERMINING THE CONFORM	12355
RONAL	INTERACTIONS IN FROG CEREBELLUM/ NEU	13936
THYROIDAL RELATIONSHIP	INTERACTIONS OF ESTROGENS AND THYROI	13660
BRAIN SLICES/ METABOLIC	INTERACTIONS OF GLUCOSE LACTATE AND	13946
PTOR	INTERACTIONS OF NITROGEN/ DONOR ACCE	12426
NFLUENCE OF SHORT RANGE	INTERACTIONS ON PROTEIN CONFORMATION	12404
SPECTROSCOPY/ COENZYME	INTERACTIONS PART 3 CHARACTERISTICS	12432
M UV SPECTRUM/ COENZYME	INTERACTIONS PART 4 MOLECULAR COMPLE	12431
ROBACTER STREPTOCOCCUS/	INTERACTIVE PHENOMENA AMONG BACTERIA	15421
EPORTS TO THE ALL-UNION	INTERCOLLEGE CONFERENCE ON PLANT MOR	16365
NE IN HELIX RANDOM COIL	INTERCONVERTING MEDIA/ SOLUTION PHAS	12386
ON AND TRANSPORT HUMAN/	INTERDISCIPLINARY INVESTIGATION OF M	12490
ALS PSYCHO THERAPY/ THE	INTERDISCIPLINARY TEAM AN EDUCATIONA	14188
ES ON THE HOST PARASITE	INTERFACE OF STRIGEOID TREMATODES VI	17339
CHLORIDE AT LIPID WATER	INTERFACES LOCAL ANESTHETIC-DRUGS PH	14437
OIDS/ THE	INTERFACIAL ACTIVITY OF CERTAIN STER	14403
MUCO POLY SACCHARIDES/	INTERFERENCE BY PROTEINS IN ALCIAN S	11762
IN THE NONCURARIZED CAT	INTERFERENCE WITH THE FACIAL NERVE A	13942
URY A D HUMAN/ SURGICAL	INTERFERENCES AT THE FORMAEN OCCIPIT	11955
E METAB-DRUGS/ CANALINE	INTERFERING WITH ORNITHINE METABOLIS	14274
SPIRATORY VIRUSES TO AN	INTERFERON INDUCER IN HUMAN CELLS IN	15805
PROCEEDINGS OF THE 3RD	INTERINSTITUTE CONFERENCE IN MEMORY	15592
THE BIO CLIMATE OF THE	INTERIOR OF A FOREST MATERIAL FROM A	12142
BLAGES FROM THE WESTERN	INTERIOR OF CANADA ALDER-D BIRCH-D S	16259
LF POLLINATED LINES AND	INTERLINE HYBRIDS OF CORN-M POLY PHE	16418
SE OF AGING/ CHANGES OF	INTERMEDIARY METABOLISM IN OBESE PER	12633
AND NUCLEIC-ACIDS BOOK	INTERMEDIARY METABOLISM STRUCTURE RE	12570
TURNOVER AND A POSSIBLE	INTERMEDIATE /N ACETYL HISTIDINE MET	12566
ER BY RATS/ SOLUTION OF	INTERMEDIATE SIZE PROBLEM AND DISCRI	12127
ON BY SELF ADMINISTERED	INTERMITTENT INHALATION HUMAN/ METHO	14522
N ANTI NEOPLASTIC-DRUG/	INTERMITTENT MELPHALAN THERAPY IN MU	14998
ONCHOSCOPIC ADAPTOR FOR	INTERMITTENT POSITIVE PRESSURE BREAT	13372
OTUBERANTIA OCCIPITALIS	INTERNA AND ITS POSITION IN RELATION	13736
A MIRROR OF	INTERNAL DISEASES HUMAN/ THE SKIN AS	12512
DISEASES OF	INTERNAL ORGANS BOOK/ CHRONIC SEPTIC	15719
IAZINE MIXTURES AGAINST	INTERNAL PARASITES OF THE HORSE STRO	16156
VARIANTS IN THE	INTERNAL THORACIC VEINS/ STRUCTURAL	13092
	DUC	13169
	AL	11671
ELECTRIC RESISTANCE OF	INTERNODAL CELLS OF CHARACEAE NITELL	16518
KALOIDS ON THE FIELD OF	INTERO RECEPTORS CAT ISOLATED INTEST	14543
D BRAIN IN CARRYING OUT	INTEROCEPTIVE GLYCEMIC REFLEXES IN A	14015
OF THE ADRENAL GLANDS/	INTEROCEPTIVE STIMULATION OF THE EPI	13517
TONOMIC NERVOUS SYSTEM/	INTEROCEPTIVE UNCONDITIONAL REFLEXES	13519
ION OF RNA SYNTHESIS IN	INTERPHASE NUCLEI OF ALLIUM-CEPA-M E	11789
ANALYSIS AND FUNCTIONAL	INTERPRETATION CAT NEURONAL INTERACT	13947
DIRECT MEASUREMENT AND	INTERPRETATION OF CONVERGENCE AND DI	12244
O DUODENAL TRACT HUMAN/	INTERPRETATION OF FULL-SIZE 70 MILLI	12755
HODS IN CANINE MEDICINE	INTERPRETATION OF PICTURES HUMAN LYM	12517
CHIRONOMUS-TENTANS/ THE	INTERPRETATION OF PUFF PATTERNS IN P	11846
HYRIN COMPLEXES AND THE	INTERPRETATION OF THE MAGNETO OPTICA	12403
PLASMA MEMBRANE TO THE	INTERPRETATION OF THE PLASMA AMINO-A	12814
STATISTICS CLIMATE/ THE	INTERPRETATION OF VARIATION IN NORTH	12203
DS RESPIRATION/ FURTHER	INTERPRETATIONS OF PHOTOSYNTHESIS VA	16404
UENCY IN BLOOD CULTURES	INTERPRETATIVE ATTEMPT/ BACTERIA MOD	15724
EROGENESIS/ THE CURIOUS	INTERRELATION BETWEEN CANCER AND DIS	14979
ITAMIN A AND SODIUM ION	INTERRELATION IN PROTEIN BIOSYNTHESI	12565
SERUM URINE/ METABOLIC	INTERRELATIONS OF CALCIUM MAGNESIUM	12556
OF BREAST CANCER HUMAN/	INTERRELATIONSHIP BETWEEN THE SEX CH	14906
OF HYPOTHALAMO CORTICAL	INTERRELATIONSHIPS IN VERTEBRATES AM	11779
NOPUS-LAEVIS/ EFFECT OF	INTERRUPTING RETINO HYPOTHALAMIC CON	13643
N RESPONSE TO RED LIGHT	INTERRUPTION OF THE DARK GROWTH PERI	16504
ZATION OF SINGLE STRAND	INTERRUPTIONS IN THE DNA OF BACTERIO	15534
ALASKA WITH COMMENTS ON	INTERSPECIFIC COMPETITION LAGOPUS-MU	12308
ND POPULATIONS OF SEALS	INTERSPECIFIC COMPETITION/ A REVIEW	12266
I MESOCRICETUS-AURATUS/	INTERSPECIFIC CROSSES BETWEEN THE RU	11865
CUS BIOLOGICAL CONTROL/	INTERSPECIFIC RELATIONS AMONG 3 HYME	17341
S/ URINARY EXCRETION OF	INTERSTITIAL CELL STIMULATING HORMON	13618
OGRAPH/ EVIDENCE FOR AN	INTERSTITIAL GLAND IN THE TESTIS OF	13429
HORMONE OF THE OVARIAN	INTERSTITIAL TISSUE IN VARIOUS TYPES	13640
/ ON THE INTER CELLULAR	INTERSTITIUM OF THE SO-CALLED MIXED	14904
S OF AGE INTER STIMULUS	INTERVAL AND INTER TRIAL INTERVAL YO	12131
ECTRODES CAT CEREBELLUM	INTERVAL HISTOGRAM AUTO CORRELOGRAM/	13941
DURATION UNDER A FIXED	INTERVAL REINFORCEMENT SCHEDULE AND	12116
NTERVAL AND INTER TRIAL	INTERVAL YOUNG/ RABBIT EYELID CONDIT	12131
F DISTRIBUTION OF PULSE	INTERVALS IN EFFERENT AXONS OF THE A	17384
ARDIAC VOLUME AFTER THE	INTERVENTION OF MITRAL SUBSTITUTION	13042
CT HUMAN PSYCHO THERAPY	INTERVIEW /VARIABILITY IN THE COMMUN	14180
N HUMAN HEXAFORM GASTRO	INTEST /GENESIS AND TREATMENT OF ENT	12039
ETION OF RABBITS GASTRO	INTEST-DRUG /EFFECT OF HISTAMINE ON	14387
TH PENTA GASTRIN GASTRO	INTEST-DRUG /GASTRIC SECRETION IN TH	14393
SS STIMULUS ACTH GASTRO	INTEST-DRUG /LIVER REGENERATION AFTE	12813
TIVE TRACT HUMAN GASTRO	INTEST-DRUG /RESEARCH ON THE USEFULN	14388
PTIC ULCER HUMAN GASTRO	INTEST-DRUG /THE EFFECT OF FUBROMEGA	14390
TIS DOG HISTALOG GASTRO	INTEST-DRUG CINE ESOPHAGOGRAM/ A MUC	12775
HUMAN ORNICETIL GASTRO	INTEST-DRUG ORNITHINE CARBAMYL TRANS	14391
DANTHRON ISTIZIN GASTRO	INTEST-DRUG PYRECOL CENT-STIM-DRUG/	15703
E SODIUM BROMIDE GASTRO	INTEST-DRUGS /CLINICAL PICTURE AND T	14394
M EGGS CORN MEAL GASTRO	INTEST-DRUGS /IN-VIVO EFFECT OF SOME	14384
INOLINE CHLORIDE GASTRO	INTEST-DRUGS /REFLUX ESOPHAGITIS AND	14382
ONIC EXPERIMENTS GASTRO	INTEST-DRUGS /THE EFFECT OF HONEYSUC	14385
E METHOPROMAZINE GASTRO	INTEST-DRUGS /THE EFFECT OF SOME NEU	14389
CT RAT HISTAMINE GASTRO	INTEST-DRUGS LIVER SKELETAL MUSCLE R	14380
MINS PROTEIN AND FAT ON	INTESTINAL ABSORPTION OF IRON IN RAT	12664
ATION PROPERTIES OF THE	INTESTINAL ALPHA GLUCOSIDASES OF NOR	12584
THE BURSA OF FABRICIUS/	INTESTINAL AND CLOACAL LYMPHO EPITHE	13200
AND THEIR EVALUATION IN	INTESTINAL AND EXTRAINTESTINAL AMOEB	15656
PELVIS HUMAN/ COMBINED	INTESTINAL AND SQUAMOUS METAPLASIA O	13371
UILINUM-P/ INDUCTION OF	INTESTINAL AND URINARY BLADDER CANCE	14983
ES/	INTESTINAL ATRESIA A STUDY OF 50 CAS	12935
D HUMAN NOBACID/ GASTRO	INTESTINAL BLEEDING AND SALICYLATES	14720

IOD 1 INHIBITION OF 1ST INTERNODE ELONGATION/ ZEA-MAYS-M SHO 16504

Page from the Subject Index of BIOLOGICAL ABSTRACTS.

[Vol. 51(3)] LIFE LIPID S119

R FISHERY BIOLOGY VOL 1	LIFE HISTORIES ECOLOGY GROWTH AGE RE	12267
US-STROBUS-G MELANOTUS/	LIFE HISTORY AND DAMAGE OF THE PINE-	17021
DA/ OBSERVATIONS ON THE	LIFE HISTORY AND HABITS OF ALNIPHAGU	17203
EER WITH STUDIES ON ITS	LIFE HISTORY AND PATHOLOGY ELAPHOSTR	17140
ES MOCHLONYX-VELUTINUS/	LIFE HISTORY HABITAT AND TAXONOMIC C	17233
DUCTION DEVELOPMENT AND	LIFE HISTORY OF BERTHELLINA-CITRINA	17167
L CAPSULE MEASUREMENTS/	LIFE HISTORY OF MICROTETRAMERES-CENT	17145
OLERANCE/ AN ECOLOGICAL	LIFE HISTORY STUDY OF SUAEDA-DEPRESS	12158
AL SCHOOL LEAVERS FOR A	LIFE IN INDUSTRY/ PREPARATION OF IMM	14113
ONARY ARTERIES IN EARLY	LIFE IN 3 DIFFERENT ETHNIC GROUPS HU	12988
DURING THE 1ST YEAR OF	LIFE METAB-DRUG/ DIURNAL VARIATIONS	16088
CONDITIONS OF MILITARY	LIFE NEURO PSYCHOLOGICAL STRESS VITA	12710
MINATIONS OF THE WAY OF	LIFE OF GLAUCOMA PATIENTS CAFFEINE A	13856
STORES/ ON THE STORAGE	LIFE OF WHITE CABBAGE-D IN REFRIGERA	12721
WORKS INDUSTRIAL URBAN	LIFE SIGNIFICANCE/ MICRO OCEANOGRAPH	12231
TION STUDY/ THE IN-VIVO	LIFE SPAN OF NORMAL AND PRE NEOPLAST	14901
IN AND LIVER DURING THE	LIFE SPAN OF THE RAT PROTEIN RNA SEN	12466
APABILITIES TO MAINTAIN	LIFE SUPPORT SYSTEMS AND CABIN HABIT	11972
ATASTROPHE PLANKTON SEA	LIFE THREAT DANGERS TEMPERATURE TURB	12247
CELLS A BASIS OF FINITE	LIFESPAN OF ANIMALS HUMAN CANCER CAR	14980
NOGENESIS AGING/ FINITE	LIFETIME OF SOMATIC CELLS A BASIS OF	14980
POSTERIOR LONGITUDINAL	LIGAMENT A CAUSE OF CERVICAL MYELOPA	14044
CHILD/ AVULSION OF THE	LIGAMENTUM PATELLAE FROM THE LOWER P	13785
PHYLOCOCCAL NUCLEASE BY	LIGANDS CALCIUM CHLORIDE DEOXY THYMI	15448
ATION ON THE BINDING OF	LIGANDS TO ACCEPTOR MOLECULES IMPLIC	12535
IDNEY HUMAN/ SUPRARENAL	LIGATION OF THE INFERIOR VENA CAVA W	13081
H PERFUSION AND PYLORUS	LIGATURE /COMPARATIVE VALUE OF 2 STU	12765
NCE OF FERROUS IONS AND	LIGHT /ACTION OF OXYGENIZED WATER ON	12359
IA AND ITS RESPONSES TO	LIGHT /ASPERGILLUS-PROTUBERUS NEW SP	16287
FANTS LETTER NATURAL	LIGHT /BLUE LIGHT FOR JAUNDICE IN IN	12901
CE SYCAMORE-D LETTUCE-D	LIGHT /DETERMINATION OF THE ORIENTAT	16515
ANA CHLOROPHYLL PIGMENT	LIGHT /DEVELOPMENT OF THE ETIOLATED	16412
ORESIS AUTO RADIOGRAPHY	LIGHT /LAMELLAR LIPO PROTEIN OF NICO	16406
O-ACID AUTO RADIOGRAPHY	LIGHT /ON THE MECHANISM OF GIBBERELL	16474
NE ISOPRENE TEMPERATURE	LIGHT /PARTICIPATION OF PLANTS IN TH	16615
RIA-MEDITERRANEA BEAN-D	LIGHT /PHOTOSYNTHESIS PHOTO RESPIRAT	16407
PERFORMANCE IN CHILDREN	LIGHT /THE EFFECTS OF DIFFERENTIAL A	14136
ER-D FRAXINUS-D DROUGHT	LIGHT /VEGETATION DEVIATIONS IN A ME	12209
UCTURES FLAVO COENZYMES	LIGHT ABSORPTION SPECTROSCOPY IR SPE	12433
ENSITY 15 SINGLE ENGINE	LIGHT AIRCRAFT HUMAN/ COCKPIT NOISE	16061
HE DAILY ALTERNATION OF	LIGHT AND DARK/ THE DEPENDENCY OF TH	13551
HE VARIABILITY OF HUMAN	LIGHT CHAINS BENCE JONES PROTEINS/ A	15564
YELOMA/ THE IDENTITY OF	LIGHT CHAINS OF MONO CLONAL IMMUNO G	14829
ENCE JONES PROTEINS AND	LIGHT CHAINS TO K AND L CLASSES/ USE	15561
ELLS IRRADIATED WITH UV	LIGHT CHINESE HAMSTER/ DNA DEGRADATI	12014
FRAGILLARIA-CROTONENSIS	LIGHT DARK EXPERIMENTS ALGAE MASS DE	12253
COPEPODS ARTEMIA-SALINA	LIGHT EFFECTS/ INVESTIGATIONS ON THE	12296
NATURAL LIGHT/ BLUE	LIGHT FOR JAUNDICE IN INFANTS LETTER	12901
SITY OF THE STIMULATING	LIGHT GREEN ALGAE LETTUCE-D PEA-D LE	16415
DARK AND ANAEROBICALLY	LIGHT GROWN RHODOSPIRILLUM-RUBRUM PH	15476
IDAE/ LATERAL SPREAD OF	LIGHT INDUCED POTENTIALS ALONG DIFFE	13831
PART 5 RELATION OF THE	LIGHT INDUCED PROTON UPTAKE TO PHOTO	15440
IC TECHNIQUE/ EFFECT OF	LIGHT INTENSITY AND CARBON DI OXIDE	16398
ESCULENTUM-D/ EFFECT OF	LIGHT INTENSITY AND PLANT SIZE ON RA	16506
ESPONSE TO CANOPY COVER	LIGHT INTENSITY AND THROUGHFALL PREC	12155
DITIONS I		
GEAN SEA		
MENT IN RESPONSE TO RED	**LIGHT INTERRUPTION OF THE DARK GROWT**	**16504**
NAPHTHALENEACETIC-ACID	LIGHT INTENSITY TEMPERATURE/ METHODS	16455
CARMINOPHILOUS GRANULES	LIGHT MICROSCOPE ELECTRON MICROSCOPE	11793
BAT PTEROPUS-GIGANTEUS	LIGHT MICROSCOPE ELECTRON MICROSCOPE	13830
AINE NARCEINE NARCOTINE	LIGHT MICROSCOPE ELECTRON MICROSCOPE	16373
LOGICAL CONSEQUENCES OF	LIGHT MICROSCOPIC AND ELECTRON MICRO	12858
ROG CEREBELLAR CORTEX A	LIGHT MICROSCOPIC AND ELECTRON MICRO	13926
RDIOPATHY WITH NECROSIS	LIGHT MICROSCOPIC AND ELECTRON MICRO	14749
NORMAL HUMAN PLACENTA A	LIGHT MICROSCOPIC AND ELECTRON MICRO	15126
II POMPES DISEASE CHILD	LIGHT MICROSCOPY ELECTRON MICROSCOPY	12581
HE FUSION TO THE PALATE	LIGHT MICROSCOPY/ FURTHER STUDIES ON	15108
I TRANSDUCTION PHAGE UV	LIGHT MUTAGEN NITROUS-ACID MUTAGEN D	15498
LARVAE COPEPODA IN THE	LIGHT OF BIOLOGICAL AND PHYSIOLOGICA	17057
TIONAL TECHNIQUE IN THE	LIGHT OF CURRENT APPROACHES HUMAN/ T	14175
ATS AND HYPNOSIS IN THE	LIGHT OF EMPIRICAL RESEARCH HUMAN/ Y	14220
S OF THE STOMACH IN THE	LIGHT OF REMOTE RESULTS OF OPERATIVE	12972
ONGO BRAZZAVILLE IN THE	LIGHT OF THEIR LIPID CONTENT PEANUT-	12617
THE ACTION OF POLARIZED	LIGHT ON BETA CAROTENE SUBJECTED TO	16513
KIN HUMAN/ REACTIONS TO	LIGHT ON THE NORMAL AND PELLAGROUS B	13805
D HOMOTHERMAL ORGANISMS	LIGHT PENETRATION CLOUD LAYER PRECIP	12142
FREDERICI-D/ EFFECTS OF	LIGHT QUALITY ON THE CIRCADIAN RHYTH	16505
ONTIANA-D UNDER VARIOUS	LIGHT REGIMES/ ROLE OF CARBON DI OXI	16512
ICINALIS/ INDICATRIX OF	LIGHT SCATTERING AND THE PERIODICAL	17352
CED ATPASE ACTIVITY AND	LIGHT SCATTERING CHANGES MEDIATED BY	12417
THIO GUANINE DEPENDENT	LIGHT SENSITIVITY OF PERITHECIAL INI	16501
ETYLENE REDUCTION ASSAY	LIGHT TEMPERATURE RESPIRATION/ THE D	16541
NG OF PLANT COMMUNITIES	LIGHT TEMPERATURE SOIL MOISTURE HUMI	12157
IN ANAGALLIS-ARVENSIS-D	LIGHT TEMPERATURE/ ADAPTIVE SIGNIFIC	12176
DIOGRAPHY OSCILLATIONS/	LIGHT TRIGGERED TRANSIENT CHANGES OF	16402
Y LINEAR PROGRAMMING ON	LIGHT TYPE LAYERS RAISED ON DEEP LIT	15335
ARBON DI OXIDE EXCHANGE	LIGHT WATER STRESS/ STOMATAL MOVEMEN	16383
F THE DURATION OF DAILY	LIGHTING ON THE LUTEINIZING HORMONE	13641
YSODES TYLOS-LATREILLCI	LIGIA-ITALICA /OBSERVATIONS ON THE U	17381
AND STORAGE OF PEARS-D	LIGNIN CELLULOSE/ CHANGES IN THE CHE	16448
LLA-D/ DEMETHYLATION OF	LIGNIN IN ALKALINE AQUEOUS SOLUTION	16829
CHLORIDE ANTIDOTE-DRUG/	LIGNOCAINE FOR ARRHYTHMIAS LETTER HU	14705
RELATION BETWEEN PLASMA	LIGNOCAINE LEVELS AND INDUCED HEMODY	14364
SSUES OF SEVERAL PLANTS	LILAC-D COTTON-D NIGHTSHADE-D HERBIC	16489
USA/ TRILLIUM-GRACILE-M	LILIACEAE-M A NEW SESSILE FLOWERED S	16308
LONGING TO THE FAMILIES	LILIACEAE-M IRIDACEAE-M AMARYLLIDACE	16309
ICAL STUDY OF CAUCASIAN	LILIES-M LILIUM-CAUCASICUM-M LILIUM-	16310
-M LILIUM-MONADELPHUM-M	LILIUM-ARMENUM-M LILIUM-GEORGICUM-M	16310
SPORA-LILII NEW SPECIES	LILIUM-CANDIDUM-M USSR/ NEW SPECIES	16290
Y OF CAUCASIAN LILIES-M	LILIUM-CAUCASICUM-M LILIUM-LEDEBOURI	16310
PHUM-M LILIUM-ARMENUM-M	LILIUM-GEORGICUM-M LILIUM-SZOVITSIAN	16310
M LILIUM-SZOVITSIANUM-M	LILIUM-KESSELRINGIANUM-M /COMPARATIV	16310
S-M LILIUM-CAUCASICUM-M	LILIUM-LEDEBOURI-M LILIUM-MONADELPHU	16310
UM-M LILIUM-LEDEBOURI-M	LILIUM-MONADELPHUM-M LILIUM-ARMENUM-	16310
UM-M LILIUM-GEORGICUM-M	LILIUM-SZOVITSIANUM-M LILIUM-KESSELR	16310
PHOTOMETRY/ PIGMENTS OF	LILY-M FLOWERS II SURVEY OF CAROTENO	16820
F THE HOSPITAL DEL NINO	LIMA PERU PARA AMINO SALICYLIC-ACID	15715
YI PARABASCUS-LEPIDOTUS	LIMATULOIDES-DUBOISI ALLASOGONOPORUS	17125
ON CAPACITY ELEMENTS OF	LIMB AND SPLEEN DOG CHLORALOSE CENT-	14786
CENT-DEPRESS-DRUG HEAD	LIMB DOPAMINE NOREPINEPHRINE METAB-D	14514
MILK FEVER OF COWS III	LIMB DYS FUNCTIONS/M/ PARTURIENT PAR	16065
LIMATED RATS INJURED BY	LIMB ISCHEMIA EPINEPHRINE NOREPINEPH	14780
CLE IN THE REGENERATING	LIMB OF THE NEWT TRITURUS/ DEVELOPME	13924
TENSION HUMAN OCCLUDED	LIMB TECHNIQUE/ SENSITIVITY OF METHO	12981
CULAR ACTIVITY DOG HIND	LIMBS TRUNK ELECTRO MYO GRAPHY/ EXPE	14474
NE QUARRIES OF SOUTHERN	LIMBURG NETHERLANDS TROGLOBIONTS HIB	12137
OSPHATE REMOVAL BY A	LIME BIOLOGICAL TREATMENT SCHEME/ PH	16011
MOLYBDATE MIXED IN THE	LIME SEED PELLET ON NODULATION NITRO	16699

S-PARRECTUS NEW SPECIES	LIMESTONE CAVES/ 2 NEW POLYDESMID MI	17198
ITIES/ THE SUBTERRANEAN	LIMESTONE QUARRIES OF SOUTHERN LIMBU	12137
GANGLION IN THE MOLLUSC	LIMNEA-STAGNALIS /IDENTIFICATION OF	17348
ATINOPECTEN-YESSOENSIS/	LIMNO BIOLOGICAL STUDIES OF LAKE SAR	12256
ROMANIA TUBIFEX-TUBIFEX	LIMNODRILUS-HOFFMEISTERI /OBSERVATIO	12258
DISTRIBUTION OF DONORS	LIMNODYNASTES-TASMANIENSIS CRINIA-TA	14782
BOSMINA CHAOBORUS PALEO	LIMNOLOGY /THE BIO STRATIGRAPHICAL H	17216
S/ RELATIONSHIP BETWEEN	LIMNOLOGY AND THE FISHING TECHNIQUE	12288
ANCES MATERIALS FOR THE	LIMNOLOGY OF LAKE SUWA III JAPAN SEC	12255
NCE LIMNOPHILUS-LUNATUS	LIMNOPHILUS-CENTRALIS ADICELLA-REDUC	17261
TERA OF BRITTANY FRANCE	LIMNOPHILUS-LUNATUS LIMNOPHILUS-CENT	17261
OF LARVAE AND PUPAE OF	LIMONIINAE TIPULA-SCHUMMELIA-ZERNYI	17231
AMPELOPSIS-VITIFOLIA-D	LIMONIUM-OTOLEPIS-D LIMONIUM-SOGDIAN	16872
A-D LIMONIUM-OTOLEPIS-D	LIMONIUM-SOGDIANUM-D CONVOLVULUS-SUB	16872
S IN THE LATERAL EYE OF	LIMULUS-POLYPHEMUS ELECTRON MICROSCO	17437
LL-OR-NONE RESPONSES OF	LIMULUS-POLYPHEMUS PHOTO RECEPTOR CE	17436
G/ EXPERIENCE IN USE OF	LINCOMYCIN DURING THE TREATMENT OF S	16127
ENETRATION OF	LINCOMYCIN HUMAN/ THE INTRA OCULAR P	16086
FECT-DRUG/ INTRA VENOUS	LINCOMYCIN IN HIGH DOSES STAPHYLOCOC	16136
FAMYCINS NEOMYCIN GROUP	LINCOMYCIN LYSOSTAPHIN CEPHALOTHIN M	16144
R-D DIELDRIN HEPTACHLOR	LINDANE INSECTICIDES/ STUDIES ON TAN	17009
ULA-LITWINOWII-D PINE-G	LINDEN-D BEECH-D HORNBEAM-D USSR/ TH	12169
LIVER AND ON A RAT CELL	LINE IN CONTINUOUS CULTURE OF THE SA	15722
TORS INFLUENCING VISUAL	LINE SELECTION IN ADZUKI BEAN-D PHAS	16809
APY/ MITSUBISHI MEDICAL	LINEAR ACCELERATOR 2ND REPORT 6 MEV	12002
N INSULIN ACTION PART 3	LINEAR DISTRIBUTION OF D GLUCOSE D R	13571
TOLMAN A COMPARISON OF	LINEAR PERFORMANCE MODELS HUMAN LEAR	14115
UBSTANCES CALCULATED BY	LINEAR PROGRAMMING ON LIGHT TYPE LAY	15335
EEDLINGS OF 4 SORGHUM-M	LINES /COLCHICINE INDUCED COMPLEX DI	11827
ED HUMAN CARCINOMA CELL	LINES /HISTOGENETIC BEHAVIOR OF TUMO	14926
YMES OF SELF POLLINATED	LINES AND INTERLINE HYBRIDS OF CORN-	16418
HUMAN/ FISSURE	LINES IN THE PEDIATRIC ROENTGENOGRAM	13413
RATING BONE MARROW CELL	LINES IN 2 DIFFERENT X IRRADATION CO	12015
ANY/ GROUND WATER GRAPH	LINES OF SOME PLANT COMMUNITIES IN T	12195
HOD OF SELECTING INBRED	LINES OF SUGAR BEETS-D FOR GENERAL C	16716
CULTIVARS AND BREEDING	LINES TO APHIDS MACROSIPHUM-EUPHORBI	17001
ESCUE OF SV40 FROM CELL	LINES TRANSFORMED AT HIGH AND AT LOW	15803
Y AUTOMOBILE TRANSVERSE	LINES TRAPS/ EXPERIENCE IN MAPPING R	12226
ORGAN BRANCHES OF THE	LINGUAL ARTERY/ THE NUMBER OF INTRA	12993
STRUCTURE OF THE MUCOUS	LINING OF THE HUMAN TONGUE/ AGE RELA	13810
LOCALIZATION TO THE 5TH	LINKAGE GROUP OF A GENE AFFECTING AN	11838
SATELLITE/ POSITION OF	LINKAGE GROUP-V MARKERS IN CHROMOSOM	11825
ELL WALL PEPTIDO GLYCAN	LINKAGE GROUPS N ACETYLMURAMYL-L ALA	15471
USE HEMO GLOBIN PART 10	LINKAGE OF DUPLICATE GENES AT THE AL	11862
F CHROMOSOME NUMBER 2 A	LINKAGE STUDY HUMAN/ INHERITED PERIC	11901
HENYL HYDRAZONE/ ENERGY	LINKED REACTIONS IN PHOTOSYNTHETIC B	15440
S SPECIES/ NAD AND NADP	LINKED SUCCINIC SEMI ALDEHYDE DEHYDR	15472
DETERMINATION OF THE 5	LINKED TERMINI OF RIBOSOMAL RNA CAUL	11781
ON OF ENZYMES WHICH ARE	LINKED TO THE NADH NAD SYSTEM L ASPA	12439
DS STARCH LINOLEIC-ACID	LINOELAIDIC-ACID OLEIC-ACID THIO GLY	16589
PA-M GAS CHROMATOGRAPHY	LINOLEIC-ACID /RESEARCH ON OIL FROM	16775
SUNFLOWER-D GREASE LARD	LINOLEIC-ACID /VARIATIONS IN THE COM	12698
TERFAT MAIZE-M/ DIETARY	LINOLEIC-ACID AND THE PLASMA PHOSPHO	12684
IFFERENT DIETS COW MILK	LINOLEIC-ACID ARACHIDONIC-ACID EICOS	12678
M EXTRACTS SEEDS STARCH	LINOLEIC-ACID LINOELAIDIC-ACID OLEIC	16589
	EED-D OIL COCONUT-M OIL SAFFLOWE	12700
	RON /EFFECT OF SOIL AND CLIMATE	16989
SIS FOR SELECTIVITY FOR	LINURON ON CARROT-D AND COMMON RAGWE	16814
UM MELOSIRA-NUMMULOIDES	LIOMOPHORA-EHRENBERGII OSCILLATORIAC	12190
R IN THE CALIFORNIA SEA	LION ZALOPHUS-CALIFORNIANUS HUMAN PO	12273
/ RECTAL PROLAPSE IN A	LIONESS AS A RESULT OF FEEDING BONES	16066
MNOCANTHUS-HERZENSTEINI	LIOPSETTA-OBSCURA PATINOPECTEN-YESSO	12256
LIP IN CONGENITAL HARE	LIP /SOME DATA ON THE INNERVATION OF	15209
TS WITH CONGENITAL HARE	LIP HUMAN CLEFT PALATE/ HEARING COND	15242
ERY IN W IN TUMORS AND	LIP HYPERTROPIES HUMAN/ PLASTIC SURG	15034
NNERVATION OF THE UPPER	LIP IN CONGENITAL HARE LIP/ SOME DAT	15209
HORESIS/ ISO ENZYMES OF	LIP OXIDASE PEA-D GREEN BEAN-D WHEAT	16589
IC RELATIONSHIP BETWEEN	LIPARID FISH CAREPROCTUS-SPP AND MAL	12217
GEL FILTRATION ESTERASE	LIPASE /CARNITINE ESTER HYDROLASE OF	12457
CHOLESTEROL FATTY-ACIDS	LIPASE /THE EFFECT OF VARIOUS KINDS	12691
M PHOSPHO LIPID PHOSPHO	LIPASE A PHOSPHO LIPASE C GLYCERO 3	15467
SM IN RAT ADRENAL GLAND	LIPASE ACTH HORMONE-DRUG/ THE LIPID	13537
ENDOGENOUS LIPO PROTEIN	LIPASE ACTIVITY FREE FATTY-ACIDS AND	13595
THE SERUM LIPO PROTEIN	LIPASE ACTIVITY IN EXPERIMENTAL ANIM	12671
E STUDY OF FAT GLYCOGEN	LIPASE AND SUCCINIC DEHYDROGENASE IN	13718
HOSPHO LIPASE A PHOSPHO	LIPASE C GLYCERO 3 PHOSPHATE DEHYDRO	15467
CID GLYCO PROTEIN/ ACID	LIPASE OF THE CASTOR BEAN-D SEED END	16523
	LIPEMIA IN PONIES/ HYPER	12588
DIABETES MELLITUS HYPER	LIPEMIA SENILE APPEARANCE/ WERNERS S	12582
URONIC-ACID VI ANTIGEN	LIPID A GLUCOSAMINE FATTY-ACIDS/ COM	15491
IT AORTA/	LIPID ACCRETION IN THE PERFUSED RABB	13084
ORONARY ARTERY DISEASE/	LIPID AND CARBOHYDRATE ABNORMALITIES	13052
DS BY HELA CELLS/ GLYCO	LIPID BIOSYNTHESIS SYNTHESIS OF GLUC	14942
CT OF TRACE ELEMENTS ON	LIPID BIOSYNTHESIS WHEN YEASTS ARE C	16391
V RADIATION HUMAN MOUSE	LIPID CHANGE CELL ORGANELLE MEMBRANE	12022
ESIS OF THE PROTEIN AND	LIPID COMPONENTS OF THE INNER AND OU	12408
AL PROTEIN TOTAL LIPIDS	LIPID COMPOSITION COPPER IRON AND MU	17036
T DEFICIENCY ON PHOSPHO	LIPID COMPOSITION IN THE CHICK BRAIN	12667
LISM AND CHANGES OF THE	LIPID CONCENTRATION IN THE VESSEL WA	13082
E IN THE LIGHT OF THEIR	LIPID CONTENT PEANUT-D GOURD-D PALM	12617
MA KWASHIORKOR/ PROTEIN	LIPID CORRELATION IN INFANTILE MAL N	12634
METHYL BIGUANIDE ON THE	LIPID DISTURBANCES CAUSED BY THE ADM	14405
IFUGATION DOG METABOLIC	LIPID EFFECTS PARENTERAL CALORIC SUP	12546
RINOLOGY EASTERN EUROPE	LIPID ESTROGENS PROTEIN THYROXINE TH	11742
/ EFFECT OF STORAGE AND	LIPID EXTRACTION ON THE PROPERTIES O	12732
H FAILURE INFANT/ TOTAL	LIPID FATTY-ACID PATTERNS OF UMBILIC	15162
Y OF INDIVIDUAL PHOSPHO	LIPID FRACTIONS OF THE RAT LIVER/ IR	12391
E PRESENCE OF A PHOSPHO	LIPID IN THE PG SITE WHICH REVEALS T	14831
S/ CYTOCHROME C PHOSPHO	LIPID INTERACTION STRUCTURAL TRANSIT	12392
STIMULATION OF PHOSPHO	LIPID LABELING IN GOLDFISH BRAIN/ TE	13974
INS AND HORMONES ON THE	LIPID LEVEL IN BLOOD AND TISSUES DUR	14297
TES MELLITUS AND PLASMA	LIPID LEVELS HUMAN/ RELATIONSHIP OF	13586
T OF ARTIFICIAL PHOSPHO	LIPID MEMBRANES IN THE PRESENCE OF U	12407
OF ARTIFICIAL	LIPID MEMBRANES/ PROTONIC CONDUCTION	12405
AT INFUSION HUMAN INTRA	LIPID METAB-DRUG CEREBRAL HEMORRHAGE	14273
Y VOL 4 DRUGS AFFECTING	LIPID METABOLISM BOOK FREE FATTY-ACI	11712
TIS/ CHARACTERISTICS OF	LIPID METABOLISM IN CHILDREN WITH DI	13370
CHANGES OF MYO CARDIAL	LIPID METABOLISM IN EXPERIMENTAL MYO	14285
MITIC-ACID 1 CARBON-14/	LIPID METABOLISM IN HELMINTH PARASIT	17338
ACTH HORMONE-DRUG/ THE	LIPID METABOLISM IN RAT ADRENAL GLAN	13537
ON THE RATE OF PHOSPHO	LIPID METABOLISM IN RAT BRAIN DURING	13510
OXAN/ IN-VITRO STUDY OF	LIPID METABOLISM IN THE UTERUS OF TH	13564
CTION/ INBORN ERRORS OF	LIPID METABOLISM PART 13 FAMILIAL HY	12596
A A GENETIC DISORDER OF	LIPID METABOLISM WITH NERVOUS INVOLV	12583
N LAYER CHROMATOGRAPHIC	LIPID METHOD OF EXAMINING CARDIAC SE	12984
O MOLECULAR ASSEMBLY OF	LIPID MICELLES IN NEGATIVELY STAINED	12415
LLUS-MEGATERIUM PHOSPHO	LIPID PHOSPHO LIPASE A PHOSPHO LIPAS	15467
ADISM OSTEO POROSIS AND	LIPID PNEUMONIA/ AN AUTOPSY CASE OF	13617
ECT OF RAW PETROLEUM ON	LIPID SYNTHESIS BY YEASTS/ SOME CHAR	16564

Page from the Author Index of BIOLOGICAL ABSTRACTS.

DANI-FARADZHEV **[Vol. 51(3)] A4**

DANI R 12914
DANIEL C W 14901
DANIEL E E 13704
DANIEL M 17041
DANIELS F JR 12022
DANIELSON L L 16821
DANILEVS KA M S 16690
DANILOV I P 13128
13198
DANILOVA T F 16214
DANKO V 16624
DARBINYAN T M 14457
DARBY C W 15015
DARDEN J H 12023
DARDIK H 13967
DARNELL J E JR 14937
DARTOIS A-M 12619
DARZYNKIEWICZ Z 13161
13163
DAS B C 16867
DAS GUPTA N N 15458
DASS H C 16787
DASTE P 15437
DAUM S J 14633
DAVENPORT D G 15289
DAVER J 15974
DAVEY K G 17408
DAVID A 12482
DAVID P 11956
DAVID P A 15925
DAVIDSON E H 11837
DAVIDSON J T 13063
DAVIES D S 14289
DAVIES H D 15908
DAVIES H T 16811
DAVIES J N P 14853
DAVIES P N 16546
DAVIS A E 14683
DAVIS G M 17042
17360
DAVIS J S 13282
DAVIS J W 13527
DAVIS L E 16155
DAVIS P H 16348
DAVIS T R A 15646
DAVIS W H 16716
DAVISON W C 15094
DAVSON J 12876
DAVYDOV O N 17346
DAVYDOVA S I 13636
DAVYS M N G 12717
DAW J C 13534
DAWE C J 11727
DAWE S T 14624
14625
DAWES A C 14178
DAWS G T 11937
DAWSON C O 15807
DAWSON J H 16946
DAYHOFF M O 11716
DAZAI M 16023
DE AZEVEDO E SILVA E
14976
15038
DE BACKER H 16745
DE BARJAC H 15367
15371
DE BERNARDI M 14284
DE BIANCHI A M D L
15742
DE BOER W G R M 15620
DE BORGER R 16745
DE BRABANDER K 12233
DE CAMARGO P N 16374
16381
DE CAMPOS J M 13150
DE CARNERI I 17056
DE CARVALHO S 14824
DE CASTRO E 13138
DE CASTRO PASZTOR Y P
16843
DE CASTRO PASZTOR Y
16844
DE CLERCK E 12737
DE D C 15204
DE FERNANDEZ E K 14900
DE GANDARIAS J M 14315
DE GAVINA-MUGICA M
16621
DE GROOT S J 17454
DE GROOTE G 15335
DE HALLER G 17319
DE HARO N B 12435
DE JONG J C 14936
DE LA BORBOLLA-Y
ALCALA J M R 12727
DE LA FUENTE P 15167
DE LA LANDE CREMER L
C N 16750
DE LA LLATA M 17070
DE LACHICA M 17092
DE LAURENZI A 12957
DE LEY J 15387
DE LORIMIER A A 12900
DE LUCA H F 12831
DE LUCA Y 17394
DE LUISE M 13607
DE MARCO S 14724
DE MARIA A 14724
DE MEIRLEIRE H 17007
DE MELLO C G 11689
DE MELLO F A F 16392
DE MENDONCA J M 17066
17149
DE MIGUEL J 12895
DE MOOR C E 15350
DE MORAIS L A 16844
DE MUCHA-MACIAS J
15939
DE MUYNCK A 16165
DE O SANTOS C F 16375
DE OLIVEIRA M A 17011
DE OLIVEIRA RODRIGUES
H 17149
DE OLIVEIRA S A 13024
DE OLIVEIRA-BASTOS F
14207
DE OME K B 14901
DE PAILLERETS F 12893
DE PAIVA I 17193
DE PALMA R G 13589
DE PALO F 13208
DE PHAN E M 16425
DE PLANQUE B A 13090
DE PORTUGAL ALVAREZ J
13096
DE POURBAIX F 13410
DE RODRIGUEZ E Q 14305
DE ROO M J K 11747
DE SA MENEZES W 17242
DE SAINT-BLANQUAT G
12765
DE SANDO T 14456
DE SEZE S 13789
DE SOUZA FERNANDES P
16855
DE STADELHOFEN C M
12254
DE TORO G 14907
DE VENANZI F 11736
DE WET P D 13961
DE WIED D 14565
DEA I C M 16316
DEAN H A 12720
17015
DEAN J 17184
DEANER R M 13723
DEBELYI G A 16792
DEBENATH A 11950
DECKER H A 15976
DECOURT L V 13024
DEFENDI V 15597
DEFLANDRE G 17499
DEGGINS B A 14971
DEGKWITZ R 14087
DEGLIN V L 14575
DEGTYAREV P A 16511
DEICHMANN W B 14990
DEIST J 16926
DEL BIANCHI S 15381
DEL CAMPO A 12756
12969
DEL PISTOIA L 14179
DEL SOLAR E 12937
DEL VECCHIO R 15865
DELAGE J 15318
DELAND F H 14801
DELAUNAY P 14361
DELBARRE F 15616
DELL AIRA A 14614
DELLA BELLA D 14458
DELORME L D 17189
DELP M H 12506
DELREZ M 12015
DELUCA H F 12572
DEMEESTER G 14324
DEMIRER M A 12716
DEMKO E B 13671
DEML F 12813
DEMPSTER R P 12293
DENARI J H 13564
DENENBERG V H 14214
DENIS C 17261
DENISOVA O P 12991
12992
12993
12994
DENLIEV P D 16872
DENNIS C J 17243
DENSEN P 15981
DENSEN P M 15991
DEO P 14154
DEODHAR K P 13110
DEPAOLI J R 13937
DEPARIS M 15724
DEPARTMENT OF
EDUCATION AND SCIENCE
AND THE BRITISH
COUNCIL 11695
DEPOE C E 16346
DEPOORTERE H 17345
DERACHE R 12765
DERNUET S 15386
DEROO J 12013
DERRINGTON A W 15962
DERYLO M 15453
DESAI A J 15486
DESAI B M 16519
16520
16539
DESAI P 12628
DESHAYES P 13789
DESHMUKH M M 15258
DESHMUKH S M 15258
DESHPANDE C K 14912
DESILETS D 14710
DESMOND M M 14636
DESMOTTES R M 13119
DESSAU J 11967
DESYATNIK G G 14813
DETAILLE J-Y 14386
DETELS R 15878
DETTLI L 16088
DETTMER N 13768
DEUMLING B 13160
DEUTSCH V 15230
DEVERALL B J 16962
DEVERILL J 15567
DEVIDZE T A 15056
DEVLIN H B 12892
DEVYNCK M A 15462
DEYO R 12030
DHALA S A 15486
DHALL D P 13178
DHOND R V 15061
DI CONZA J J 15666
17139
DI GIUSEPPE F 17104
DI JESO F 17444
DI LA GRUTTA V 13942
DI MATTEO G 12762
DI PALMA J R 15045
DI PAOLO N 12347
DI PRISCO G 12446
DIAMOND E I 15141
DIAMOND H 13291
DIANA S 14426
DIAZ H 13363
DIAZ-AMADOR C 12635
DIAZ-COLON J D 16985
DIAZ-FIERROS F 16744
DIBOBES I K 16047
DICK D A T 13597
DICK R I 16014
DICK S 16483
DICKES R 14187
DICKINSON D B 11815
DIEGOLI PIRES E V
17011
DIERSCHKE H 12195
DIETZ A 15402
DIETZ S B 12569
DIETZ T H 15137
DIETZMANN G B 14399
DIGEON M 15546
DIKSHITH T S S 11888
DILGER B 13136
DILGER K 13136
DILLEY W G 13464
DIMA V F 15383
DIMOPOULLOS G T 17037
DINGENDORF W 12005
DINGWALL D 13426
DINIZ O JR 12496
DINN J J 13333
DINNEN A 14856
DINTER Z 15701
DINU M 15304
DIOP B 15903
DIOT M-F 16261
DISKO R 17033
DISSANAYAKE S A W
13649
DITTRICH A 14603
DITTRICH P 15645
DIX R L 12151
DIXON B D 12742
12744
DIXON R L 15184
DIXON W J 14066
DIXSON S 15760
DIZON J J 15878
DJANKOV I 15754
DJURIC D 14090
DMITROVSKAYA I P 11932
DNYANSAGAR V R 11808
DO NASCIMENTO KRONKA
F J 16846
DOANE W A 14922
DOBBEN G D 13644
DOBBS C 15122
DOBOS I 14679
DOBOS P 15557
DOBOSZ-CYK
DOBROKHOTC
DOBROSSY L
DOBROVOL SKII N M
12078
DOBRYANS KYI V S 16683
DOCIU I 15814
DOCKE F 13550
DODSON M G 12842
DOEHRING D G 13883
DOEPFMER R 12582
DOERING C H 13528
DOERNER G 13637
DOETSCH R N 15423
DOGADINA T V 16020
DOGLIA AZAMBUJA R
14641
DOGU T S 14532
DOHOTARU V 15837
DOLABCHYAN Z L 12977
DOLBY A E 15682
DOLD U 14744
DOLE J W 12218
DOLEZAL V 14673
DOLGIKH Y R 16754
DOLGODVOROV A F 12706
DOLGOPOL SKAYA M A
17295
DOLL E 13392
13395
DOLL R 14235
DOMAN N G 16565
DOMER F R 14294
DOMINGUEZ-ASENSIO J
14465
DOMINIK T 16248
DOMINO E F 14549
DONACHIE W D 15502
DONALDSON A W 15913
DONALDSON D M 15573
DONATI L 13664
DONGO R A 13383
DONNER L 11859
13189
14163
DONNET V 13258
DONNOU J 13379
DONTSOV V V 16366
DOOB A N 14125
DOOLEY S 15988
DOPPMAN J L 14800
DORFMAN A 11714
DORHOUT MEES E J 13090
DORIGUZZI T 14573
DORNER G 13550
DOROFEICHUK V G 15547
DORR P 13457
DORSCHEID T 16400
16401
DOSHOYANTS M S 12795
DOSKOCIL M 15187
DOSSETOR J 13307
DOSTALEK M 16209
DOTT R H 16736
DOUDOROFF M 15406
DOUGHTY C C 16522
DOUGLAS B H 13475
15180
DOUGLAS S D 13195
DOUGLAS V 14572
DOUROV N 15244
DOVGALEV S I 13128
DOVGALOV S I 13198
DOWDLE W R 15586
DOWLER C C 16988
DOWLING P C 14036
DOWNARD T R 12213
DOWNES J J 13034
DOWNING S E 13004
DOYLE A E 13074
DRAKE B J 13349
DRAKE R L 11938
DRAPER S A 13657
DRASH A 13584
DREGOL SKAYA I N 17356
DREIER G A 12910
DREVINA A I 13144
DREVVATNE T 13234
DREW J V 16989
DREW W A 17254
17416
DREXEL H 13402
DREXLER K 16893
DREYFUS J 11909
DREYFUS J-C 13122
DREYFUS P 13789
DROUILLAT M 14325
DROZDOVA I V 16528
DRUCKER H 15479
DRUDGE J H 16156
DRY J 14099
DRY L 12867
DRYFOOS J G 15970
DRYLL A 13789
DU PLESSIS S F 16742
DU PRE C 15372
DUBININA G A 15379
DUBINS KA H M 16680
DUBITZKY M 15977
DUBOIS J 13312
DUBOIS R 15125
DUBOS R 15744
DUBOURDIEU M 15450
DUBREUIL R 16178
DUBROV A P 16514
DUBROVINA M I 17205
DUBROVSKY N 13960
DUBUIS A 16354
DUBYNIN T L 13275
DUCHEN L W 14081
DUCKMAN S 15147
DUCROT R 14386
DUDGEON G 15211
DUDOREV M A 16830
DUERRE J A 12463
DUESBERG P H 15555
DUFLOT J C 13258
12326
14321
12292
DUKE S O 16504
DULIC B 11869
DULL H B 15879
DUMA D 11865
DUMAS J-C 13630
DUMAZERT C 14293
16580
DUMMETT C O 13818
DUMNOVA A G 13326
DUNAEVSKII G A 12708
DUNBAR J S 13225
DUNCA I 15304
DUNCAN H 12019
DUNCAN W H 16345
DUNCHIK V N 13313
DUNEA G 13351
DUNHAM E W 13068
DUNJIC A 15623
DUNKELBERG W E JR
15389
DUNKERLEY J 12740
DUNLOP W R 15063
DUNN J 13645
DUNNHILL R M 17166
DUNSTAN M K 15155
DUPERRAY B 12812
DUPEYRAT M 12411
DUPUY J-M 15647
DUPUY-COIN A M 12524
DUQUE O 17078
DUQUENOIS P 16866
DURAN P V M 14357
DURAND C 13344
DURE L III 16548
DURLAKOWA I 15396
DURMISHIDZE S V 12724
DURNEV V I 14330
DURNIN R E 15245
DUSSART B 17377
DUTREIX J 12042
DUTT A K 14860
DUTTA G P 17101
DVORACKOVA I 12813
DVORAK M 13448
DWURAZNA M M 16918
16919
DYBALL R E J 14402
DYBKAER R 12338
DYE C L 14352
DYK V 17100
17274
DYKAN S O 16216
DYKEN P 14067
DYKES M H M 14541
DYKHOVICHNAYA V A
16140
DYLIS N 12159
DYLIS N V 12141
DYSHKANT M G 16606
DYUISALIEVA R G 15948
DZAMOEVA E I 13919
DZHABAROV K A 13672
13675
DZHAFAROV S M 17222
DZHAKSON L S 14495
14497
14498
DZHUMABAEV T 14876
DZURIK R 13192
DZWILLO M 11881

E

EAKINS D 12852
EASTERLING S B 15521
EASTWOOD D P 16016
EAVES A 11794
EBENEZER L N 11928
EBERMANN R 13126
EBERT M 12026
EBERT P S 15000
ECKER J A 14922
ECKERT P 13131
ECKERT R E JR 16984
ECKHARDT A L 11917
ECKHARDT R D 13502
ECKHOFF D W 16010
ECLANCHER B 17385
EDDINGER C R 17479
EDEL H H 13284
EDLIN H L 11680
EDMONDS P 15459
EDMUNDS L K 11818
EDMUNDS L N JR 16502
EDMUNDSON A B 15561
EDU T 15338
EDWARDS F J 17399
EDWARDS H M JR 12654
EDWARDS H W 16029
EDWARDS J E 13049
EDWARDS J H 14133
EDWIN E 13777
EE-SIRIPORN V 17062
EECKHOUT W 15311
EFEIKIN A K 16433
EFREMOV M A 13682
EFREMOVA G P 16044
EGAMI F 15452
17368
EGEA CORTES A 14360
EGGERMONT E 12584
EGOROV S A 13965
EGOROVA L I 13368
EGOROVA N D 12032
12068
EHNHOLM C 11920
EHRHARDT J-P 12240
EHRLICH S 11775
EHRLICH W 12132
EIBL H 12468
EICH E 16205
EICHENBERGER-FAVARGER
12548
EIDSON C S 15071
15072
15074
EIFERT A 16785
EIFERT J 16785
EILER J J 12326
EIMAS P D 14132
EIMHJELLEN K E 17324
EISENBACH G-M 14656
EISENBERG H 14836
EISENBRAUN E J 17254
17416
EISENBUD M 16052
EISENMANN V 17491
EISNER G M 13247
EITEL L 15145
EITZMAN D V 12998
EKKEL MAN V I 12972
EKSTEIN E 15196
EKZEMPLYAROV O N 16095
EL-GAMAL A A 16194
16195
EL-KADY I A 16192
16193
EL-NAGEH M M 13264
EL-REFAI A H 16192
16193
ELBERG K 17226
ELDER R A 11707
11707
ELENTUKH M 16726
ELIAS L G 12600
ELINEK O 12397
ELION G B 15047
ELISEEV V A 16449
ELLEMAN T S 11999
ELLER L 12264
ELLIOT O S 11866
ELLIOTT H C 13246
ELLIOTT R B 12606
ELLIOTT R C 15293
ELLIS C J 17145
ELLIS D V 17166
ELLIS H 12632
ELLIS V L 17329
ELLNER P D 15769
ELORZA M V 16477
ELSTON H R JR 15359
ELSTON H R SR 15359
ELWOOD P C 13118
ELYASHEVICH B L 14481
ELZINGA W E 12983
EMEL YANOV V B 17352
EMEL YANOVA O S 15374
EMERICK R M 11992
EMERSON K 13562
EMERSON P A 14419
EMERSON T R 16031
EMERSON W K 17157
17158
EMLEN J T 12107
EMMELIN N 14018
EMMERICH W 16844
ENDE V 14852
ENDO R 16030
16032
16035
16036
16038
ENDO T 13460
ENDRES O 14804
ENDTZ L J 11927
ENEMAR A 11702
ENG C P 14939
ENGBRING H 12878
ENGEL P 14217
ENGELMANN F 17425
ENGLESBERG E 15516
15533
ENIGK K 17097
ENIKEEV D G 14568
ENIKEEV K K 16822
ENOMOTO M 15524
ENOMOTO S 17054
ENSER M 12429
ENTIN G M 14232
EPSTEIN N A 17088
EPSTEIN S M 14981
EQUI A 15381
ERASMUS D A 17339
ERBENOVA Z 12813
ERBSLOEH F 13731
ERCOLI A 14475
ERDEI M 15329
15330
15331
15333
15339
EROSTEIN S 15742
EREMINA I M 12670
ERHARDOVA-KOTRLA B
17050
17061
ERICHSEN A W 16716
ERIKSSON A W 11920
ERIKSSON G 13802
ERNEK E 15797
15798
ERNST J M 16430
ERPINO M J 13470
ERSHOFF B H 14651
ERTINGHAUSEN G 12318
ESAKOVA G G 13993
ESANU C 12629
ESCH G W 17333
ESCHWEGE F 12042
ESCHWEILER W 16190
ESGUERRA O 15243
ESKES T 15188
ESMANN V 15264
ESMOND W G 13219
ESNOUF M P 14708
ESPADA S 16668
ESPANA PINETTA O 16164
ESPARRACH E 13760
ESPOSITO M S 16544
ESPOSITO R E 16544
ESQUIVEL HERNANDEZ F
14329
ESTES J W 12771
ESTEVES M B 15835
ETIENNE M 15736
ETILI L 13595
ETO H 12027
ETTALA E 15306
EVANS A D 16145
EVANS E E 17363
EVANS J F 14303
EVANS J H 14612
EVANS J I 14576
EVANS M B 14158
EVANS R A 16694
16984
EVANS R G 13352
EVDOKIMOV V F 12879
EVELOFF H H 14713
EVENS R G 14800
EVERETT C 13723
EVERETT M A 11802
EVJE M 14160
EXLEY E M 17246
17247
17248
EXTON J H 13573
EYAL Z 13063
EYE J D 16016
EYLAR E H 15674
EYRING H 12395
EZCURDIA M 15167

F

FAARUP P 13271
FAASCH H 17206
FABREGA H JR 14198
FABRI Z I 13688
FABRICIUS J 13060
FABRIS N 13612
FADDA G 15566
FADEEVA E A 16140
FADIGA E 13943
FADNESS L 15583
FAGAN J F III 14242
FAGIANI M B 14577
FAGUNDES J O 14207
FAHMY M H 15324
15325
FAHRENBACH W H 12483
FAILLACE L A 14602
FAINBOIM M M 16202
FAIRBAIRN A S 15928
FAIRBAIRN D 17338
FAIRNIE I J 14770
FALBRIARD A 13267
FALCONER I R 13657
FALCONI L 12969
FALLON J 15403
FALLSTROM S P 15146
FALTIN J 14729
FAMILIARI M 15842
FANGE R 14447
FANGHANEL J 13927
FANTALIS I A 12489
FARADZHEV G R 17196

Page from the Biosystematic Index of BIOLOGICAL ABSTRACTS.

B12 [Vol. 51(3)]

MONOCOT	GRAMINEAE	ECOL PLANT	12195
MONOCOT	GRAMINEAE	ECOL PLANT	12208
MONOCOT	GRAMINEAE	ECOL PLANT	12209
MONOCOT	GRAMINEAE	WILDLIFE TER	12309
MONOCOT	GRAMINEAE	WILDLIFE TER	12310
MONOCOT	GRAMINEAE	WILDLIFE TER	12311
MONOCOT	GRAMINEAE	NUTR GEN	12598
MONOCOT	GRAMINEAE	NUTR GEN	12599
MONOCOT	GRAMINEAE	NUTR GEN	12600
MONOCOT	GRAMINEAE	NUTR GEN	12611
MONOCOT	GRAMINEAE	NUTR GEN	12613
MONOCOT	GRAMINEAE	NUTR GEN	12615
MONOCOT	GRAMINEAE	MALNUTRITION	12630
MONOCOT	GRAMINEAE	NUTR MINERAL	12664
MONOCOT	GRAMINEAE	NUTR DIET	12684
MONOCOT	GRAMINEAE	FD TECH GEN	12715
MONOCOT	GRAMINEAE	FD TECH GEN	12717
MONOCOT	GRAMINEAE	FD MILL TECH	12725
MONOCOT	GRAMINEAE	FD CER CHEM	12728
MONOCOT	GRAMINEAE	FD CER CHEM	12729
MONOCOT	GRAMINEAE	FD CER CHEM	12730
MONOCOT	GRAMINEAE	FD CER CHEM	12731
MONOCOT	GRAMINEAE	FD CER CHEM	12732
MONOCOT	GRAMINEAE	PHARM DIGEST	14384
MONOCOT	GRAMINEAE	PHARM REPROD	14629
MONOCOT	GRAMINEAE	TOXIC GEN	14657
MONOCOT	GRAMINEAE	TOXIC PHARM	14749
MONOCOT	GRAMINEAE	AN PROD FEED	15289
MONOCOT	GRAMINEAE	AN PROD FEED	15290
MONOCOT	GRAMINEAE	AN PROD FEED	15291
MONOCOT	GRAMINEAE	AN PROD FEED	15293
MONOCOT	GRAMINEAE	AN PROD FEED	15295
MONOCOT	GRAMINEAE	AN PROD FEED	15298
MONOCOT	GRAMINEAE	AN PROD FEED	15299
MONOCOT	GRAMINEAE	AN PROD FEED	15301
MONOCOT	GRAMINEAE	AN PROD FEED	15303
MONOCOT	GRAMINEAE	AN PROD FEED	15304
MONOCOT	GRAMINEAE	AN PROD FEED	15306
MONOCOT	GRAMINEAE	AN PROD FEED	15310
MONOCOT	GRAMINEAE	AN PROD FEED	15311
MONOCOT	GRAMINEAE	AN PRCD BRED	15320
MONOCOT	GRAMINEAE	POULT FEED	15327
MONOCOT	GRAMINEAE	GENET BAC VI	15493
MONOCOT	GRAMINEAE	FD MICR FOOD	16180
MONOCOT	GRAMINEAE	FD MICR ANTB	16192
MONOCOT	GRAMINEAE	FD MICR ANTB	16204
MONOCOT	GRAMINEAE	FD MICR SYN	16212
MONOCOT	GRAMINEAE	FD MICR SYN	16214
MONOCOT	GRAMINEAE	PALYNOLOGY	16259
MONOCOT	GRAMINEAE	ALGAE SYST	16267
MONOCOT	GRAMINEAE	MONOCOT SYST	16303*
MONOCOT	GRAMINEAE	MONOCOT SYST	16304
MONOCOT	GRAMINEAE	MONOCOT SYST	16305
MONOCOT	GRAMINEAE	MONOCOT SYST	16306
MONOCOT	GRAMINEAE	MONOCOT SYST	16307
MONOCOT	GRAMINEAE	MONOCOT SYST	16311*
MONOCOT	GRAMINEAE	FLOR DISTRIB	16354
MONOCOT	GRAMINEAE	FLOR DISTRIB	16356
MONOCOT	GRAMINEAE	BOT GEN SYST	16358
MONOCOT	GRAMINEAE	PL MORPH	16366
MONOCOT	GRAMINEAE	PL MORPH	16378
MONOCOT	GRAMINEAE	PL TEMP	16385
MONOCOT	GRAMINEAE	PL TEMP	16386
MONOCOT	GRAMINEAE	PL MINL NUTR	16390
MONOCOT	GRAMINEAE	PL MINL NUTR	16394
MONOCOT	GRAMINEAE	PL PHOTOSYN	16397
MONOCOT	GRAMINEAE	PL PHOTOSYN	16398
MONOCOT	GRAMINEAE	PL PHOTOSYN	16411
MONOCOT	GRAMINEAE	PL PHOTOSYN	16413
MONOCOT	GRAMINEAE	PL RESP FERM	16418
MONOCOT	GRAMINEAE	PL GROWTH	16426
MONOCOT	GRAMINEAE	PL GROWTH	16435
MONOCOT	GRAMINEAE	PL GROWTH	16438
MONOCOT	GRAMINEAE	PL REPROD	16460
MONOCOT	GRAMINEAE	PL GROW SUB	16466
MONOCOT			
MONOCOT	**GRAMINEAE**	**PL RAD EFFCT**	**16504**
MONOCOT	GRAMINEAE	PL GROW SUB	16478
MONOCOT	GRAMINEAE	PL RAD EFFCT	16508
MONOCOT	GRAMINEAE	PL ELEC MAG	16516
MONOCOT	GRAMINEAE	PL ENZYMES	16530
MONOCOT	GRAMINEAE	PL ENZYMES	16534
MONOCOT	GRAMINEAE	PL METB	16551
MONOCOT	GRAMINEAE	PL METH APP	16589
MONOCOT	GRAMINEAE	PL METH APP	16590
MONOCOT	GRAMINEAE	PL METH APP	16595
MONOCOT	GRAMINEAE	PL PHYSL GEN	16607
MONOCOT	GRAMINEAE	PL PHYSL GEN	16610
MONOCOT	GRAMINEAE	ECON BOT GEN	16622
MONOCOT	GRAMINEAE	CROPS GEN	16623
MONOCOT	GRAMINEAE	CROPS GEN	16624
MONOCOT	GRAMINEAE	CROPS GEN	16625
MONOCOT	GRAMINEAE	CROPS GEN	16633
MONOCOT	GRAMINEAE	CROPS GEN	16634
MONOCOT	GRAMINEAE	CROPS GEN	16636
MONOCOT	GRAMINEAE	CROPS GEN	16637
MONOCOT	GRAMINEAE	CROPS GEN	16638
MONOCOT	GRAMINEAE	CROPS GEN	16639
MONOCOT	GRAMINEAE	CROPS GEN	16641
MONOCOT	GRAMINEAE	CROPS GEN	16643
MONOCOT	GRAMINEAE	CROPS GEN	16644
MONOCOT	GRAMINEAE	CROPS GRAIN	16645
MONOCOT	GRAMINEAE	CROPS GRAIN	16646
MONOCOT	GRAMINEAE	CROPS GRAIN	16647
MONOCOT	GRAMINEAE	CROPS GRAIN	16648
MONOCOT	GRAMINEAE	CROPS GRAIN	16649
MONOCOT	GRAMINEAE	CROPS GRAIN	16650
MONOCOT	GRAMINEAE	CROPS GRAIN	16651
MONOCOT	GRAMINEAE	CROPS GRAIN	16652
MONOCOT	GRAMINEAE	CROPS GRAIN	16653
MONOCOT	GRAMINEAE	CROPS GRAIN	16654
MONOCOT	GRAMINEAE	CROPS GRAIN	16655
MONOCOT	GRAMINEAE	CROPS GRAIN	16656
MONOCOT	GRAMINEAE	CROPS GRAIN	16657
MONOCOT	GRAMINEAE	CROPS GRAIN	16658
MONOCOT	GRAMINEAE	CROPS GRAIN	16659
MONOCOT	GRAMINEAE	CROPS GRAIN	16660
MONOCOT	GRAMINEAE	CROPS GRAIN	16661
MONOCOT	GRAMINEAE	CROPS GRAIN	16662

MONOCOT	GRAMINEAE	CROPS GRAIN	16663
MONOCOT	GRAMINEAE	CROPS GRAIN	16664
MONOCOT	GRAMINEAE	CROPS GRAIN	16665
MONOCOT	GRAMINEAE	CROPS GRAIN	16666
MONOCOT	GRAMINEAE	CROPS GRAIN	16667
MONOCOT	GRAMINEAE	CROPS GRAIN	16668
MONOCOT	GRAMINEAE	CROPS GRAIN	16669
MONOCOT	GRAMINEAE	CROPS GRAIN	16670
MONOCOT	GRAMINEAE	CROPS GRAIN	16671
MONOCOT	GRAMINEAE	CROPS GRAIN	16672
MONOCOT	GRAMINEAE	CROPS GRAIN	16673
MONOCOT	GRAMINEAE	CROPS GRAIN	16674
MONOCOT	GRAMINEAE	CROPS GRAIN	16675
MONOCOT	GRAMINEAE	CROPS FORAGE	16676
MONOCOT	GRAMINEAE	CROPS FORAGE	16679
MONOCOT	GRAMINEAE	CROPS FORAGE	16680
MONOCOT	GRAMINEAE	CROPS FORAGE	16681
MONOCOT	GRAMINEAE	CROPS FORAGE	16682
MONOCOT	GRAMINEAE	CROPS FORAGE	16683
MONOCOT	GRAMINEAE	CROPS FORAGE	16685
MONOCOT	GRAMINEAE	CROPS FORAGE	16686
MONOCOT	GRAMINEAE	CROPS FORAGE	16687
MONOCOT	GRAMINEAE	CROPS FORAGE	16688
MONOCOT	GRAMINEAE	CROPS FORAGE	16690
MONOCOT	GRAMINEAE	CROPS FORAGE	16691
MONOCOT	GRAMINEAE	CROPS FORAGE	16692
MONOCOT	GRAMINEAE	CROPS FORAGE	16694
MONOCOT	GRAMINEAE	CROPS FORAGE	16695
MONOCOT	GRAMINEAE	CROPS FORAGE	16696
MONOCOT	GRAMINEAE	CROPS FORAGE	16697
MONOCOT	GRAMINEAE	CROPS FORAGE	16698
MONOCOT	GRAMINEAE	CROPS FORAGE	16700
MONOCOT	GRAMINEAE	CROPS SUGAR	16713
MONOCOT	GRAMINEAE	WEED CONTROL	16728
MONOCOT	GRAMINEAE	WEED CONTROL	16729
MONOCOT	GRAMINEAE	SOIL GEN	16734
MONOCOT	GRAMINEAE	SOIL FERTIL	16752
MONOCOT	GRAMINEAE	SOIL FERTIL	16754
MONOCOT	GRAMINEAE	SOIL FERTIL	16756
MONOCOT	GRAMINEAE	SOIL FERTIL	16757
MONOCOT	GRAMINEAE	SOIL FERTIL	16761
MONOCOT	GRAMINEAE	PL DIS FUNG	16896
MONOCOT	GRAMINEAE	PL DIS FUNG	16897
MONOCOT	GRAMINEAE	PL DIS FUNG	16901
MONOCOT	GRAMINEAE	PL DIS FUNG	16906
MONOCOT	GRAMINEAE	PL DIS FUNG	16907
MONOCOT	GRAMINEAE	PL DIS FUNG	16909
MONOCOT	GRAMINEAE	PL DIS FUNG	16913
MONOCOT	GRAMINEAE	PL DIS AN	16916
MONOCOT	GRAMINEAE	PL DIS VIRUS	16917
MONOCOT	GRAMINEAE	PL DIS OTHER	16926
MONOCOT	GRAMINEAE	PL DIS OTHER	16930
MONOCOT	GRAMINEAE	PL DIS RESIS	16932
MONOCOT	GRAMINEAE	PL DIS RESIS	16940
MONOCOT	GRAMINEAE	PL DIS RESIS	16943
MONOCOT	GRAMINEAE	PL DIS CONT	16948
MONOCOT	GRAMINEAE	PL DIS CONT	16949
MONOCOT	GRAMINEAE	PL DIS CONT	16954
MONOCOT	GRAMINEAE	PL DIS MISC	16968
MONOCOT	GRAMINEAE	PL DIS MISC	16969
MONOCOT	GRAMINEAE	PL DIS MISC	16971
MONOCOT	GRAMINEAE	PL DIS MISC	16972
MONOCOT	GRAMINEAE	PL DIS MISC	16973
MONOCOT	GRAMINEAE	PL DIS MISC	16980
MONOCOT	GRAMINEAE	PEST CONT	16981
MONOCOT	GRAMINEAE	PEST CONT	16984
MONOCOT	GRAMINEAE	PEST CONT	16985
MONOCOT	GRAMINEAE	PEST CONT	16987
MONOCOT	GRAMINEAE	PEST CONT	16990
MONOCOT	GRAMINEAE	PEST CONT	16991
MONOCOT	GRAMINEAE	PEST CONT	16992
MONOCOT	GRAMINEAE	ENT FLD CROP	17004
MONOCOT	GRAMINEAE	ENT FLD CROP	17006
MONOCOT	GRAMINEAE	ENT FLD CROP	17009
MONOCOT	GRAMINEAE	ENT FLD CROP	17013
		DIPTERA SYST	17230
		HEM-HET SYST	17238
MONOCOT	GRAMINEAE	LEPIDOP SYST	17257
MONOCOT	GRAMINEAE	PROTOZ EXPT	17313
MONOCOT	GRAMINEAE	PROTOZ EXPT	17316
MONOCOT	GRAMINEAE	PLATYHM EXPT	17336
MONOCOT	GRAMINEAE	ARACHNO EXPT	17441
MONOCOT	HYDROCHARITA	ECOL PLANT	12163
MONOCOT	HYDROCHARITA	PL RAD EFFCT	16499
MONOCOT	HYDROCHARITA	ECON BOT GEN	16618
MONOCOT	IRIDACEAE	MONOCOT SYST	16309
MONOCOT	IRIDACEAE	PHARMAC BOT	16872
MONOCOT	JUNCACEAE	CROPS FORAGE	16683
MONOCOT	LEMNACEAE	PL GROW SUB	16487
MONOCOT	LEMNACEAE	PL RAD EFFCT	16499
MONOCOT	LEMNACEAE	PL METH APP	16600
MONOCOT	LILIACEAE	CYTOL PLANT	11789
MONOCOT	LILIACEAE	CYTOL ANIMAL	11801
MONOCOT	LILIACEAE	GENET PLANT	11810
MONOCOT	LILIACEAE	GENET PLANT	11819
MONOCOT	LILIACEAE	ECOL PLANT	12173
MONOCOT	LILIACEAE	ECOL PLANT	12210
MONOCOT	LILIACEAE	LYMPH RE SYS	13201
MONOCOT	LILIACEAE	PHARM METAB	14283
MONOCOT	LILIACEAE	PHARM CV	14366
MONOCOT	LILIACEAE	PHARM NEURO	14499
MONOCOT	LILIACEAE	PHARM REPROD	14629
MONOCOT	LILIACEAE	NEOPLM THERA	15035
MONOCOT	LILIACEAE	FUNGI SYST	16290
MONOCOT	LILIACEAE	MONOCOT SYST	16308*
MONOCOT	LILIACEAE	MONOCOT SYST	16309
MONOCOT	LILIACEAE	MONOCOT SYST	16310
MONOCOT	LILIACEAE	FLOR DISTRIB	16353
MONOCOT	LILIACEAE	VEGETAB HORT	16795
MONOCOT	LILIACEAE	VEGETAB HORT	16804
MONOCOT	LILIACEAE	VEGETAB HORT	16810
MONOCOT	LILIACEAE	FLOWER HORT	16820
MONOCOT	LILIACEAE	PL DIS FUNG	16905
MONOCOT	LILIACEAE	PL DIS AN	16916
MONOCOT	LILIACEAE	ENT FLD CROP	17003
MONOCOT	MUSACEAE	BEHAV ANIMAL	12101
MONOCOT	MUSACEAE	PL DIS AN	16916
MONOCOT	ORCHIDACEAE	PL PHOTOSYN	16408
MONOCOT	PALMAE	NUTR GEN	12617
MONOCOT	PALMAE	NUTR PATH	12700

MONOCOT	PALMAE	FD DAIRY PRO	12743
MONOCOT	PALMAE	PHARM REPROD	14629
MONOCOT	PALMAE	FRUIT TROP	16774
MONOCOT	PALMAE	FRUIT TROP	16775
MONOCOT	PALMAE	HOMOP SYST	17242
MONOCOT	PONTEDERIA	PL REPROD	16454
MONOCOT	PONTEDERIA	ECON BOT GEN	16618
MONOCOT	PONTEDERIA	INVERT SYST	17110
MONOCOT	POTAMOGETONA	WEED CONTROL	16728
MONOCOT	POTAMOGETONA	WEED CONTROL	16729
MONOCOT	POTAMOGETONA	PEST CONT	16986
MONOCOT	TYPHACEAE	PL REPROD	16462
MONOCOT	ZINGIBERACEAE	ANGIOSP SYST	16301*
MONOCOT	ZINGIBERACEAE	PHARMAC BOT	16862
MONOCOT	ZINGIBERACEAE	PHARMAC BOT	16894
MONOCOT	ZINGIBERACEAE	HOMOP SYST	17241
DICOT		ECOL PLANT	12164
DICOT		ECOL PLANT	12192
DICOT		NUTR DIET	12682
DICOT		MONOCOT SYST	16308
DICOT		DICOT SYST	16317
DICOT		PL GROWTH	16432
DICOT		HORT MISC	16822
DICOT		PL DIS BAC	16914
DICOT	ACANTHACEAE	ANGIOSP SYST	16301
DICOT	ACANTHACEAE	PL DIS MISC	16966
DICOT	ACERACEAE	ECOL PLANT	12207
DICOT	ACERACEAE	ECOL PLANT	12209
DICOT	ACERACEAE	ECOL PLANT	12210
DICOT	ACERACEAE	PALYNOLOGY	16257
DICOT	ACERACEAE	PL MORPH	16369
DICOT	ACERACEAE	PL TEMP	16386
DICOT	ACERACEAE	PL RAD EFFCT	16515
DICOT	ACERACEAE	FLOWER HORT	16821
DICOT	ACERACEAE	PEST CONT	16983
DICOT	AIZOACEAE	DICOT SYST	16342
DICOT	AIZOACEAE	PL METB	16565
DICOT	AMARANTHACEAE	GENET PLANT	11826
DICOT	AMARANTHACEAE	CROPS GRAIN	16645
DICOT	AMARANTHACEAE	PEST CONT	16987
DICOT	ANACARDIACEAE	FORESTRY	16847
DICOT	ANACARDIACEAE	PEST CONT	16985
DICOT	ANACARDIACEAE	HOMOP SYST	17242
DICOT	ANNONACEAE	PL DIS AN	16916
DICOT	APOCYNACEAE	GENET PLANT	11808
DICOT	APOCYNACEAE	CV PHYSL	13008
DICOT	APOCYNACEAE	CV HRT PATH	13031
DICOT	APOCYNACEAE	MUSC PHYSL	13704
DICOT	APOCYNACEAE	NERV PHYSL	13965
DICOT	APOCYNACEAE	PHARM GEN	14255
DICOT	APOCYNACEAE	PHARM CV	14314
DICOT	APOCYNACEAE	PHARM CV	14323
DICOT	APOCYNACEAE	PHARM CV	14326
DICOT	APOCYNACEAE	PHARM CV	14327
DICOT	APOCYNACEAE	PHARM CV	14338
DICOT	APOCYNACEAE	PHARM CV	14347
DICOT	APOCYNACEAE	PHARM CV	14360
DICOT	APOCYNACEAE	PHARM NEURO	14435
DICOT	APOCYNACEAE	PHARM NEURO	14459
DICOT	APOCYNACEAE	PHARM NEURO	14510
DICOT	APOCYNACEAE	PHARM NEURO	14517
DICOT	APOCYNACEAE	PHARM NEURO	14543
DICOT	APOCYNACEAE	PHARM NEURO	14552
DICOT	APOCYNACEAE	TOXIC VET	14769
DICOT	APOCYNACEAE	NEOPLM THERA	15010
DICOT	APOCYNACEAE	NEOPLM THERA	15015
DICOT	APOCYNACEAE	DEV BIO EXPT	15169
DICOT	APOCYNACEAE	PL GROW SUB	16473
DICOT	APOCYNACEAE	PHARMAC BOT	16881
DICOT	APOCYNACEAE	PHARMAC BOT	16887
DICOT	APOCYNACEAE	PHARMAC BOT	16892
DICOT	AQUIFOLIACEAE	PALYNOLOGY	16257
DICOT	ARALIACEAE	PHARM CV	14332
DICOT	ARALIACEAE	PHARMAC BOT	16864
DICOT	ARISTOLOCHIA	PHARM CV	14331
DICOT	ASCLEPIADA	PHARMAC BOT	16891
DICOT	BALSAMINACEAE	ECOL PLANT	12162
DICOT	BEGONIACEAE	PL REPROD	16456
DICOT	BEGONIACEAE	HORT MISC	16825
DICOT	BEGONIACEAE	PL DIS OTHER	16929
DICOT	BERBERIDACEAE	DICOT SYST	16337*
DICOT	BERBERIDACEAE	PHARMAC BOT	16882
DICOT	BETULACEAE	BIOCLIMATOL	12142
DICOT	BETULACEAE	ECOL PLANT	12169
DICOT	BETULACEAE	ECOL PLANT	12207
DICOT	BETULACEAE	ECOL PLANT	12210
DICOT	BETULACEAE	PALEOBOTANY	16255
DICOT	BETULACEAE	PALYNOLOGY	16257
DICOT	BETULACEAE	PALYNOLOGY	16259
DICOT	BETULACEAE	PALYNOLOGY	16265
DICOT	BETULACEAE	PL TEMP	16385
DICOT	BETULACEAE	PL METB	16541
DICOT	BETULACEAE	PL METH APP	16604
DICOT	BETULACEAE	FORESTRY	16829
DICOT	BETULACEAE	ENT TREE WD	17018
DICOT	BIGNONIACEAE	PL MORPH	16374
DICOT	BIGNONIACEAE	PL WATER REL	16381
DICOT	BIGNONIACEAE	PHARMAC BOT	16884
DICOT	BORAGINACEAE	ECOL PLANT	12172
DICOT	BORAGINACEAE	TOXIC RESID	14688
DICOT	BORAGINACEAE	DICOT SYST	16329*
DICOT	BORAGINACEAE	PHARMAC BOT	16872
DICOT	BORAGINACEAE	PL DIS MISC	16970
DICOT	BURSERACEAE	DICOT SYST	16338
DICOT	BUXACEAE	HEM-HET SYST	17236
DICOT	CACTACEAE	PHARM GEN	14253
DICOT	CACTACEAE	DICOT SYST	16318
DICOT	CACTACEAE	PL REPROD	16457
DICOT	CAMPANULACEAE	PHARM NEURO	14505
DICOT	CAMPANULACEAE	DICOT SYST	16340*
DICOT	CAPPARACEAE	DICOT SYST	16337*
DICOT	CAPPARACEAE	PHARMAC BOT	16860
DICOT	CAPRIFOLIA	ECOL PLANT	12162
DICOT	CAPRIFOLIA	PHARM DIGEST	14385
DICOT	CAPRIFOLIA	PHARMAC BOT	16868
DICOT	CARICACEAE	NUTR GEN	12599
DICOT	CARICACEAE	DIGEST PHYSL	12823
DICOT	CARICACEAE	DIGEST PATH	12949
DICOT	CARICACEAE	LYMPH RE SYS	13199
DICOT	CARICACEAE	TOXIC PHARM	14700

FIGURE 15. BIORESEARCH INDEX is published monthly, and BIOLOGICAL ABSTRACTS is published twice a month. The "GUIDE" is an aid to efficient use of the Indexes.

CROSS Index: Additional information pertinent to your interest can be located in the *CROSS Index* found in the pink pages of *Biological Abstracts*. Most research papers deal with more than one biological subject. However, BIOSIS assigns each abstract under only one heading for publication in *Biological Abstracts*. Nevertheless, BIOSIS enables you to identify materials related to your subject, although printed under another subject heading. This system makes it possible to find cross references in every subject area relating to your research.

In 1965 BIOSIS added a second publication, *BioResearch Index*, which is published once each month. This is a bibliographic index and refers the investigator to publications that may contain information pertinent to his interest. The papers announced in *BioResearch Index* are in addition to those found in *Biological Abstracts*. A year's volume contains more than 90,000 citations to papers from symposia, trade journals, semi-popular journals, government reports, and bibliographies. The same indexing system is used for *BioResearch Index* as is used for *Biological Abstracts*.

As the scientific literature grows, the problems of abstracting and indexing all articles published becomes increasingly great. Without modern data-processing and computer techniques, such as those used by BIOSIS, it would be almost impossible to provide this important service to scientists.

Personal Records. Scientists are engaged in a continuous search for papers relating to their own work. With the help of the abstracting services, an investigator can search the literature for information about a particular topic or problem. Many different systems are used for keeping personal files of significant papers. Some scientists keep a file of titles and authors on 3″ x 5″ or 4″ x 6″ cards, alphabetized by author and date, so that they can be referred to quickly. Others have worked out a system on "punched cards," which permits them to retrieve in a short time all the papers written either by a particular person, or between certain years, or on a particular subject, and so on. Each system is quite personal, since in most cases it is designed to meet the needs of an individual scientist.

Until you work out your own system, use index cards to record information about the papers you read. This method is similar to that used by many scientists and students. The card below is one that a student might make if he were to come across the paper by Duke and Wickliff that you read earlier in this section.

The person who made this reference card noted certain details about *Zea* shoot development and finally wrote a brief summary of the paper. In other words, each person makes out a card that contains somewhat different information, depending on his interests.

File card used to record information about a single scientific paper.

Duke, S.Q. and Wickliff, J.L. 1969
Zea shoot development in response to red light interruption of the dark-growth period. 1. Inhibition of first internode elongation.   Plant Physiol. (1969) 44:1027-1030

1. ORGANISM: seeds of zea maya 1. cu. f. m. cross
2. LIGHT SOURCE: cinemoid filters
3. STATISTICAL INTERPRETATIONS: standard error of the mean and t-test.
4. METHODS GIVEN IN DETAIL.

SUMMARY —
    Brief low energy red light interruption of the early growth period of zea induced inhibition of the first internode elongation. The coleoptile tip showed →

It is important, at this time, to realize that the responsibility for reading in the literature and for keeping adequate records of what you have read is your own. This is the way scientists work. You can easily gain the habit of browsing through scientific journals, reading and making notes on articles of interest or of relevance to work you are doing. The value of the library depends entirely on how much use you make of it.

INVESTIGATIONS FOR FURTHER STUDY

As a rule, larger colleges and universities have extensive collections of scientific journals. Public libraries in the larger cities also subscribe to many of these journals. Smaller public libraries and high school libraries, on the other hand, may subscribe to a few journals only.

1. If possible, visit a large public library or a university library. It receives journals that deal with a science in general; make a list of them. Then list some journals dealing primarily with one or more of the biological sciences.

2. Using the card catalog, determine how these journals are cataloged in the library.

3. In one issue of a journal, scan each paper and determine its structure. Construct a chart which compares all the articles in this one issue. Now do the same thing for an issue of an entirely different journal. Do all papers have about the same organization (structure), or is this determined in part by the journal in which they appear?

4. Listed below are the names of some of the journals which are published in foreign countries. Can you translate the titles and determine the biological subject matter with which each journal deals?

La Cellule	*Embryologia*
Zeitschrift für Botanik	*Crustaceana*
Cytologia	*Chromosoma*
Hereditas	*Insectes Sociaux*
Hydrobiologica	*Zoologischer Anzeiger*

5. In what field of science would you place each of the following journals?

The Journal of Biological Chemistry
The Bulletin of Mathematical Biophysics
Biochimica et Biophysica Acta
Journal of Molecular Biology

SUMMARY

In the preceding sections of this book, we have discussed the meaning of science and the processes of biological investigation. By this time you have had considerable experience in working with these processes, and it may be profitable now to take a second look at them. With your added experience, it may be possible to grasp better the ways in which ideas and experiments interact to produce new explanations and understandings of biological phenomena. In other words, your experience can help you to understand better the way in which biological science grows.

Problems

Questions which could form part of a successful experimental design occur to each of us nearly every day. Which route is the shortest distance between my home and my school? What brand of soap in our store is the most destructive to microorganisms on my skin? Which of the gasolines in my neighborhood will give the best mileage to my car over a given road test? Many of these questions you probably will leave unanswered. Science, however, if it is to progress, must constantly seek the answers to its questions.

How does the research worker select the questions which he will attempt to answer? Occasionally, he may experiment merely *to discover* whether or not something happens, with little previous experience, either of his own or of others, to guide him. More often, though, the scientist draws from previous experience with either the current or related problems in order to make an educated guess (hypothesis) that a predicted event can be confirmed by new data.

Competent research workers gain the insight necessary to make logical hypotheses through a constant study of the problems in which they are interested. This study began in some small sense on the day they were born, acquired momentum during their formal schooling, and finally approached maturity when they assumed the role of active researchers. The fascination of science is that complete mastery of knowledge is never attained. The more a scientist learns, the more he realizes how much he does not know. It is this realization, plus the desire to probe the unknown, which has paved the way for human progress.

While a laboratory investigator draws heavily on his personal experience to frame hypotheses, an even *more important* part of his total experience is his familiarity with the knowledge of others. Knowledge of the contributions of other scientists can be gained in several ways; however, it is most important that the scientist be familiar with scientific literature, especially as it pertains to his particular problem. Perhaps someone has already answered his question. Perhaps someone asked his question before but came to a conclusion that is different from the one which now seems plausible. Perhaps new insights to his question can be gained by study of the experiments of others concerning related questions. Let there be no under-emphasis of this point: the scientist, if he is to frame useful, effective hypotheses and avoid wasting time in testing them by collecting new data from nature or by experiments, must be familiar with the literature pertaining to his problem.

What kinds of questions can be answered or evaluated by new data which are collected? How does one frame a particular hypothesis in order that it may be most successfully tested? The common sense born of experience has shown that the most essential element of a good question for experimental design is *keeping it as simple and direct as possible.*

Suppose you intend to experiment with mileage from different gasolines. Don't ask what is the best gasoline to use in cars. To work a question stated in this manner into an experimental design would involve the impossible task of testing all the effects of all gasolines upon all cars. Rather decide what you want to know about gasoline and cars and stick to it. You may, for instance, wish to ask which of the gasolines available in your neighborhood will give the best mileage to your car at 60 miles per hour over a given stretch of road. Now you have something you can work with. If you wish to expand the findings of this experiment in future experiments, you may, but keep your current question a *workable one.*

It must be stressed, however, that the simplicity of the questions, which have proved to be good ones for experimentation, is a very misleading simplicity. They are not simple in the sense that you bite off one little piece of a problem and ignore all the other aspects that surround it. This can only lead to unrelated little pieces of knowledge that nobody could really consider good science.

Let us consider again the value of different gasolines for automobiles. Of course, what you really want to know is all the different effects of all the different gasolines on an automobile. How much mileage you can get from a gallon of fuel is an important aspect of this question, but you wouldn't want to get more mileage from a gallon of fuel if that fuel was one which burned up your engine. However, when you design experiments, you want to break up your large question into smaller, workable ones, and you ask each of these, one at a time. Remember, however, that the answer to any one of these questions is only a part of the answer to the initial important question, "Which is the best gasoline?" So, in the end, what is wanted is a way of putting all the questions and their answers together again.

You will find many opportunities to express the questions being asked in the remainder of the investigations in this course. Express them well. They are the reasons for doing the experiments. Without an understanding of its purpose, one can neither profitably design nor evaluate the results of an experiment.

Design of Experiments

It has become commonplace to consider science and experiments reasonably close synonyms. This is not true because, as it was pointed out earlier, many scientists do little or no experimentation, but it is true that in most areas of biological science today, experimentation is a major tool of the scientist.

Experimentation is a broad term that covers what scientists do in both the field and the laboratory during the process of inquiry: asking questions, gathering data, and evaluating these data. An experiment is not a haphazard trial; it is a well-planned attack on a specific problem. Experiments are designed to ask one or more questions within a framework which will yield possible answers. To design an experiment is to prepare an investigation from which data may be gathered and critically evaluated.

In former days, when modern science was just beginning, much scientific work was done in the field where nature could be observed firsthand; and, indeed, much biological work must still be done in the field. As time passed, however, it became more and more apparent that more precise measurements and more accurate observations could be made in the laboratory, because the scientific worker could then so control conditions that a comparison could be made between situations which differed in only one significant respect. That was the beginning of experimentation (the experiment of the single variable) which made possible such logical certainty in drawing a conclusion that it has become the preferred method of science wherever possible.

Conclusions. The technique of drawing conclusions from a set of data is seldom as obvious as it may seem at first. An important consideration in scientific procedure is to recognize that no results are ever completely unequivocal. No matter how closely the conditions of an experiment are controlled, variations do occur; these variations can lead to error in the conclusions. Thus, the conclusions of the scientist must take into account the ego-deflating fact that his deductions may be wrong. To minimize this possibility, the scientist often repeats his experiments to be sure he can get similar results from trial to trial.

Often, too, the difference in the results from the control and the treated specimens may be small, causing one to wonder if a difference actually exists or if the apparent difference is just chance variation between the two samples. These problems in drawing conclusions have been greatly alleviated by the techniques of data analysis developed by statisticians

If the data obtained are adequate, the scientist may be able to discard his hypothesis or to say that the data are consistent with it. It may never be possible to *prove* a hypothesis, but, at least until other data indicate that we should not do so, we may *accept* the hypothesis. Then we can look for another, which may be a refinement of the earlier hypothesis, or one which combines several hypotheses. More data may be necessary for the proper evaluation of this new hypothesis. It may call for a new experiment, and thus the cycle continues, not as a circle, but as a spiral, leading to greater and greater understanding of nature.

EXPERIMENTS AND IDEAS IN BIOLOGICAL INVESTIGATIONS

PART 2

GROWTH AND INTERACTIONS OF POPULATIONS

SECTION **9** POPULATION DYNAMICS

Study of Populations by the Use of Microbes

Unicellular forms of life are convenient for studying population growth. Each cell division results directly in an increase in the population of individuals, and many cells reproduce themselves every few hours or even every few minutes. Such short generation times permit the production of millions or billions of individuals in a few hours or days.

Numerical changes in populations are *best* estimated by actually counting the individuals in a population at different times. The sociologist often does this by measuring the population growth of a country over ten-year intervals. This information is available from the national census. From these data, he can plot a population-growth curve such as that seen in Figure 16. Understanding the nature of this curve helps the sociologist to make certain hypotheses about future populations: what their needs will be, and how these needs may be met.

Microbiologists use two methods of counting the individuals in a population. The choice is determined by whether one wishes to measure only *viable* microbes (those which are able to reproduce) or all the organisms in the culture (both *viable* and *nonviable*). The inability

113

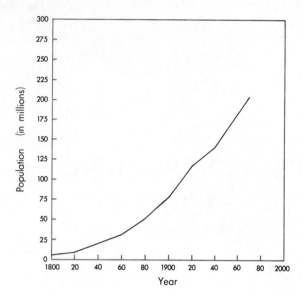

FIGURE 16. Human population in the United States from 1800 to 1968.

to reproduce is, in practice, the criterion by which microbes are said to be "dead."

Viable microbes may be counted by distributing a suspension of the organism in a suitable, liquefied, agar-culture medium. When the agar hardens, each of these microscopic individuals is then separated in space by the agar. During incubation, each of the viable microbes produces enough cells to form a visible colony. Each colony is then assumed to have originated from a single viable cell. Of course, the nonviable cells fail to reproduce in the agar and therefore cannot be detected.

The individuals in a microbial population can also be counted by direct microscopic observation of the culture. Most methods of direct microscopic counting of the cells in a microbial suspension measure *both* viable and nonviable cells. Hence, a direct microscopic enumeration of microbes is sometimes called the "total count." Ideally, then, the difference between the total count and the viable count would equal the nonviable count. The curve obtained by plotting the increase of a population of microbes during its growth will differ, depending on the technique used for counting.

Understanding the growth of populations is simplified if the mathematics of their growth is understood. The growth of microbes can demonstrate the application of these mathematics to population growth in general.

Exponential or Logarithmic Growth. A microbial cell grows to a certain size and divides to become 2 cells; the 2 new cells repeat this process to become 4; the 4 become 8, and so on. Each doubling of the population by cell division is known as a *generation* of the microbe. During periods of maximum growth, the time required for each successive doubling of a population remains constant; this time is called the *generation time*.

Since the population doubles with each generation, its numbers can be expressed by exponents of the number 2. Thus, 1 cell can be expressed as 2^0 cells, 2 cells can be written as 2^1 cells, 4 cells as 2^2 cells, and so on. Such progressions in populations are called either *exponential growth* or *logarithmic growth*.

Table 11 illustrates exponential growth during the development of a population from a single microbe with a generation time of one hour. Note the following four points in this table:

(1) The time between successive generations (generation time) is constant during exponential growth.

(2) With each generation the population doubles. Compare the constant factor of increase in *geometric progressions*—1, 2, 4, 8, 16, and so on—with the constant difference of increase in *arithmetic progressions*—2, 4, 6, 8, 10, and so on.

(3) Exponents may also be expressed as *logarithms*. A logarithm is an exponent to which power a fixed base must be raised in order to yield a certain number. Since microbes double with each generation, a handy base to use with our table is 2. This relationship may be expressed mathematically as:

$$\log_2 \chi = n \text{ (read as "log to the base 2}$$
$$\text{of } x \text{ equals } n\text{")}$$

where χ = actual number of cells produced
in n generations, and
n = the number of generations.

(4) The number of generations (n) between any two measurements in an exponentially growing population is the difference between the exponents of the two measurements. By following the arrows, it can be seen that it took two generations for 4 cells to become 16 cells. Quick arithmetic shows that the difference in exponents of $4 = 2^2$ and $16 = 2^4$ is $4 - 2$, or 2. In like manner, the number of generations which occurred between 8 or 2^3 cells and 128 or 2^7 cells is $n = 7 - 3 = 4$.

TABLE 11. EXPONENTIAL GROWTH OF MICROORGANISMS

Population age (hours)	0	1	2	3	4	5	6	7	n^*
Population growth	o →	o o →	oo oo →	oooo oooo →	oooo oooo oooo oooo oooo → etc. →		→		→
Actual number of cells	1	2	4	8	16	32	64	128	χ
Number of generations	0	1	2	3	4	5	6	7	n
Number of cells expressed as powers of 2	2^0	2^1	2^2	2^3	2^4	2^5	2^6	2^7	2^n
Number of cells expressed as logarithms of the base 2	0	1	2	3	4	5	6	7	n

*In this example we are assuming that the generation time is 1 hour. As a result, the population age in hours is equal to the number of generations; $\chi =$ number of cells in n generations. For example, the logarithm to the base 2 of 16 is 4; that is, the base 2 must be raised to the 4th power (2^4) to equal 16. "The logarithm to the base 2 of 16 is 4" may be stated in shorthand as follows: $\log_2 16 = 4$.

Logarithms and the Calculation of Growth. Since logarithms are also exponents, the number of generations can also be computed by the difference in logarithms to the base 2. Using the data of the last problem,

$$n = \log_2 128 - \log_2 8$$
$$\log_2 128 = 7 \text{ and } \log_2 8 = 3$$
$$n = 7 - 3 = 4 \text{ generations between 8 cells and 128 cells.}$$

This relationship between logarithms and the number of generations can be expressed as a generalized mathematical formula. If B is the number of microbes present after a certain period of exponential growth and b is the number present at an earlier age of the population, then

$$\text{number of generations} = n = \log_2 B - \log_2 b.$$

The logarithms to the base 2 of the foregoing examples can be computed mentally; however, it would be helpful to have a table of logarithms for any number. Tables for $\log_2$ are not readily available, but fortunately there are tables for logarithms to the base 10 ($\log_{10}$). Logarithms based upon 10 are called common logarithms. A biologist can convert from $\log_{10}$ to $\log_2$ if he is familiar with common logarithms.

A logarithm usually is not a whole number; rather, it consists of two parts—an integer, called the *characteristic*, and a decimal, called the *mantissa*. To find the logarithm of a number, we determine the characteristic by noting the position of the number's decimal point. The following are three rules for determining the characteristic of a logarithm:

(1) If the decimal point of a number immediately follows its first digit, the characteristic of its log is 0. Thus, the characteristic of the log of any number from 1 to 10, but not including 10, is 0.

(2) If the decimal point appears after the second digit, the characteristic of its log is 1; if it appears after three digits, it is 2; if it appears after four digits, it is 3, and so on. Thus, the characteristic of 711.58 is 2.

(3) If the decimal appears immediately before the first nonzero digit, the characteristic of its log is -1, usually written $\bar{1}$; if there is one zero between the decimal point and the first digit, the characteristic is $\bar{2}$, and so on. Hence, the characteristic of 0.008 is $\bar{3}$.

Using common logs, state the characteristics of the following numbers:

a. 100

b. 274

c. 1000

d. 0.4790

e. 7,456,132

f. 0.000002

g. 8.561

h. 0.0479

In this list, the characteristic of the logs of both 100 and 274 is 2; the characteristic of the $\log_{10}$ 1000 is 3. Note that the exponential expression 10^2 exactly equals 100 and that 10^3 exactly equals 1000, but that neither the digit 2 nor the digit 3 can alone express the logarithm (exponent), to the base 10, which represents 274. This logarithm is a number somewhere between 2 and 3 and is expressed as 2 plus a decimal. This decimal is called the mantissa of the logarithm.

Therefore, a logarithm is incomplete until its mantissa is determined. This is done most easily by consulting the table of common logarithms in Appendix F. The first two digits of the number for which you want to find the mantissa are listed in the left-hand column; the third digit is listed in the top horizontal column.

As an example of how the logarithm of a number is found, follow through the calculation of the common logarithm of 274.

characteristic of 274 $= 2$

mantissa $= .4378$ (found in the table at the junction of the row for the number "27" with the column for the number "4")

$\log_{10} 274 = 2.4378$ or in exponential terms,

$10^{2.4378} = 274$

Our second example will be to calculate the log of 1,378,000. You may have some difficulty in finding the logarithm of this number. Notice that 1,378,000 lies between 1,370,000 and 1,380,000. The logarithm we wish to find lies somewhere between the logarithms of these two numbers in our table. We will adopt the convention of rounding our numbers off to 3 significant digits before finding the logarithm. (There are methods of finding the logarithm more accurately, but we do not need this much accuracy for our data.) To find the logarithm of 1,378,000, we would first round off to 3 significant digits (1,380,000) and then look in the tables where we would find the logarithm 6.1399.

PROBLEM

What is the logarithm of 0.00458?

The table of logarithms can also be used to convert a logarithm to its original number. The original number is called the *antilogarithm*. As an example, find the antilogarithm of $\overline{2}.6812$. The logarithm table shows that the mantissa (0.6812) represents the digits 480. The characteristic is $\overline{2}$, so the antilogarithm of $\overline{2}.6812$ is 0.0480. Expressed as a power of 10, $10^{\overline{2}.6812}$ is 0.0480.

Suppose we were asked to find the antilogarithm of $\overline{2}.6818$. We can determine that our antilog lies between 0.0480 and 0.0481, but $\overline{2}.6818$ is closer to the logarithm of 0.0481 than it is to the logarithm of 0.0480. The value 0.6818 is only .0003 from the logarithm of 0.0481, while it is 0.006 from the logarithm of 0.0480. Therefore, we shall take as the antilog the value 0.0481. As stated before, it is possible to get a more accurate value, but this is not considered necessary for our purposes.

PROBLEM

What is the antilog of 6.9243? of $\overline{4}.7634$?

We can use our knowledge of logarithms and antilogarithms in a handy method for doing multiplication and division. As an example, consider a procedure for finding the product of 100 and 100,000. We could write 100 as 10^2 and 100,000 as 10^5. Therefore, $\log_{10} 100 = 2$ and $\log_{10} 100,000 = 5$. We know that $10^2 \cdot 10^5 = (10 \cdot 10)$ $(10 \cdot 10 \cdot 10 \cdot 10 \cdot 10) = 10^{2+5} = 10^7$. Notice that we have added the exponents of our factors to get the exponent of our product. Because these exponents are logarithms, we have actually added the two logarithms to get the logarithm of our product. Finally, if we can find out what number is represented by 10^7—in other words, what number has a logarithm of 7 when the base is 10—we will have our answer. This is exactly what we have done above, except that this time we do not need a table to determine that the number which is the antilogarithm of 7 is 10,000,000.

Using a similar argument, we could show that division involves the subtraction of logarithms. Carefully review the following two rules:

(1) To multiply, add the logs of the numbers and take the antilog of this sum. Thus, to multiply 339 $\times$ 864, add $\log_{10} 339$ to $\log_{10} 864$; the antilog of this sum is the product of 399 $\times$ 864. Compute the answer.

(2) To divide, subtract the logs of the numbers and find the antilog of this difference. Thus, to divide 2557 by 450, round 2557 to 2560 and then subtract $\log_{10} 450$ from $\log_{10} 2560$; the antilog of this difference will be the quotient. Compute the answer.

A knowledge of logarithms permits one to perform computations involving the logarithmic nature of growth. You will observe this phenomenon in the next Investigation. To illustrate, use 10 as a base to find the answer to the following problem.

Suppose a culture starts with 10 cells/ml which reproduce without interruption until 1,000,000 cells/ml are present. How many generations (n) will have elapsed? Remember from the previous discussion of generations that n may be expressed by:

$$n = \log_2 B - \log_2 b$$
$$B = 1,000,000 \text{ and } b = 10$$

Remember also, however, that from generation to generation, microbes double rather than increase by a factor of 10. Thus, we must convert our common logs to $\log_2$. To do this, we divide the common logs by $\log_{10} 2$.

As another illustration, consider the following problem. Suppose B represents the total number of cells present at a given time all having originated from a single cell. If we can determine the exponent of 2 so that $B = 2^n$, then, as before, n is the number of generations required to produce B cells. Therefore, we must devise a method for finding n if we know B.

If $B = 2^n$, then certainly $\log_{10} B = \log_{10} 2^n$, because if two numbers are equal they can both be represented by 10 raised to the same power. For example, suppose $B = 10^{2.3142}$. If $B = 2^n$, then 2^n would also have to be equal to $10^{2.3142}$. In other words, their logarithms to the base 10 are equal. For our next step we will show that $\log_{10} 2^n = n \cdot \log_{10} 2$. To do this, first look at $\log_{10} 2^3$. Recall that $\log_{10} 2 = .3010$; that is, $10^{.3010} = 2$. We can then write 2^3 as $(10^{.3010})^3$, but $(10^{.3010})^3 = 10^{.9030}$. (Recall from algebra that $(2^3)^4 = 2^{12}$ and so on.)

We now have $2^3 = 10^{.9030}$ or, in other words, $\log_{10} 2^3 = .9030$. Since $\log_{10} 2 = .3010$, we observe that $\log_{10} 2^3 = 3 \log_{10} 2$. This is a *specific* example, but it is *always* true that $\log_{10} 2^n = n \cdot \log_{10} 2$.

We have previously shown that if $B = 2^n$, then $\log_{10} B = \log_{10} 2^n$. We can now state that $\log_{10} B = n \cdot \log_{10} 2$. Dividing both sides of the equation by $\log_{10} 2$, we have

$$\frac{\log_{10} B}{\log_{10} 2} = n$$

where n is the number of generations. This is the method we will use to find the number of generations (n) if we know the number of cells present (B) at a given time.

For example, suppose we start with a single cell. How many generations will have elapsed when we have 10,000,000 cells? We want to express 10,000,000 as 2^n where n is the number of generations. As we explained,

$$\frac{\log_{10} 10,000,000}{\log_{10} 2} = n.$$

$\log_{10} 10,000,000 = 7.0000$ and $\log_{10} 2 = .3010$ so that,

$$\frac{\log_{10} 10,000,000}{\log_{10} 2} = \frac{7.0000}{.3010} = 23.3 \text{ generations.}$$

Of course, .3 generations is meaningless since generations are represented by whole numbers.

Apply this reasoning to another problem. Suppose a culture begins with 10 cells/ml and reproduces without interruption until 1,000,000 cells/ml are present. How many generations (n) will have elapsed?

Remember that n may be expressed as

$$n = \log_2 B - \log_2 b$$

where $$B = 1,000,000 \text{ and}$$
$$b = 10.$$

Now, $$\log_2 B = \frac{\log_{10} B}{\log_{10} 2} \text{ and } \log_2 b = \frac{\log_{10} b}{\log_{10} 2}$$

therefore, $$n = \frac{\log_{10} B}{\log_{10} 2} - \frac{\log_{10} b}{\log_{10} 2} = \frac{\log_{10} B - \log_{10} b}{\log_{10} 2}$$

$$n = \frac{\log_{10} 1,000,000 - \log_{10} 10}{\log_{10} 2}$$

$$= \frac{6.0000 - 1.0000}{.3010} = \frac{5.0000}{.3010}$$

$$= 16.6 \text{ generations}$$

The *growth rate* of the population is a measure of its increase in numbers during a given time and can be expressed as $r = \frac{n}{t}$, where r is the growth rate, n is the number of generations, and t is the time required to produce that number of generations. The time required for the population to produce a new generation is called the *generation time* (g) and can be expressed by:

$$g = \frac{\text{total time between } B \text{ and } b}{\text{number of generations between } B \text{ and } b}$$

If in our example the lapsed time between 10 and 1,000,000 cells was 10 hours, then

$$g = \frac{10}{16.6} = 0.6 \text{ hour, or 36 minutes.}$$

In other words, this population doubled in number every 36 minutes. Its growth rate is $r = \dfrac{16.6}{10}$ or 1.66 generations per hour.

Plotting the Microbial Growth Curve. Now consider the graph of a microbial population of 100 microbes having a generation time of 1 hour and immediately starting to multiply by cell division. The graph of this population may be plotted by using the actual number of organisms (y-axis) as a function of culture age (x-axis). However, as the plot in Figure 17A shows, the actual number of microbes soon attains immense values requiring a huge sheet of graph paper for accurate plotting. More important, the population increase is really not a direct arithmetic function of culture age. The resulting line is a curve rather than a straight line.

However, population growth will become a straight-line function of culture age if the data are plotted on a semilogarithmic graph. In this graph, the population is plotted by logarithms of the number of organisms (see Figure 17B, page 123), while the culture age remains on an arithmetic scale. By the proper scaling of these logarithms, enormous numbers of individuals may be plotted on ordinary semilogarithmic graphing paper. By extrapolation, it can be seen that up to 10,000,000 (or 10^7) organisms involving 14 generations can be plotted on semilog graph paper. The arithmetic scale, on the other hand, will not permit us to graph even the fifth generation of 3200 organisms.

Characteristics of the Microbial Growth Curve. Population growth cannot continue indefinitely. It is limited by space, food, and other factors. A "typical" microbial culture develops in a way remarkably similar to other expanding populations of living things. Figure 18

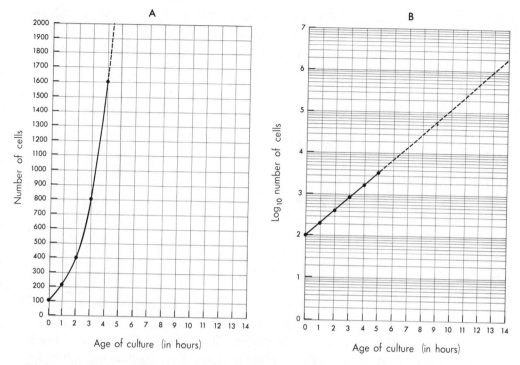

FIGURE 17A and B. Comparison of plots of microbial growth using arithmetic and logarithmic scales to express the number of individual organisms.

(page 124) shows the rise and fall of a population of microbes in a defined environment.

Some rather definite phases (A, B, C, and D) can be recognized in this idealized plot of the growth curve. Do not be surprised if your results fail to reproduce this plot exactly. If deviations from this "typical" growth curve are found, they will be understood better by a recognition of those factors which influence population development.

First, note Phase A. This period, called the *lag* period, is often found in populations. It appears before newly developing populations enter into the period of logarithmic (exponential) growth (Phase B). One explanation for the static lag period is that the cells must adjust to their new environment before division begins. Perhaps new enzymes must be synthesized before growth occurs. Another explanation for the lag phase in certain cases is that a mutant cell is naturally selected from the population at large as the dominant organism of the culture. In this case, most of the cells inoculated at the beginning (zero time) fail to reproduce in the new environment of the culture medium, while the mutant strain may be highly successful. Some investigators suggest that although cell division may lag in Phase A, syntheses of several important constituents in the individual cells do occur at this time. Thus, individual cells may grow during the lag period, but there is little or no increase in the number of individual organisms.

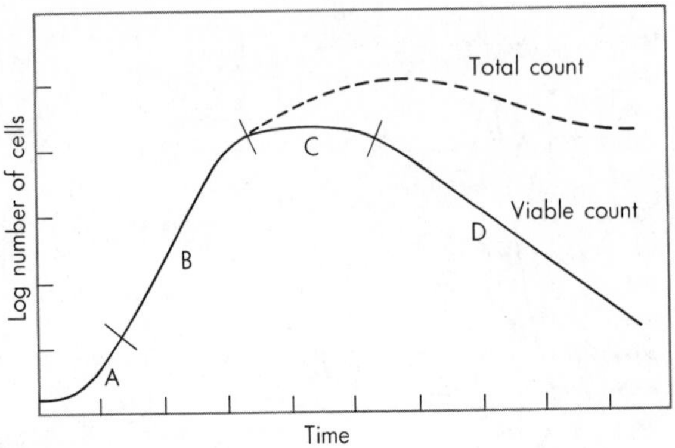

FIGURE 18. Idealized microbial growth curves as determined by viable and total counts of the same population.

Now note Phase B in Figure 18. This is the exponential or logarithmic *growth* phase. It represents the time during culture development when a constant rate of increase in the number of microorganisms is maintained. Nearly all the cells are viable and reproducing. This constant increase in numbers of individuals per unit of time eventually leads to the maximum population of the culture. When sufficiently large numbers of microbes have accumulated, food shortages, lack of space, the accumulation of toxic products, and so on, cause the population to enter the next phase.

Phase C is known as the *maximum stationary* phase of a culture. A population may remain in this phase for some time. Note that the total count (viable plus nonviable cells) continues to increase beyond the beginning of the stationary growth phase for viable cells. The explanation for this is that the production of new cells in the viable-cell curve is counterbalanced by the death of old cells.

Phase D is called the *death* phase. The ratio of viable to nonviable cells becomes less and less. The death rate, like the logarithmic growth rate, attains a constant value. Later, the curve for the death phase may undergo considerable change. Mutant cells, resistant to the harmful effects of the old environment, may appear and give rise to new populations. Old cells may disintegrate, causing the total count and the viable count to decrease. On the other hand, the products of cell breakdown may serve as the substrate for forming new cells. (Watch for these phenomena when plotting your results for Investigation 8 which deals with the growth of a yeast population, *Saccharomyces cerevisiae*.)

PROBLEMS

1. Starting with one bacterium with a generation time of one hour and assuming that all cells remain viable, how many individuals would be present after 24 hours? after 4 days? Express these numbers of cells as exponents of 2.

2. Assume that an increase in the temperature of incubation for the culture described in question 1 reduced the generation time to 30 minutes. How would this affect your answers to the question?

3. Examine the family of curves in Figure 19 demonstrating parts of the growth curves from four cultures, A, B, C, and D. Which of the following statements are correct and which ones are incorrect? Give the reasons for your answers.

   a. Culture A has a more rapid rate of growth (generation time) than does Culture C.

   b. Culture A has a more rapid rate of growth than does Culture B.

   c. Culture D may be in either the initial lag or maximum stationary phases of the growth curve.

   d. Cultures A and C will both attain the same populations in the maximum stationary phase.

   e. Cultures A, B, and C are each in the logarithmic phase of growth.

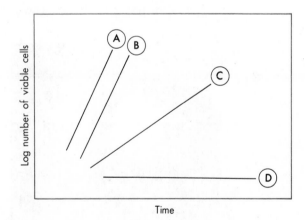

FIGURE 19. Family of growth curves.

Microbial Methods and Techniques

Biologists had to solve many problems in order to grow microbes for study in the laboratory. One such problem was the nutritional requirements of bacteria. Different sorts of organisms require different foods. You would not think of feeding the same food to a fish, a dog, and a cow; each requires a special food. The different species of microbes are just as demanding in their nutritional requirements: some will grow in a medium containing simple inorganic molecules; others require complex organic molecules for growth.

Culturing Microorganisms. If a pinch of soil and all the substances needed for growth of microbes were placed in a container, the result would be a tremendous hodgepodge of species. Many kinds of bacteria, yeasts, and molds would grow and form colonies, and it would be impossible to learn much about any single species. Serious work in microbiology was difficult until methods were developed for growing a single species in pure culture.

During the earlier career of Louis Pasteur (Figure 20) and German microbiologist Robert Koch (Figure 21), it was difficult to obtain pure cultures. The standard practice was to grow microbes in a culture solution containing all the substances necessary for their growth. When Pasteur was interested in the microbes involved in wine production, he would place a tiny drop of wine in a culture solution. The original drop would frequently contain more than one species of microbe. It was difficult, therefore, to obtain a pure culture.

The problem of separating one species of microbe from its neighbors so that it could be grown in pure culture was finally solved

FIGURE 20. Louis Pasteur.

FIGURE 21. Robert Koch.

in the laboratories of Robert Koch. Koch and several other biologists discovered that a solid, rather than a liquid culture medium, could be used to isolate pure cultures of microbes. As is often the case in science, the discoveries of Koch and his group were based on the observations of another investigator, Karl Schroeter.

In 1872, Schroeter noted that bacteria would grow on the surface of several materials such as that of a cut potato. The bacteria formed large masses, called *colonies*. When these individual colonies were examined under a microscope, each was found to consist of great numbers of individuals. More important, however, all organisms in a single colony were usually of the same species. Apparently single cells had landed on the surface of the potato and had then multiplied into millions of cells. These colonies thus became visible to the naked eye, and because all the individuals in a single colony were derived from one cell, they formed a pure culture.

Koch then tried spreading dilute suspensions of several species of microbes over a solid culture medium. The individual cells would stick at different locations on the surface, and the invisible single cells would start to multiply. Eventually, as Schroeter had noted, millions of cells resulted to produce a colony from each original cell. Each colony was a pure culture.

Koch, however, was dissatisfied. The potato is not an adequate food for all species of microbes; in fact, only a few kinds will grow on a slice of potato. The potato slice had another serious disadvantage: it was difficult to keep sterile. When it was sliced, microbes from the knife, or even from the air, would become attached to its surface. These, too, would grow and form colonies. One could not be sure whether the colonies came from accidental contamination of the potato or from the microbes placed on the potato by the investigator. It was necessary, therefore, to develop a culture medium that would (1) have all the substances necessary for the growth of microbes and (2) be sterile.

Koch had already made a culture medium known as *nutrient broth*. His nutrient broth contained meat extract and partially hydrolyzed protein (peptone). These products served as sources of amino acids, vitamins, carbohydrates, and other essential nutrients. To solidify the broth, he added gelatin. Then he boiled the mixture in order to kill any microbes that were present.

To solidify the liquid mixture, Koch poured it into sterile, specially constructed dishes, known as *petri dishes* or *petri plates,* so named because one of Koch's students, Petri, designed them. Each half of a petri dish is a round, flat-bottomed glass dish with vertical sides. The culture medium is poured into the bottom half of the dish; the other half serves as a lid to prevent microbes in the air from getting on the culture medium. The closed dishes can be sterilized by heating in an oven or sterilizer.

Koch's nutrient medium contained materials necessary for the growth of many species of microbes. It was also possible to keep the medium sterile. He could smear microbes across the medium, a process known as *streaking*. The streaking separated individual cells, and from each cell produced a pure colony.

Koch discovered, however, that gelatin has its disadvantages. It melts at 28°C; this is below the best temperature for the growth of many microbes. In addition, many microbes synthesize an enzyme, *gelatinase,* which digests gelatin, causing the medium to liquefy. Whenever the medium liquefied, the separate, pure colonies would run together and mix.

It required the experience of Frau Hesse, the wife of one of Koch's students, to solve Koch's problems with gelatin. On her suggestion, Koch replaced gelatin with an extract of seaweed known as *agar*. Agar proved to be ideal for the special purpose of making a solid culture medium. It is attacked by only a few microbial species, and, in a solid state, it does not soften until it reaches a temperature of about 90°C. Liquid agar does not solidify until it is cooled to 42°C. The fact that it remains liquid until cooled to this temperature makes possible a refinement of the methods of obtaining pure cultures. Instead of streaking mixed cultures to isolate pure ones, the researcher can mix the culture with melted, but cooled, agar and pour the mixture into sterile petri dishes. When the agar sets, the individual bacteria are trapped at different locations in the agar where they develop into separate colonies.

Today, the pure-culture techniques of Koch are still used in microbiology laboratories. When cultivating microbes in the laboratory, microbiologists still use his old but completely adequate techniques. Koch was the first scientist to demonstrate convincingly that a particular disease is produced by infection with a specific bacterium. He could never have done this without first having developed a method of obtaining pure cultures.

When it became possible to grow microorganisms in pure culture, it also became possible to observe the effects of the environment (the physical and chemical surroundings) on the growth of microbes. Not all species of microbes were able to grow in the first medium developed by Koch. It was found that if different substances were added to the medium, different species would be able to grow. One species might grow well on a medium that contained only a few types of molecules; another species might require a medium with many kinds of molecules.

Other environmental conditions similarly affect each microorganism's ability to grow and multiply. Some species grow best at 20°C and others at 30°C. Some require acid conditions; others require basic conditions. Some need oxygen; others do not.

Laboratory Rules. The microbial attributes of small size and rapid reproduction are an advantage to the researcher who is using these organisms to study populations of living things. These attributes are, however, also of advantage to the microbe and help explain its success in the world. A trifling inoculation with microbes, inconspicuous though it may be at the start, may cause serious infections to develop. Fortunately, the vast majority of microbes is not harmful to human beings.

If work with microbes is to proceed successfully and safely, the general rules which apply to all laboratory work (see page 17) must be carefully followed. These rules are largely to prevent contamination and to protect you from dangerous organisms. Although in this course you will not intentionally be working with dangerous organisms, the chance of inadvertent contamination must be guarded against. Therefore, a few additional rules which apply to the Investigations in this section are:

1. *Never lay contaminated and used equipment (pipettes, inoculating loops, and so on) on desk tops—use pipette jars, trays, or other provided receptacles.*
2. *If a living culture is spilled, immediately notify the instructor and then disinfect your hands and the contaminated table area. (A 2% aqueous Lysol solution is convenient for this purpose.)*
3. *Liquid waste, agar, and contaminated materials should be discarded in special containers.*

Transfer of Agar Cultures. Before agar slant cultures of living organisms are used, each person should practice the following manipulations several times. Instead of using the culture of *Bacillus subtilis* suggested in step 1, use two sterile slants. Make a transfer from one tube to the other as if you were actually inoculating the second tube from the first. After using these sterile slants, incubate them for 3 days and observe. If any growth occurs, your techniques were not aseptic.

To transfer live cultures, use the operations that are outlined in the following steps and illustrated in Figure 22.

1. Hold the test tube containing an agar slant culture of *B. subtilis* and a tube containing a sterile agar slant in your left hand. Hold the inoculating loop in your right hand.
2. Heat the transfer loop in the burner until it glows orange. Allow it to cool for a period of approximately 30 seconds.
3. While the loop is cooling, open both test tubes by placing the plugs between the fingers of the right hand. Flame the mouths of the test tubes by passing them slowly through the flame 3 times.
4. Remove a small bit of bacterial growth from the surface of the agar culture and transfer it immediately to the surface of the sterile nutrient agar. Do not dig into the agar surfaces. The pressure of the needle is sufficient to remove and to apply the inoculum.
5. Flame the needle again and pass the mouths of the test tubes through the flame as before.
6. Replace the cotton plugs and stand the test tubes up in the rack provided for this purpose.

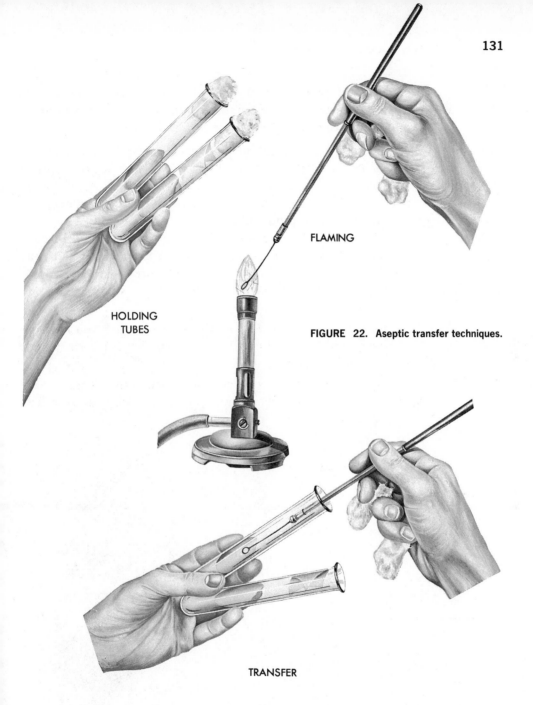

FLAMING

HOLDING
TUBES

FIGURE 22. Aseptic transfer techniques.

TRANSFER

7. This transfer practice may be repeated by using a tube of sterile nutrient *broth* and another tube of nutrient *broth* containing a culture of the bacterium, *B. subtilis.*

8. Incubate your culture for 48 hours at either 22°C or room temperature and observe it for growth.

Volumetric Transfer of Broth Cultures. A pipette will be used to transfer a known volume of a liquid culture of *B. subtilis* to a tube of sterile nutrient broth. Use the following procedure:

1. Unwrap a sterile 1.0-ml pipette. Take care that the 2 or 3 inches nearest the tip touches nothing that is not sterile. If you touch any part of the pipette except that small portion held in your fingers, to any object in the environment, the pipette will become contaminated and cannot be used for transfer. Hold the pipette between your thumb and second finger. This will leave your index finger free for easy manipulation over the end of the pipette (see Figure 23). Practice drawing up and measuring quantities of pure water before proceeding.

2. In your left hand, hold the tube of broth which you have just inoculated and a tube of sterile broth. Take a sterile pipette in your right hand.

3. Open both test tubes as you did when transfers were made with a loop. Flame the mouths of the test tubes by passing them through the flame 3 times.

FIGURE 23. Proper method of holding the pipette.

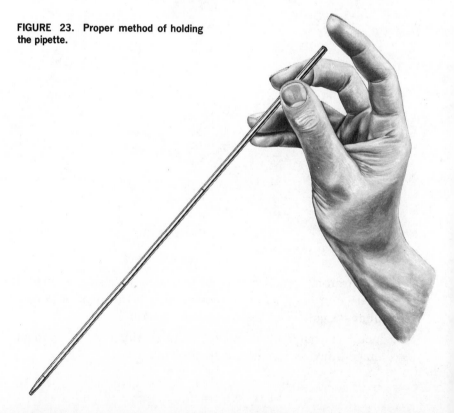

4. Insert the sterile end of the pipette into the inoculated broth culture. Use *careful* mouth suction to fill the pipette to the 1.0 ml mark. Remove the end of the pipette from the inoculated broth to a position about 1 inch above the surface of the sterile broth. Keep your index finger firmly over the end of the pipette so that *no dripping occurs.*

5. Aseptically allow 0.1 ml of the inoculated culture to fall on the surface of the medium in the tube of sterile broth.

6. Immediately place the pipette and its remaining contents into a disposal cylinder containing 2% aqueous Lysol.

7. Flame the mouths of the test tubes and replace the cotton plugs. Stand both tubes of inoculated broth in the rack provided for this purpose.

8. Incubate cultures at room temperature or 37°C for 48 hours and observe for growth.

The Hemocytometer. A direct census of microbes in a given volume of liquid must necessarily be taken by microscopic examination. A common device for providing known small volumes of fluid for microscopic examination is the counting chamber or *hemocytometer* (so-called because it was developed originally for counting cells in the blood). The counting chamber is a microscope slide with regularly ruled chambers, each holding a known volume of liquid. A glass cover is placed over the slide before counting.

By counting the cells in samples of known volumes taken from a culture of a known volume, the population of the whole culture or any part thereof can be estimated. To justify such estimates, every effort must be made to be sure that the tiny volumes which are being counted are random samples of the larger culture volumes.

The figures on page 134 show two views of the rulings in the counting chamber. Figure 24A shows the entire ruled area of the chamber. Note that it consists essentially of nine squares of one square millimeter (mm^2) each. Each of these is subdivided into smaller units. Only one complete square millimeter area can be observed at a time in the low-power ($100\times$) field of your microscope. Such an area is indicated by the circle on Figure 24A. Since many of the populations of yeast and fungal spores you are to count will be counted in this center area, we should concentrate on it for a moment.

Figure 24B shows an enlargement of the rulings of the center, large square. Note that this 1 mm^2 area is subdivided into 25 squares which are 0.2 mm on a side (0.04 mm^2 in area). With these pictures

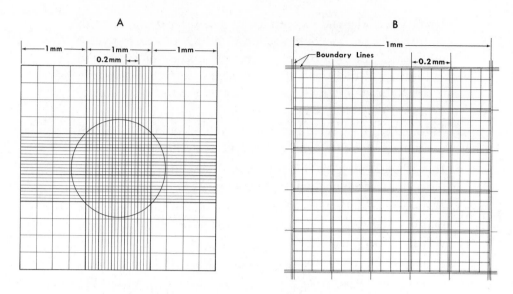

FIGURE 24. Two views of the rulings in the counting chamber.

clearly in mind, examine the counting chambers with the 100-power lens system of the microscope. Note the squares and rulings as outlined in Figure 24.

Each large square has an area of 1 mm^2; however, counting techniques are concerned with the number of individuals in a given volume rather than in a given area. Here then is the problem: How can we calculate the volume of a liquid over a flat area of a known size? To do this, the third dimension—the depth of the liquid over the ruled area—must be considered. The distance between the bottom of the counting chamber and the cover glass overlaying the chamber determines the depth, which in this case is 0.1 mm. The volume (length times width times depth) of a single large square can now be calculated as $1.0 \times 1.0 \times 0.1 = 0.1$ mm^3.

Counts of microorganisms are commonly recorded as the number of individuals per cubic centimeter (also cm^3, cc, or ml). To estimate the number of cells per cm^3 from those present in a counted sample of 0.1 mm^3, one must know how many 0.1 mm^3 there are in 1 cm^3. Remember, 1 cm = 10 mm, or, stated in another way, 1 mm = 0.1 cm. The 1 mm^2 area of the large squares can then be converted to cm^2 as follows:

$$1 \text{ mm}^2 = 1 \text{ mm} \times 1 \text{ mm} = 0.1 \text{ cm} \times 0.1 \text{ cm} = .01 \text{ cm}^2$$

Next, to obtain the volume in cm^3 of the area represented by a large square, one may convert its depth of 0.1 mm to 0.01 cm. Then the volume represented by a large square becomes, as before, length times width times depth, or

$$0.1 \text{ cm} \times 0.1 \text{ cm} \times 0.01 \text{ cm} = 0.0001 \text{ cm}^3$$
$$= (1/10,000 \text{ cm}^3)$$

Thus, if the number of organisms in one large square is counted, the number in 1.0 cm³ of the culture can be estimated by multiplying the count by 10,000. Figure 25 compares the volume of 1 mm³ with that of 1 cm³.

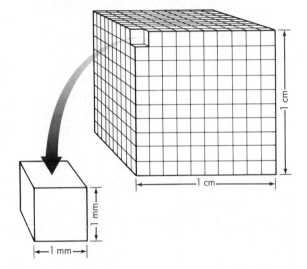

FIGURE 25. A comparison of the volume of a one-millimeter cube with that of a one-centimeter cube.

As an example, suppose that 210 yeast cells were counted in the large center square. Expressed in numbers of cells per cm³, the original population contains

$$210 \times 10,000 = 2,100,000 \text{ cells per cm}^3,$$

or, expressed as an exponential quantity, 2.10×10^6 cells per cm³.

You should count about 200 to 300 cells in a specified area in order to obtain a reasonably accurate count. Some cultures will be populations which do not contain 200 to 300 cells in a single large square. In these cases, the other large squares or the smaller squares may be used to count volumes with approximately 200 to 300 cells.

Counting the Cells in a Yeast Suspension. Use the following steps for practice in counting:

1. Shake an aqueous suspension of yeast cells well and, before it settles, use a sterile pipette to remove 0.1 cc of the suspension.

2. Replace the first two drops of this suspension from the pipette into the original suspension. Work rapidly but carefully.

3. Immediately transfer enough of the cell suspension from the pipette to the counting chamber so that no air bubbles are formed. If air bubbles are formed, remove the cover glass and wash the slide before repeating the procedure.

4. Mount the counting chamber on your microscope and observe it under low power.

5. After the lines of the counting slide are in focus, find the large center square in your microscopic field and reduce the light until the small, oval, yeast cells are in focus.

6. When you start counting the cells in the large center square, use as a guide not the smallest but the medium-sized squares (0.2 mm × 0.2 mm). Start counting from the top row of these medium-sized squares and continue to the bottom row. Some cells will probably touch the lines forming these squares. Count these cells only if they are on the lines forming either the top or the right sides of the square. If a cell touches either the bottom or the left sides of a medium-sized square, do not count it; it will be counted with another square.

7. Count approximately 200 to 300 cells before determining the number of cells per cm³. Four different situations may arise in your counting:

   a. There may be about 200 to 300 cells per large square (0.1 mm³ or 1/10,000 cm³). In this case, multiply the cell count by 10,000 to obtain the number of cells per cm³. (Recall from the discussion of exponents that 10,000 may also be expressed as 1×10^4.)

   b. There may be fewer than 200 cells visible in the large center square. It will then be necessary to count cells in several of the same size corner squares to obtain a total of about 200 cells. Divide the total number of cells counted by the the number of squares involved to obtain the average for one square. Then multiply this number by 10,000 to obtain the number of cells per cm³. For example, 240 cells were counted in four large squares. This is an average of 60 cells per square. Sixty cells per large square $\times 10^4 = 6 \times 10^5$ cells/cm³.

c. There may be a much larger number of cells than 200 per large square. If all cells in a large square were counted, too much time would be spent. In this case, count 200 to 300 cells in the medium-sized (0.2 mm × 0.2 mm) divisions of the large square. (Remember that the large center square is divided into 25 medium-sized squares.) To get the average number of cells per medium-sized square, divide the total number of cells counted by the number of medium-sized squares counted. Multiply this average by 25 to estimate the number of cells per large square. This number times 10^4 equals the number of cells per cm^3.

Example: There was a total of 210 cells in three medium-sized squares. How many cells were there per cm^3 in the original culture?

$$\frac{210}{3} = 70 \text{ cells per medium-sized square}$$

$$70 \times 25 = 1750 \text{ cells/0.1 mm (1750 cells/large square)}$$

$$1750 \times 10^4 \text{ (10,000)} = 17,500,000 = 1.75 \times 10^7 \text{ cells/}cm^3$$

d. The cell numbers may be too dense to count. The suspension must then be diluted before a count is made. Do not forget to multiply the final cell number by the amount with which your suspension was diluted.

Example: 1 cm^3 of a yeast culture was diluted with 99 cm^3 of water (1/100 dilution) before using the counting chamber. When counted, this dilution yielded 220 cells per 0.1 cm^3. How many cells were there per cm^3 of the original culture?

$$220 \times 10^4 \times 100 = 220,000,000 \text{ } (22 \times 10^7) \text{ cells/}cm^3$$

PROBLEMS

1. Suppose the center large square of your chamber contains 53 yeast cells. You count the four large corner squares and find that they contain 51, 54, 55, and 60 cells, respectively. What is the best estimate of the total number of yeast cells per cm^3 in your original culture? the best estimate of the total number of cells per liter in your original culture? (Use the relationship, 1000 cm^3 = 1 liter.)

2. You have a suspension of yeast cells suspected to contain approximately 300,-000,000 cells per cm³. How many cells is this per 0.0001 cm³ (the volume of one large square)? How much should a sample of this culture be diluted before counting in order to obtain about 300 cells per large square?

INVESTIGATION 8
Growth of a Yeast Population

The yeast *Saccharomyces cerevisiae* has an oval shape and measures 50 microns (μ) in diameter. Asexual reproduction in this yeast comes about by a process of cell division known as budding. In budding, the mother cell forms a small outgrowth or bud at its surface. The bud enlarges until it is about the size of the mother cell. Then nuclear division occurs, and one new nucleus goes to each cell. Finally, a cross wall is laid down between the two cells and they separate into two daughter cells. Both of these may reproduce a third generation, and so on. A photograph showing the different steps involved in a single generation of a yeast cell is shown in Figure 26.

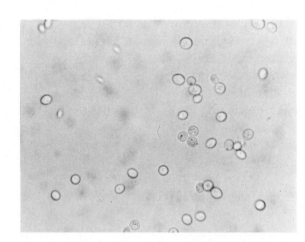

FIGURE 26. Budding and reproduction by the yeast, *Saccharomyces cerevisiae*.

Some of the cells in our population studies will be in various stages of budding. This leads to the question, "Should budding cells be counted as two cells or as one cell?" Think this question over and make your decision before the experiment starts. Discuss the reasons for your decisions until the class reaches an agreement as to the best course of action for the group.

MATERIALS (per class)

1. One package dry yeast

MATERIALS (per team)

1. One microscope
2. One hemocytometer
3. One 250-ml flask

4. Two 150-ml flasks
5. One volumetric pipette,
   1-ml accuracy

PROCEDURE

1. Suspend 1 g of dried yeast in 99 ml of sterile water.
2. After shaking the liquid to suspend the yeast cells uniformly, transfer 1 ml of this stock to a flask containing 99 ml of sterile water.
3. Shake the second dilution of yeast cells to form a uniform suspension. Transfer 1 ml of the second suspension to each of two flasks containing 49 ml of sterile *culture medium*.
4. Each student should determine the number of yeast cells per ml in the second (99-ml) dilution flask by use of the hemocytometer counting chamber. How many cells per ml are to be found in each of your two culture flasks immediately after inoculation? This number represents the population count per ml in your cultures at the zero (starting) time of your study. Record it on a data chart similar to Chart A.
5. Each team now has two cultures of *S. cerevisiae*. Weigh each flask to the nearest g and record this weight. Incubate one of these at 12°C and the other at 22°C.

CHART A. POPULATION GROWTH IN YEAST CULTURES MAINTAINED AT 12°C AND 22°C

Date	Time of day	Age of culture	Team average of cells per ml at	
			12°C	22°C
		0		
		24		
		48		

6. Each student should count and record the number of yeast cells in each of the two cultures daily. Each count should be recorded in the notebooks of the individual students and the average number entered in the chart. For later ease in computations and graphing, these counts should be recorded as exponents of 10. For example, 1,400,000 cells per ml may be expressed 1.4×10^6 cells per ml. Continue the daily counts until the cultures are about 10 days old. Weigh each flask to the nearest g each time a count is made and record the weight.

7. Graph your own and your team average on semilogarithmic graph paper. The exponential scale on the y-axis is used to indicate the numbers of organisms present at a particular time. The time is shown on the x-axis with an arithmetic scale.

8. The data from each team should now be pooled with data from the rest of the class. The mean values for the class at each temperature ($12°C$ and $22°C$) may be recorded in charts similar to Chart B.

CHART B. INDIVIDUAL COUNTS BY TEAMS PERFORMING YEAST POPULATION GROWTH STUDIES AT 12°C

Age of culture (hours)	Number of cells per ml team number 1, 2, 3, 4, 5, etc.	Total	Mean ($\bar{x}$)	Standard deviation (s)
0				
24				
etc.				

9. Plot the yeast population for the class on the same semilogarithmic graph paper which was used for plotting individual and team data.

10. Plot the weight loss for each flask. Correct the observed weight loss for the weight loss due to water evaporation. Your teacher has set up a flask similar to yours but containing only water and has weighed it over a similar period of time. You will be provided with the information so obtained. On the last day, shake the flask vigorously to remove as much CO_2 as possible.

11. The yeast (=alcoholic) fermentation is 1 sucrose→4 ethanol + 4 CO_2. The molecular weight of CO_2 is 44; therefore, for each 44 grams of CO_2 lost (represented by the weight lost by your flasks), there would be one molecular weight of ethyl alcohol (ethanol) formed. Since this has a molecular weight of 46, for each gram of CO_2 formed you will have $46/44 = 1.04$ gram of alcohol formed. At what % alcohol did the fermentation stop?

INTERPRETATIONS

1. Compare the charts and graphs of the pooled-class data with those of your own team. Calculate the standard deviations of your team's data on different days. Compare these values with the standard deviations from the data of the class. Are there variations in the two? Discuss the reasons for any such variations.

2. Concentrate on a 48-hour period during the logarithmic period of growth. Calculate the following from your team's data and from the data of the class.
   a. How many generations occurred during this 48-hour period in the cultures grown at 12°C and 22°C?
   b. What was the generation time in each culture?
   c. Considering generation time as a *growth rate*, how much faster (or slower) was the rate of growth at 22°C than at 12°C?
   d. Do you think your answer for part c would hold for any 10°C difference in temperature? Explain.

3. Observe the period beyond the phase of logarithmic growth.
   a. Did the total number of individuals in your culture remain constant each day during this period?
   b. What are some explanations for the particular constancy or variations which your graphs show in this period? Did your results agree with those of the class?
   c. If you had determined viable cells, what may have been the general trend of the growth curve in its last few days? What procedures could be used to check this hypothesis? If time permits, check your hypothesis with results from another set of cultures.

PROBLEMS

1. Figure 16 (page 114) presents a population-growth curve for the United States (1800–1968). From what you have seen in your yeast-population studies, in what phase of growth do you think our national population is now?

2. Extrapolate the U.S. population curve to the year 2000. What are some possible events that may cause your extrapolation to be invalid?

3. The common and most reliable method for determining "live" microbes is by the somewhat cumbersome and lengthy plate-count method. A principle of staining is that only "dead" cells take stain, a factor which could provide a rapid method of distinguishing viable cells from nonviable ones. How would

you design an experiment to test how different stains A, B, C, and D might correlate with the plate-count method in determining the number of viable cells in a culture?

4. How do you think temperatures of 5°C, 35°C, 45°C, and 55°C might affect the growth of yeast populations? Test your hypothesis.

5. The criterion for calling a microbe "living" is more commonly based on its ability to reproduce. This criterion is not used in judging many other "living things" as live or dead. Why is it a convenient one for microbes?

6. What probably happens to the shapes of yeast protoplasts when dry yeast is placed in pure water? Why?

PATTERN OF INQUIRY 5
Factors Affecting Population Development

A microbiologist was investigating the effects of an antibiotic on the inhibition of growth of a species of bacteria. He made a series of dilutions of the antibiotic in a culture medium and added the same amount of inoculum to each. After 24 hours, he counted the number of viable cells remaining in each concentration of the antibiotic. He obtained the following results:

EXPERIMENT 1

Concentration of antibiotic in micrograms per liter	Number of viable cells per ml after 24 hours of incubation
none	6×10^8
0.1	3×10^7
0.2	5×10^4
0.4	2×10^2
0.8	none
1.6	none
3.2	none

A few days later, he wished to repeat the experiment, but he found that his original culture of bacteria had become contaminated. He still had the plates which he used in making the counts, and he made a new isolation from one plate which had only a few colonies on it. Using routine procedures, he found that he had managed to isolate a pure culture of the original species. Using this new culture, he repeated the first experiment and obtained the results in Experiment 2.

What explanations can you give for the different results of the two experiments?

EXPERIMENT 2

Concentration of antibiotic in micrograms per liter	Number of viable cells per ml after 24 hours of incubation
none	8×10^8
0.1	6×10^8
0.2	2×10^8
0.4	9×10^7
0.8	4×10^7
1.6	7×10^6
3.2	8×10^3

Populations may be composed of many individual cells, or they may be composed of cells with a degree of relationship and control over each other. In both cases the cells may be all of one kind, a "pure culture," or the cells may be different. In the populations composed of individual cells, the cells may interact with each other, and, if there are mixed populations, there may be antagonisms between them; but while the cells do interact, the individual cells are not dependent on the interaction, and each cell could grow by itself without the presence of the others. But populations of cells also exist in which there is a relationship among the cells. You and I, for example, are a population of cells, but should we cut off a finger, it could no longer remain a finger; and even if we cultivated the cells from it, they would become a mass of cells, not a structure. In general, when we think of populations, we are thinking of the "uncontrolled" kind where each individual may go his own way. This is the type of population we shall study first. Later we shall look at the more complex populations represented by those dependent on cell interaction—even as you and I.

INVESTIGATION 9
A Mixed Population

Soil represents an environment in which many organisms live and thrive. Among the microorganisms are yeasts, molds, actinomycetes, and bacteria. The latter two exist in the soil in amounts of a million or more individual organisms per gram, and it is these which we will examine.

MATERIALS (per team)

1 g of soil

5 sterile 9-ml water blanks

6 sterile petri dishes

1 flask sterile soil agar (approximately 50 ml)

1 flask sterile nutrient agar (approximately 50 ml)

1 spreading rod

70% alcohol

Sterile 1-ml pipettes graduated in 0.1-ml units

1 Bunsen burner

PROCEDURE

1. Prepare a dilution of soil to 1/1,000,000 as shown in the diagram. If your pipettes are not graduated in 0.1-ml units, you can use 2 drops to equal 0.1 ml.

FIGURE 27.   Serial dilution of soil.

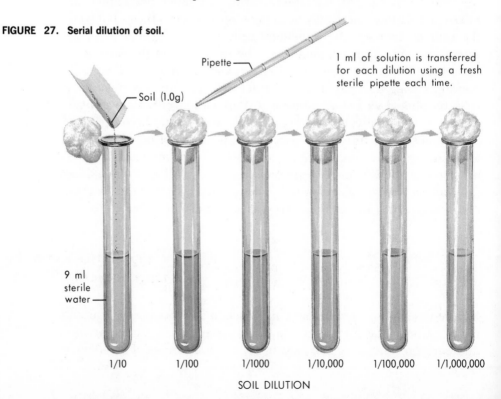

Pipette

1 ml of solution is transferred for each dilution using a fresh sterile pipette each time.

Soil (1.0g)

9 ml sterile water

1/10   1/100   1/1000   1/10,000   1/100,000   1/1,000,000

SOIL DILUTION

2. Under sterile conditions, pour 3 petri dishes with "soil" agar and 3 petri dishes with nutrient agar; allow to solidify.

3. When the agar is solid, add 0.1 ml of the 1/10,000 dilution to one plate of soil agar and to one plate of nutrient agar. Spread over the surface with a spreading rod so that the entire surface is moistened. Label these the 1/HT plates. The soil plate contains 1/100,000 gram (0.01 mg or 10 μg) of soil.

The spreading rod is a sterile glass rod as shown in Figure 28.

FIGURE 28. Spreading rod.

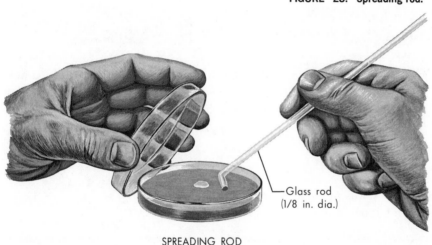

Glass rod
(1/8 in. dia.)

SPREADING ROD

It is sterilized by storing in 70% alcohol. Just before use, it is removed, air-dried, and touched to a flame which will burn off residual alcohol. Use this rod to spread the organisms on the surface.

4. Add 0.1 ml of the 1/HT dilution to the second plate of soil agar and to one of nutrient agar. Spread and label as 1/M plates (one to a million; the soil plate contains 1 μg of soil).

5. Add 0.1 ml of the 1/M dilution to the third plate of soil agar and to one of nutrient agar. Spread and label as 1/10M plates. How much soil does the soil plate have added to it?

OBSERVATIONS

1. After 1 day and after 1 week, examine the plates. Count the number of colonies on a plate that has about 50 colonies on it.

2. Note especially the differences in the appearances of the plates of soil agar and nutrient agar.

3. Save the plates for future use.

INTERPRETATIONS

1. How many bacteria were there per g of soil, based on counts found at 1 day and at 1 week? Do they differ depending on the medium? Why?

2. Do you think that you have cultured all the organisms in the soil sample? Why?

INVESTIGATION 10
Isolation of a Pure Culture

A pure culture is one containing organisms of only one kind—a field with only pea plants in it, for example. In microorganisms the best way to obtain a pure culture is to have all the cells in a culture come from a single cell.

MATERIALS

2 slants nutrient agar (sterile)
1 inoculating needle

PROCEDURE

1. In looking over the soil plates made in Investigation 9, you are almost sure to find a colony—especially on nutrient agar which looks like this:

FIGURE 29. Colonies of *Bacillus mycoides*.

Bacillus mycoides, sometimes called *Bacillus subtilis, var, mycoides,* is the organism which we are to isolate.

It is, of course, not at all necessary that you select *B. mycoides*. You may select another organism if you wish. It probably will not act exactly the same as *B. mycoides* in some of the subsequent experiments, but it might tell you even more than *B. mycoides*. We will use the term *B. mycoides* to apply to the organism you have isolated since it is more convenient to say "*B. mycoides*" than to say "*B. mycoides* or whatever organism you isolated."

2. Flame a needle, pick up a small portion of the colony of *B. mycoides,* and transfer it as a streak to an agar slant as shown in Figure 30.

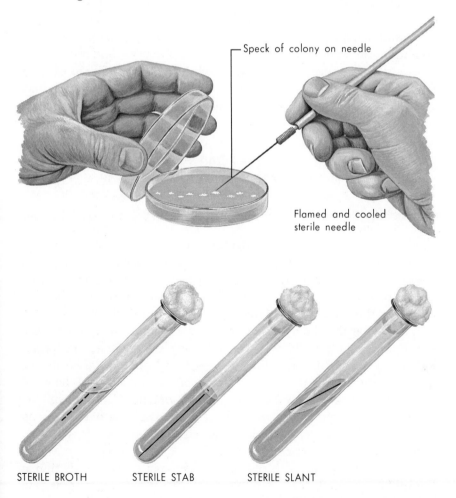

—Speck of colony on needle

Flamed and cooled
sterile needle

STERILE BROTH   STERILE STAB   STERILE SLANT

FIGURE 30. Picking a colony.

3. Prepare a second slant from another portion of the same colony. Try to pick a colony which is not mixed with other organisms—that is, one which is well isolated. These cultures should be saved and used in later investigations.

4. Incubate the streak cultures for 24 hours or until they show good growth; then hold them in the refrigerator. These will be your stock cultures to be used in future experiments.

PARALLEL READING: Microbial Ecology

During centuries of development, there has been abundant opportunity for living forms to investigate the possibilities of growth and development under a wide variety of conditions in nature, and some have been successful in exploiting energy sources of varied types or environmental conditions far from "ideal." The net result is that we find microorganisms, and higher organisms as well, living in many different environments, some of which shelter them from competition with "normal" forms. Some organisms obtained this advantage by boldly attacking their competitors—for example, antibiotic-producing organisms. Others withdrew to the physiological arctic or the biochemical tropics, to the mines, and to the mountain tops.

These "adapted" organisms are not usually the most adaptable. Frequently, in order to invade the hostile environment, it was necessary not only to add a given property, but also to lose some "normal" attribute. So long as the external environment remains constant, the specialized forms possess a considerable advantage in the specialized environment, but when the environment changes, when the hot spring grows cold, they are likely to be less able to cope with "the slings and arrows of outrageous fortune." Among the lower organisms, the more primitive, the less specialized, the more jack-of-all-trades type of organism, the better the chances of survival in a changing environment. And the environment does change—not only the climate but also the organisms living within it. Organisms grow old and die, new kinds invade, mutants outgrow the unmutated form, while fire, flood, and the wrath of the elements all slowly but surely—and sometimes in the twinkling of an eye—change the environment. A climax environment is not exactly a stable matter. It is rather an equilibrium moving slowly toward another environment.

These ecological principles are evident not only in early human history but also in the record of the mighty forest, the waving grass, and the lonely sea. But we need not study these vast and complex interactions. We can see the same cause and response, as well as the same change and reaction, in populations of microorganisms. The soil, for example, has so many micro-environments, even within a gram of it, that an exceedingly wide variety of different kinds of microorganisms can live in it; and their varied populations, the ebb and flow of their numbers, and their success or failure as a population determinant can be seen quite readily. Further, we need not wait a century for change to be evident; this microcosm can be studied and the effect which only centuries could impose upon the larger world may be observed in days.

The principles which may be deduced from microbial societies are not entirely applicable to animals for these have the added faculties of brain and tool to varied degrees. Here animal and man are capable of changing the environment—the beaver and his dam no less than man and his furnace. Man, indeed, has changed his environment to such an extent, and within such a short period, that it is neither a stable equilibrium nor a change that is entirely to his advantage as a species hoping to survive. Expand he has—at the expense of pure air and water, of space and room, and of some of the qualities of life which were available even in the recent past. Whether what he has gained is greater than what he has lost remains to be seen.

INVESTIGATION 11
A Population Sequence

Any natural population is dependent on the conditions of environment, the availability of food stuffs, and so forth. As one type of organism uses up food, others capable of using other foods come in and these, too, are followed by others. We can follow such a sequence among microorganisms rather readily.

MATERIALS

Microscope slides
Bunsen burners
Methylene blue stain (other stains can be substituted)
Milk (fat-free or skim milk; powdered milk at 5 gm/100 ml can be used)
Soil
Microscope
Flat-sided bottle
Inoculating loop

PROCEDURE

1. Pour about 10 ml of milk into a medicine bottle and place the bottle on its flat side. More milk may be added so that it is about a half inch deep over the entire surface.

2. Add about $\frac{1}{2}$ gram of soil, shake, and place the bottle on its flat side.

3. Make a smear of the milk; stain and examine under the high-power lens.
   (a) This is done in the following way. A loop is sterilized by putting it in the Bunsen flame until it is red hot. Cool in air.
   (b) The sterile loop is dipped into the milk, removed, and its contents spread thinly on a microscope slide.
   (c) The milk film is allowed to dry.
   (d) The milk film is then heat-fixed to the slide by passing the slide through the Bunsen flame (film side up) for three passes. The slide should feel hot to your hand but should not burn it. Bake a little; don't fry.
   (e) The film is then stained with methylene blue for 1 minute.
   (f) The stain is poured off, the film washed in water (dipped in a jar or tumbler or held under a *gently* flowing tap), dried, and examined under the high-power lens.

4. A smear should be made, stained, and examined each period for the next 7 days. Note also the change in the appearance of the milk.

OBSERVATIONS

What changes occur in the appearance of the milk? What organisms do you see in the stained smear? Do they differ from one observation time to another? Is there a sequence of organisms, one type coming first, then replaced by another? If you have the facilities, it is of interest to determine the pH at each observation. You will see not only bacteria but also molds and yeasts which are much larger than bacteria.

INTERPRETATIONS

If an organism grows on a given substance, it produces particular products from it. These products, then, are suitable for the growth of some other organism, and so forth. The products formed, however, will select from the mixed population those organisms that can grow on them. The products these produce again select the third type of organism, and so forth. The first organism to become the dominant population thus controls, to a large extent, the subsequent sequence of populations. From a single pail of milk, a cheese maker can make any of 300 different cheeses depending on which organisms he chooses to start with and how he manipulates the environment subsequent to this start. Where else does the ability to control population sequence benefit man? How is it possible to control the population sequence?

INVESTIGATION 12
Interaction in Populations

In a mixed population, organisms interact with each other. These are antagonisms and also instances of mutual assistance.

MATERIALS

Plates from Investigation 10 (at 1/*HT* dilution)

2 sterile slants (5-ml each) nutrient agar

Culture of *B. mycoides* (isolated in Investigation 9)

Inoculating loops

PROCEDURE

1. Look at the plates prepared in Investigation 9, especially the crowded ones, to see if you can find any cases where the growth of one organism has inhibited others. If you find one or more, transfer those colonies onto the same media; that is, if the organism is growing on a plate of soil agar, transfer to a soil agar (method of transfer described, Investigation 10, Procedure No. 2).

2. Melt two tubes (5-ml each) of nutrient agar and cool to 45°C. Inoculate each by transferring a loopful of culture from the pure culture of *B. mycoides* you isolated in Investigation 10; pour the tube, while still melted, onto the 1/*HT* plates made from soil in Investigation 9. Allow the agar to solidify and incubate.

OBSERVATIONS

1. At the next period look for any clear zones in the upper layer of agar. This agar had been seeded with *B. mycoides* which should grow abundantly throughout it.

If, however, there is an organism on the *1/HT* plate (from Investigation 9) that produced a substance inhibiting to *B. mycoides*, there would be no growth above that colony. If there are such inhibiting zones, transfer the colony responsible to nutrient agar.

2. At a subsequent period you may use the antagonistic cultures (if any) isolated as follows: In paragraph 1 of Procedures, you may have been able to isolate one or more organisms producing substances antagonistic to other organisms. Similarly, in paragraph 2, you may have been able to isolate an organism antagonistic to *B. mycoides*. If this is the case, you would first wish to know whether you can demonstrate such antagonism again (an important part of the value of any experiment is its reproducibility) and then determine what organisms may be inhibited by the substance produced by your antagonist. If you do not have antagonism demonstrated in paragraphs 1 or 2, naturally you can do no more; this ends the experiment.

INTERPRETATION

1. This is obviously a way which is used to search for antibiotics. Your chances of finding a useful antibiotic are just as good as those of anyone else. If you have found an antagonistic organism, what studies should you do next?

INVESTIGATION 13
Isolation of Mutants from Bacteria

Cell populations of 10-million individuals normally contain 3–10 (or more) mutants. These occur "spontaneously." Your slant culture of *B. mycoides* (Investigation 10) will have about 10^{12} to 10^{13} organisms on its surface. If we suspend this growth in 0.4 ml of water and use 0.1 ml on each plate, we should have 2.5×10^{11} to 2.5×10^{12} organisms per ml. We can expect about one organism in each 10^9 to have mutated, and possibly 1 in 10^{11} (one out of every 100 mutations) may have mutated to streptomycin resistance. We are therefore adding enough organisms so that we could expect anywhere from 2 to 250 streptomycin-resistant mutants.

MATERIALS

    2 sterile nutrient agar slants

    3 test tubes of sterile nutrient agar (20 ml/test tube)

    3 sterile petri dishes

    3 sterile 1-ml pipettes

    2 ml of streptomycin solution (sterile, 1 mg/ml)

1 glass rod spreader

70% alcohol

1 inoculating loop

Sterile water

PROCEDURE

1. Transfer each of your two stock cultures of *B. mycoides* to new slants of nutrient agar.

2. Melt three tubes of nutrient agar, each containing 20 ml.

3. With a sterile pipette add 1 ml of streptomycin solution (1 mg = 1000 μg/ml) to each of two sterile petri dishes. Label them *S* and *SG*. A third sterile petri dish (empty), label *C*.

4. When the agar has melted, pour 1 tube of nutrient agar into petri dish *C* and another into petri dish *S*. Pour only half of the third tube into petri dish *SG* and return the remaining portion to the melting bath. Tilt this dish (*SG*) so that the agar is thin on one side and thick on the other. This is easily done by resting one edge on a pencil, small rod, or block as shown in Figure 31.

FIGURE 31. Preparation of a gradient plate.

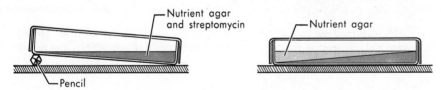

The plate labeled SG is called a "gradient plate." The agar you have just poured contains streptomycin, at 100 μg (millimilligrams)—that is, you had 1000 μg of streptomycin in the petri dish, and you added 10 ml of agar. In Section 5, you will add agar containing no streptomycin. At the side of the dish where the streptomycin agar is thickest, there is much more streptomycin (when it has diffused into the other agar to reach the surface) than at the opposite side. For a few hours, at least, the end containing the deep layer of streptomycin agar has 100 μg/ml, the opposite end has 0 μg/ml—and a linear concentration exists between them; that is, half way across the plate the concentration is 50 μg/ml. When you inoculate the entire surface (as in Section 7), the organisms are exposed to a variety of streptomycin concentrations depending on where they happen to be located on the plate. We can therefore obtain an estimate of the concentration of streptomycin required to inhibit the growth of the majority of the population. But we may also see mutants with differing degrees of resistance.

5. Allow the agar in all dishes to solidify. Then setting dish SG on a flat surface, pour in the remainder of the agar and allow the last portion to solidify.

6. Add about 0.4 ml of sterile water with a sterile pipette to one of the old stock slants of *B. mycoides* and, with a sterile loop, suspend the growth in the water.

7. With a sterile pipette add 0.1 ml of the suspended cells of *B. mycoides* to each of the plates C (control), S (streptomycin), and SG (gradient plate with streptomycin) and spread the suspension evenly over the entire surface of the plate.

OBSERVATIONS

1. Plate C was your control; it should be heavily overgrown. Plate S had 50 μg streptomycin per ml. Assuming that they were sensitive in the first place, this procedure should have killed all the organisms you put on the plate. (Some cultures of *B. mycoides*, isolated from soil, may not be sensitive to streptomycin.) Count the number of resistant mutants (if any). Transfer one resistant colony onto a nutrient agar slant. This will be your resistant culture (*B. mycoides R*) to be used in subsequent work.

2. Plate SG should have growth at one side, with a rather sharp line where growth ceases. Calculate the concentration of streptomycin which inhibits the growth of the sensitive strain. Are there mutants having different sensitivity?

INTERPRETATIONS

1. Why does a gradient plate work?

2. Assuming that in the SG plate you found several mutants, how could you determine whether these resistant mutants had differing sensitivity?

INVESTIGATION 14
Resistance and Sensitivity to Chemical Agents

You have isolated a pure culture of *B. mycoides* from soil (Investigation 10). You have also obtained a mutant resistant to streptomycin from this pure culture (Investigation 13). Does this streptomycin resistant mutant differ from its parent only in its resistance to streptomycin, or does it differ in other ways as well? Is it still as sensitive to other inhibitory materials?

MATERIALS

2 sterile petri dishes

3 tubes of sterile nutrient broth

2 tubes of sterile nutrient agar (15 ml/tube)

Sterile pipettes

Filter-paper disks (from paper punch or cuts of paper approximately 5–10 mm square)

Solutions of streptomycin (25 μg/ml), penicillin (10 μg/ml), mercuric chloride (10 μg/ml) and phenol (10 μg/ml)

Glass-spreading rod

70% alcohol

PROCEDURE
Preliminary work

Inoculate one broth tube of nutrient broth with the streptomycin-sensitive strain of *B. mycoides*. This is the strain you originally isolated from soil. After growth it will be held in the refrigerator and used as inoculum for Investigations 14, 16, and 17.

Inoculate 2 tubes of nutrient broth with the streptomycin-resistant mutant which you isolated in Investigation 13 (*B. mycoides R*). After growth, one tube will be held in the refrigerator and used as inoculum for Investigations 14, 16, 17. The other tube, after growth, will be used in Investigation 15.

After the inocula are ready, proceed with Investigation 14.

1. Pour two sterile petri plates with nutrient agar. When they have solidified, use a sterile pipette to add 0.2 ml (4 drops) of the broth culture of the sensitive strain to one petri dish and spread over the surface with a sterile glass-spreading rod (see Investigation 9 for details). Label this plate *S*. Similarly prepare the second plate with the resistant strain—after sterilizing the spreading rod in 70% alcohol and flame. Label this plate *R*.

2. Divide each plate into four sections by marking on the bottom. With a sterile forceps (flame), place a disk or square of filter

paper (which has been dipped in a solution of streptomycin containing 25 µg/ml) in Section 1 of each plate. Place a similar disk dipped in the specified solution as indicated below. (Naturally you may use materials other than those specified here; for an unknown material, a concentration of 10 µg/ml is frequently used.)

Section 2—Penicillin, 10 µg/ml

3—HgCl₂, 10 µg/ml

4—Phenol, 10 µg/ml

3. Incubate the plates until the next period.

4. After incubation, examine plates for clear zones (= zones of inhibition) around the disks. Measure their diameter with a mm rule and record the data.

INTERPRETATIONS

1. Do the R and S strains differ in ways other than in their sensitivity to streptomycin?

2. You will undoubtedly recognize that the method used here can be used to evaluate (and in many cases to quantitatively measure) toxic materials, antibiotics, and so forth. Before using it as an "assay" for penicillin, for example, what would you need to establish experimentally?

INVESTIGATIONS FOR FURTHER STUDY

1. Survey other substances, possibly home remedies, for their inhibitory action.

2. Set up an assay for an antibiotic, or other toxic substance; determine the amount in an unknown solution, possibly even in a medium in or on which an organism producing the toxic substance has grown.

Note: In setting up the following three Investigations, we are taking a possible chance because they may not "work." You will use the bacteria which you have isolated. These may differ in some subtle ways from the laboratory strains for which the methods are described. For example, in Investigation 15, after the cell is dissolved by an enzyme and a detergent, the DNA remains. The actual cell wall of different strains of even the same bacterium may differ a little in their chemical composition, and thus these agents may not dissolve your cell. And naturally you cannot do Investigations 16 and 17 unless you can get out the DNA in 15. If this happens—that is, your strain does not dissolve (and release DNA) in the procedure recommended—then you will need to try other things. For example: increase the concentration of either the detergent or lysozyme and let them act longer; try different detergents ("Dreft," "Tide," "All," etc.); try cultures of different ages and cultures grown on different media, and so forth. You may not have time to finish all the Investigations but go as far as you can.

INVESTIGATION 15
Isolation of High Molecular Weight DNA

High molecular weight DNA may be isolated from microorganisms; it is physiologically active and may, when conditions are proper, enter another bacterial cell and "transform" it—that is, add a genetic character which the second strain did not originally possess. We shall attempt to isolate the DNA from the streptomycin-resistant strain, and, in Investigation 16, we shall attempt to use this DNA to transform streptomycin-sensitive cells into streptomycin-resistant cells.

If the necessary equipment, necessary enzymes and reagents, and even the necessary skills are not available for your performance of these experiments, at least you can see how they are done.

MATERIALS (per team)

50 ml of sterile nutrient broth

50 ml of saline citrate buffer (0.6 g NaCl plus 3.0 g sodium citrate ($2H_2O$), adjusted to pH 7)

0.5 ml of 0.1 M EDTA

0.5 ml of lysozyme (10 mg/ml)

0.2 ml of 45% alcohol saturated with sodium lauryl sulfate pH 8.

6 ml chloroform

0.3 ml of isoamyl alcohol

6 ml of 95% ethanol

2 ml of sterile 10% NaCl in small test tubes

MATERIALS (per class)

Centrifuge

Centrifuge tubes

PROCEDURE

1. Inoculate 50 ml of nutrient broth with 1 ml of the broth culture of the streptomycin-resistant strain (*B. mycoides R*) which you made in Investigation 14, providing this culture is less than 7 days old. Incubate at room temperature (in a cool place, if possible, 20–25°C) for 24 hours. (Higher temperatures require shorter incubation; you do not want to achieve maximum growth.)

2. In the late lag phase, centrifuge the culture (need not be aseptic centrifugation) to sediment the cells, pour off the clear supernatant, and resuspend the cells in 5 ml saline-citrate buffer. If it is available, add 0.5 ml of 0.1M EDTA—that is, disodium ethylene diamine tetraacetate, 3.36 gm/100 ml adjusted to pH 7. (The citrate and EDTA bind Mg^{++} and thus inhibit DNAse.) To this cell suspension add 0.5 ml (or 10 drops) of a solution of the enzyme lysozyme (10 mg/ml). This enzyme hydrolyzes the cell wall of most bacteria, especially gram-positive bacteria like *B. mycoides*. Mix and incubate at 37° until the suspension clears up entirely, becomes almost a gel, or becomes very viscous. The latter is due to the release of DNA which sometimes occurs at this point. This should take 10–30 minutes.

3. When the cell wall is removed in many gram-positive bacteria, the remaining portion (called the "protoplast") will swell up and burst. In others there is still sufficient cell wall material left; therefore, detergent treatment is necessary. This may be accomplished by adding 0.2 ml of a saturated solution (in 45% alcohol) of sodium lauryl sulfate. Shake the lysozyme treated suspension with the detergent for 10 min. Vigorous shaking should be employed. Stopper the tube and really shake it.

4. DNA and other cell constituents, released by the detergent treatment, are still attached to protein. The next step is to "denature" the protein—that is, change its physical structure so that it will not readily unite with DNA. Add 6 ml of chloroform and 0.3 ml of isoamyl alcohol. (The chloroform will settle to the bottom of the tube because it is heavier than water; the isoamyl alcohol will break up an emulsion which may form, but it may be omitted if unnecessary.) Shake vigorously for 2 minutes and centrifuge down the debris. The chloroform will be at the bottom, a layer of denatured protein will be at the chloroform-water layer, and a water solution of DNA (which

may be turbid) will be above this. Pipette this layer into a clean test tube.

5. Hold the test tube containing the supernate from step 4 at a 45° angle and carefully layer 6 ml of 95% ethanol over the top. With a glass rod SLOWLY stir in the alcohol, twisting the rod on which a gummy, sticky precipitate will gradually accumulate. This is the DNA at a very high molecular weight. It should not be pipetted since this will break up the long strands.

6. After all the alcohol has been stirred in, twirl the rod slowly to pick up all the DNA on it and remove the rod containing the DNA to a test tube containing 2 ml of sterile 10% NaCl. Stir slowly until the DNA dissolves. There may be a small insoluble residue which may be ignored. Remove the glass rod and use the DNA solution in Investigation 16.

INVESTIGATION 16
Transformation

When an organism acquires DNA from its environment and thereby undergoes a hereditary alteration, it is said to be "transformed." When an organism is capable of being transformed, it is said to be "competent." There seem to be certain periods during growth or cell division when cells are "competent," but the factors involved are not clearly understood. There seem to be three stages involved, each evidently subject to some environmental requirements. First, the physiologically active (usually high molecular weight) DNA must be bound to the cell—first transiently, then permanently. Second, it must penetrate the cell or at least become involved with the cell membrane. Third, the external DNA must somehow become integrated with the internal DNA so that a copy is made of the new genetic factor as well as of the DNA of the cell. Since these processes are rather complex and are not under exact experimental control, it is probable that experiments to demonstrate transformation will not be successful for all individuals or at every attempt.

MATERIALS (per team)

    1 tube of 2-ml sterile 10% NaCl

    10 ml of enriched broth

2.5-3 ml of streptomycin solution (75 mg/100 ml)

2 sterile petri dishes

2 tubes containing 15 ml of sterile nutrient agar

1 5-ml sterile pipette

4 1-ml sterile pipettes

Broth culture of sensitive *B. mycoides* less than 7 days old

DNA saline from Investigation 15

Sterile glass-rod spreader

70% alcohol

PROCEDURE DAY I

1. In Investigation 15 you prepared 2 ml of a saline solution in which was dissolved the DNA you recovered from a 50-ml culture of your streptomycin-resistant strain. Label this tube *DR* (for resistant DNA).
2. To tube DR and to another tube containing 2 ml of sterile 10% NaCl, aseptically add 5 ml of broth. Label this second tube *SN* (sensitive strain, no DNA). We shall use as rich a broth as possible; brain-heart infusion with yeast extract— plus peptone, if possible. (You will be told what your broth contains.)
3. Inoculate each tube (DR and SN) with 1 ml of a broth culture of *B. mycoides* strain sensitive to streptomycin. Incubate for 24 hours.

PROCEDURE DAY II

4. With a sterile pipette, aseptically add 1 ml of a streptomycin solution (75 mg/100 ml) to each of two sterile petri dishes.
5. Melt and pour into each dish 15 ml of sterile nutrient agar. (It is convenient to have two sterile tubes of agar each containing 15 ml of agar.) Mix well, allow to solidify. Label one plate *SN*, the other *DR*.
6. Remove, aseptically, 0.2 ml from the SN culture prepared in step 3—after 24 hours of incubation—to plate SN and spread evenly over the surface with a sterilized bent-glass-rod spreader. Similarly, remove 0.2 ml from culture DR prepared

in step 3 to plate DR and spread over surface. Incubate both plates (inverted) for 24 and 48 hours.

PROCEDURE DAYS III and IV

7. Examine plates SN and DR for colonies at 24 and 48 hours. Count the number of colonies on each plate at each interval.

INTERPRETATIONS

1. What would you expect to happen should transformation have taken place?
2. Suppose that, in Investigation 15, you had done all the manipulations using pipettes. Would you still expect transformation?
3. How do you know that the organisms growing on plates SN and DR are resistant to streptomycin?

INVESTIGATION FOR FURTHER STUDY

1. If transformation did not occur, one might add a further 5 ml of broth to tubes SN and DR (steps 2 and 3) plus a small crystal of lysozyme (or 0.1 ml of a 10 mg/ml solution, sterilized by filtration, if possible) and then repeat steps 4 through 7 on subsequent days.

INVESTIGATION 17
Sensitive, Resistant, and Transformed Strains

You now have a sensitive strain of *B. mycoides* which you orginally isolated from soil, a streptomycin-resistant mutant of that strain, and, hopefully, a mutant which you obtained by the process of transformation. Are these alike—that is, do the two mutants differ from the parent in only one character (streptomycin resistance) or in other characters as well?

Using the same procedure as in Investigation 14 with 3 petri plates instead of two, design an experiment to answer the above question.

OBSERVATIONS AND INTERPRETATIONS

What you are looking for is some difference between the three strains with respect to their reaction to some other challenge. Do all of them ferment the same sugars? Are they all equally resistant (as sensitive) to other inhibiting materials? If their response is the same to all such challenges, we could conclude that only a small part of the DNA has been affected. If they differ we know that more than one change has taken place.

INVESTIGATION FOR FURTHER STUDY

1. Interaction Without Cell Contact Between Two Mutants of a Bacterium.

The methods by which various kinds of organisms are able to communicate has long been a subject of intense investigation. Social organization depends on communication. Porpoises appear to have a kind of language, bees communicate via sound vibrations and "dance" patterns, and termites have been demonstrated to possess some form of communication.

While it is extremely unlikely that any form of "language" exists between individual plants or microorganisms, there are various forms of complex chemical *interactions* which affect the existence of populations of countless thousands of individual organisms. The subject of chemical interaction between cells is enormously complex. One can approach this subject only by a careful study of interactions that can be investigated in the laboratory.

This Investigation centers on one general problem: the nature of an interaction between cells that results in visible changes within a population of a specific microbe—*Serratia marcescens*.

When a wild, prodigiosin-producing strain of *S. marcescens* is plated on a proper medium and incubated at 25–30°C, most of its colonies will have a red color due to the presence of the compound prodigiosin. Often, however, a few colonies are white, indicating that the cells in these colonies can no longer produce the complete pigment molecule. This could be due to a mutated gene (or genes) in the cells of these white colonies. If the mutation resulted in the loss of the ability of some gene to direct the production of any of the series of enzymes involved in the conversion of medium constituents to prodigiosin, one could account for the loss in pigmentation. Occasionally, colonies of two white mutants can interact in some way (without physical contact) so that one or both of them become pigmented. If you are interested in pursuing this subject, see your teacher for additional information.

INVESTIGATION 18
Preparation of a Scientific Paper

This Investigation will differ from those that you have done previously. Instead of concentrating on a specific laboratory investigation, you will be asked to study the results of certain Investigations and to combine your interpretations into the construction of a scientific paper that meets the criteria discussed in the literature section of Part I.

You should feel free to develop your own ideas; exact details for the structure of the paper will not be given. It should, however, reflect the following points:

1. The nature of your problem and the hypothesis you set out to test should be clearly stated.
2. You should explain in very clear terms how you went about performing various aspects of your investigation so that someone else could repeat the same series of investigations.
3. You should state as clearly as possible the actual results obtained including data and such analyses of these data as you may have performed.
4. An interpretation of the experimental results based on an evaluation of your data should be clearly stated.
5. A concluding statement should be made that reflects on your original hypothesis and gives as much of a conclusion as you are able to give in light of your interpretations.

It might be well to study various articles in the literature as well as those reproduced in this book to help you to find out how to structure this paper. In essence, what you are asked to do is as follows: state a hypothesis that could account for the results which you have observed; describe the kind of experimentation you did in seeking to verify your hypothesis; and finally, express in your own words and with your own ideas, based upon your experimental work, what you believe to be the best possible explanation for these results.

This is not going to be an easy task, but it is typical of the problem that faces the research worker; he must be able to communicate his work in as clear and concise a fashion as possible so that other investigators might carry on or verify his work by their own experimental techniques.

If your work has been careful, if you have been meticulous in keeping notes on each of these past Investigations, and if you have very carefully thought through each of the Investigations, you might be able to write a paper that would be of significant interest to those people who are now conducting research on these subjects.

Use either the data from Investigations 9, 10, 11, and 13, or the data from Investigations 10, 12, 14, 15, 16, 17, and construct a single paper from either group.

GROWTH, DEVELOPMENT, AND BEHAVIOR OF INDIVIDUALS

SECTION **10** GROWTH AND DEVELOPMENT
IN PLANTS

Living things can reproduce their kind; perhaps this is their most characteristic property. Reproduction in most organisms is sexual; it involves the production of gametes, their fusion to form a zygote or fertilized egg, and the development of an embryo from the zygote. But how does this embryo develop from the single-celled zygote to a complex bean plant or to an even more complex organism, such as a frog or a man? This is one of the most baffling questions in biology. Attempts have been made to explain these extraordinary events since the early history of man.

Meiosis

Basic to sexual reproduction in any species is the formation of gametes. Gametes have half as many chromosomes as are found in the zygote. The reduction of the number of chromosomes from the diploid number (2n) found in the zygote to the haploid number (n) found in the gametes is accomplished during *meiosis*.

Meiosis may occur at various places in the life history of the many species of organisms. In the moss plants and in ferns, the stage which produces sperms or eggs is an independent, haploid plant, and, in some cases, it may live for many years. In some plants and in nearly all animals, the formation of sperms and eggs is the direct result of meiosis, although some change in cell structure may follow the second meiotic division before the sperms or eggs are mature.

FIGURE 32. Diagrammatic representation of the meiotic divisions of a diploid cell with three pairs of chromosomes. One set of three is shaded; the other is unshaded.

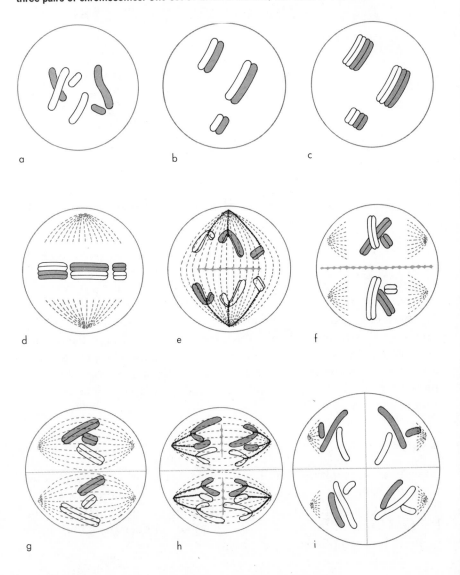

Meiosis results in two divisions of the nucleus accompanied by a single division of the chromosomes. You will find this process illustrated in Figure 32 on the opposite page. In this diagram the diploid cell has six chromosomes, and each of the four daughter cells has three chromosomes.

Diploid cells have two sets of chromosomes: one set came from the male parent through the sperm and the other from the female parent through the egg. Early in the first meiotic division, the members of each homologous pair of chromosomes synapse (come together along their entire lengths). Each chromosome of each pair then duplicates itself, yielding a quartet or tetrad of chromosomes. This process is followed by two successive divisions of the nucleus, with each of the four daughter nuclei receiving one chromosome from each quartet. We call these two divisions Meiosis I and Meiosis II.

Life History of a Flowering Plant

Our study of the growth and development of a flowering plant will be aided by a review of the life history of such a plant, as shown in Figure 33. The essential parts of a flower are the stamens and the pistil. The most important part of the stamen is the anther, in which the pollen grains are produced. Similarly, the most important part of the pistil is the ovulary (often called ovary) which produces ovules.

An anther usually produces four more-or-less cylindrical regions of tissue in which most of the cells produce microspores (little spores). Each microspore mother cell undergoes meiosis producing four microspores, each with one haploid nucleus which then undergoes one mitotic division. At the same time, the outer wall of the microspore thickens, often becoming sculptured in a pattern unique for the species. This two-nucleated structure is a pollen grain.

At the same time that pollen grains are produced in the anther, a series of events which may lead to the production of eggs is taking place in the ovulary. Depending on the species, from one to many ovules are produced in a cavity (or cavities) in the ovulary. The ovules can first be seen as small protuberances which arise usually from definite regions of the ovulary wall. In each young ovule a diploid megaspore mother cell differentiates, divides by meiosis, and produces four haploid megaspores. Usually three of these spores disintegrate; the remaining megaspore may divide by mitosis and give rise to an eight-nucleated microscopic plant which lives within the ovule as a parasite. This is a megagametophyte, a plant which produces large gametes. In the flowering plants, it is often called the *embryo sac*. One of the eight nuclei with some surrounding cytoplasm becomes an egg. Two other nuclei, called the polar nuclei, come to rest near the middle of the megagametophyte.

FIGURE 33. Shown in sequence for a generalized flowering plant are: (a) the flower; (b) pollination by an insect; (c) the interior of the pistil at the time of fertilization (a pollen tube has grown to the ovule, and double fertilization is about to take place); (d) the seed (the black arrows indicate that the seed coat is produced by the integuments of the ovule, the endosperm in the seed by the fertilized fusion nucleus, and the embryo by the fertilized egg); (e) the seedling; (f) the mature plant with flower.

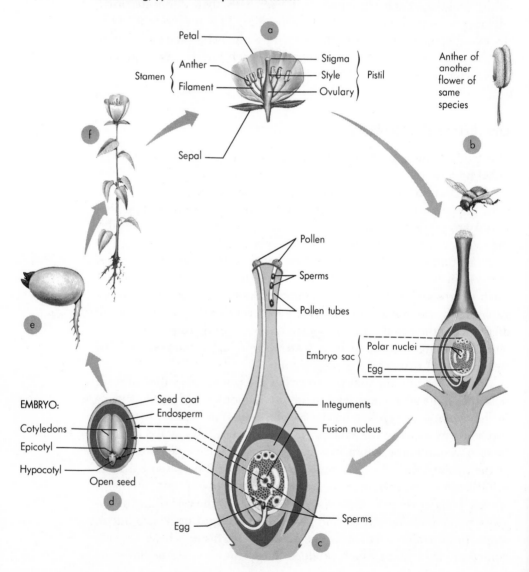

The development of the egg in the megagametophyte takes place at about the same time as the development of the pollen grains in the anther. When the pollen grains are mature, the anther opens, and pollen grains may be carried to the stigma of the pistil by gravity, wind, or insects. There the pollen grains may germinate, forming a pollen tube which grows into the stigma and down through the style to the ovulary where it may enter one of the ovules.

Soon after germination of the pollen grain, one of its two nuclei divides to form two sperm nuclei. The other nucleus moves with the growing tip of the pollen tube, and no doubt it controls the synthesis of enzymes which digest the cells of the stigma and style in the pathway of the pollen tube. Since the pollen grain or pollen tube is a plant which produces sperms, what would you call this phase of the life cycle? The pollen tube grows into the megagametophyte. Here the end of the tube disintegrates, and the two sperms move into the megagametophyte where one of them fuses with the egg nucleus and the other fuses with the two polar nuclei. Fusion of sperm and egg results in a zygote; fusion of the other sperm with the polar nuclei results in what is called a triple-fusion nucleus. The triple-fusion nucleus then divides by mitosis and forms what is called an endosperm, which is important as a food source for the developing embryo.

The zygote divides and forms a two-celled embryo which continues its development within the endosperm. In some species, by the time the seed is mature, the embryo will have completely digested the endosperm, using it as food. This happens in the common bean. In other plants, such as the castor bean or corn (the corn grain is a fruit consisting of an ovulary with an enclosed seed), most of the food stored in the seed will be in the endosperm. In these species, the embryo will have used only a small part of the food present in the endosperm.

The development of the endosperm and embryo in the ovule of shepherd's purse (*Capsella bursa-pastoris*) is shown in Figure 34. An ovule containing an embryo is called a seed. The structures present in some mature seeds are shown in Figure 35.

It is often said that all the parts of each embryo develop by cell division and cell growth. If we consider growth as simply an increase in size, it is obvious that something else has happened. The cells of leaves, stems, and roots are not all alike. What is it that causes them to differentiate? This is a question which biologists have not yet answered. Some of the processes involved in the germination of seeds and the growth of plants are well understood; others are not. We will look at some of the problems in each of these two categories on the following pages.

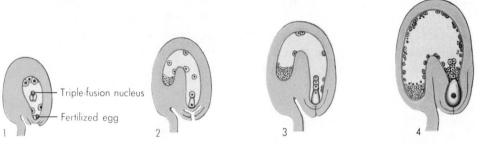

FIGURE 34. Diagrams illustrating growth of embryo and endosperm in shepherd's purse.

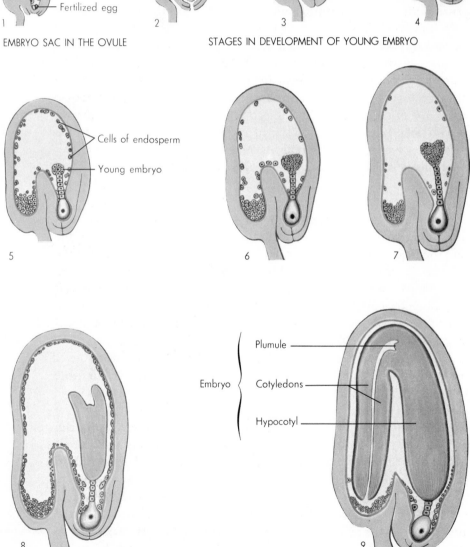

Triple-fusion nucleus

Fertilized egg

1   2   3   4

EMBRYO SAC IN THE OVULE   STAGES IN DEVELOPMENT OF YOUNG EMBRYO

Cells of endosperm

Young embryo

5   6   7

Plumule

Embryo { Cotyledons

Hypocotyl

8   9

SEED

FIGURE 35. Seeds and their parts. Starting from the top: corn, Alaska pea, Kentucky Wonder bean, and castor bean.

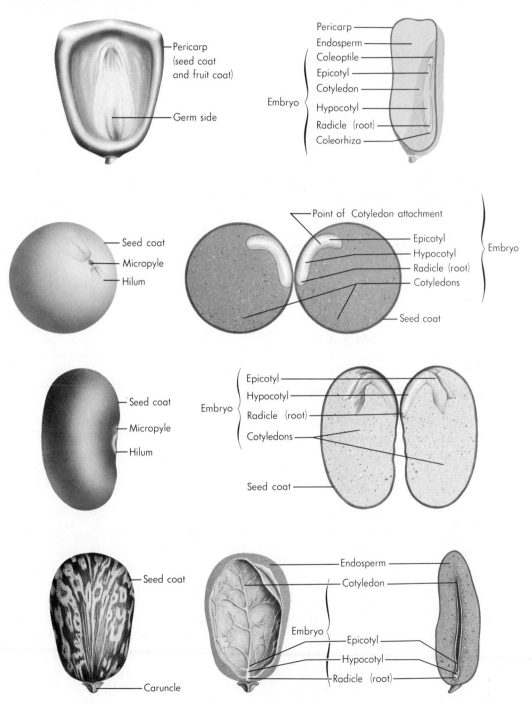

Pericarp (seed coat and fruit coat)

Germ side

Pericarp
Endosperm
Coleoptile
Epicotyl
Cotyledon
Embryo
Hypocotyl
Radicle (root)
Coleorhiza

Seed coat
Micropyle
Hilum

Point of Cotyledon attachment
Epicotyl
Hypocotyl
Radicle (root)
Cotyledons
Embryo
Seed coat

Seed coat
Micropyle
Hilum

Epicotyl
Hypocotyl
Embryo
Radicle (root)
Cotyledons
Seed coat

Seed coat

Caruncle

Endosperm
Cotyledon
Embryo
Epicotyl
Hypocotyl
Radicle (root)

PATTERN OF INQUIRY 6
Developing Seedlings

It has been observed that, as the shoot and root of the corn and castor bean grow, the endosperm becomes smaller. As the shoot and root of the bean plant grow in size and dry weight, there is a decrease in the dry weight of the cotyledons. What explanation can you give for such data? What was it that moved?

INVESTIGATION 19
Enzyme Activity in Germinating Seeds

MATERIALS (per team)

1. Corn grains
2. Three petri plates
3. Starch agar
4. Razor blades
5. Iodine solution
6. Tes-Tape

PROCEDURE, PART ONE (per team)

1. Obtain 5 dry corn grains and 10 corn grains which have been soaked for 24 hours.

2. Prepare 3 culture dishes by pouring not more than 15 ml of melted starch agar into each of 3 petri plates or similar containers. Allow the agar to cool, then cover the dishes. Label completely as to team, class, experiment, date, and treatment.

3. With a sharp razor blade or scalpel, carefully cut both the dry grains and the soaked grains lengthwise, as in the diagram (see Figure 36). Place 5 of the soaked grains (10 halves) in boiling water and boil for at least 20 minutes.

4. Place 5 half-grains from each group—dry, soaked, and boiled —in the appropriate dish so that the cut surface of the grain is in contact with the surface of the agar. Actually the embryo will be in contact with the agar. Space them wide apart but not too close to the edge of the dish (see Figure 37).

5. Leave the dishes at room temperature for 24 hours.

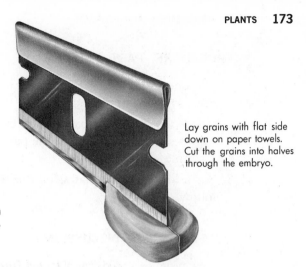

Lay grains with flat side
down on paper towels.
Cut the grains into halves
through the embryo.

**FIGURE 36. Method for cutting corn
grains longitudinally through the embryo.**

6. At the end of 24 hours, remove the grains from the agar and
pour a few drops of iodine solution on the surface of the agar
in each dish. Swish it around and immediately pour it off.
Rinse the surface of the agar carefully with tap water.

Place half-grains on starch agar
with cut surface of embryo in
contact with agar.

**FIGURE 37. Position of some split corn
grains on the starch agar.**

QUESTIONS FOR DISCUSSION

1. What change, if any, has taken place in the starch agar where the corn embryos
were placed?
2. How was this experiment controlled?

PROCEDURE, PART TWO

1. Dip one end of a Tes-Tape paper into the agar zone where the corn embryos were placed.

2. Allow the Tes-Tape paper to remain in the agar for 5 to 10 minutes. A color change in the paper indicates the presence of a simple sugar.

3. Set aside the agar plates for the tetrazolium test which will follow.

QUESTIONS FOR DISCUSSION

1. What might be a good control for the Tes-Tape investigation?
2. Explain the importance of formulating "if . . ., then . . ." statements to guide experimental design.

INVESTIGATION 20
Isolation of an Enzyme

The test just completed indicates that starch has been changed into sugar. Now consider an experiment which will give more evidence for the hypothesis that an enzyme is responsible for this change. If the enzyme diffused into the agar and changed the starch, then the agar found under the germinating embryo should contain some of the enzyme.

Design and carry out an experiment which will demonstrate the presence of an enzyme in the agar.

QUESTIONS FOR DISCUSSION

1. Do the results of this test indicate that you have been able to isolate the enzyme which is responsible for changing starch into sugar?
2. What further test would be necessary before you could say the substance is an enzyme?

INVESTIGATION FOR FURTHER STUDY

How can you test the rate of diffusion of the enzyme?

Another sugar test which can be applied to the clear area is the tetrazolium test. Colorless 2,3,5-triphenyltetrazolium chloride, when in the presence of reducing sugars, will be reduced to a pink color. Apply a solution of tetrazolium to the clear areas on the plates just tested and see if it indicates the presence of reducing sugars.

When a seed begins to sprout, we say it germinates, and we know it is viable. Imagine the disappointment and financial loss if a farmer or florist were to plant a batch of seeds and very few of them sprouted. To minimize such a possibility, many seed growers and the departments of agriculture of various states routinely test samples of seeds to determine what percentage will germinate. The germination test consists of planting a sample of seeds under standardized conditions and then determining the percentage which germinate. One drawback of this method is that it takes several days for the seeds to germinate. We may approach the problem from another angle, however, using the "if . . ., then . . ." statement. If a seed is alive, then enzymes should be changing starch to sugar in the seed. If reducing sugars are present, then the tetrazolium test should give a positive result. Thus the tetrazolium test should be an accurate indication of the percentage of live seeds in a sample.

INVESTIGATION 21
Testing for Seed Viability

You are asked to provide the procedure for this Investigation. Using the "if . . ., then . . ." reasoning, design an experiment to test the hypothesis that there is no difference in the percentage of viable seeds as determined by the germination test and by the tetrazolium test. If you use pea seeds for the tetrazolium test, split them as shown in

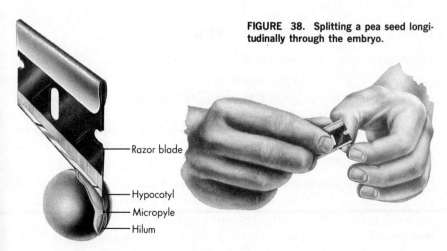

FIGURE 38. Splitting a pea seed longitudinally through the embryo.

Razor blade

Hypocotyl

Micropyle

Hilum

Figure 38. You may find a chart similar to Chart C useful in recording your data.

CHART C. COMPARISON OF GERMINATION AND TETRAZOLIUM TESTS AS MEASURES OF SEED VIABILITY

	Germination test	Tetrazolium test	
		Unboiled seeds	Boiled seeds
Total number of seeds			
Number of "viable" seeds			
Percent of "viable" seeds			

QUESTIONS FOR DISCUSSION

1. How does the percentage of viability, as determined by the tetrazolium test, compare with the percentage of viability, as determined by the germination test?
2. If you wished to know whether the data showed any significant difference between the two tests, what test for significance would you use? Why?
3. Apply the test of significance which you have selected. Are the differences significant?
4. What does the tetrazolium test actually indicate?
5. In what part of the seeds examined was the chemical action greatest?

INVESTIGATIONS FOR FURTHER STUDY

Design experiments to answer the following questions:

1. Would similar results be obtained from dry seeds?
2. What is the minimum time that seeds must be soaked to bring about the tetrazolium reaction?
3. Would similar results be obtained with a piece of raw potato cut from a point near a growing sprout? How near the growing sprout must it be cut so that the reaction may be observed?

The tetrazolium test (by making use of our knowledge of some of the chemical reactions that take place in active cells) serves as a basis for a seed-viability test, which is much quicker than the germination test. There are, however, many cases in which neither the tetrazolium test nor the germination test will tell us whether or not the seeds are capable of germinating. Many species produce seeds that will not germinate even though they are placed in an environment which we would expect to be favorable for them and in which, at some other time, they may germinate well. We call this condition in which seeds will not germinate *dormancy*. In many cases dormancy of seeds is a distinct advantage to the species. Why?

Many different factors cause dormancy in seeds. In some species, the embryo apparently must continue a part of its development under environmental conditions which are different from those in which the seeds were formed or in which they may eventually germinate. For example, some seeds may germinate well only after having been stored in a cold moist room for several months. Other seeds contain chemicals which inhibit their germination. If these are leached out, the seeds may grow immediately. Some seeds have hard coats which may prevent the entrance of water or of oxygen. They also may physically prevent growth until the coat is weakened or broken either by cracking or by the action of bacteria or fungi.

There are many other mechanisms which control dormancy in seeds. We will study only one of these in any depth. Our problem is introduced by the next Pattern of Inquiry.

PATTERN OF INQUIRY 7
Factors Affecting Seed Germination

A new home owner in Iowa was faced with the problem of establishing a lawn. His lot was on the side of a hill, and erosion was a problem, especially in the front of the house. He carefully prepared the ground and seeded it with Kentucky bluegrass, which is a very successful lawn grass in that region. Preparation of the ground and seeding of the front and back lawns were done in a similar manner, except that after scattering the seed the owner thoroughly raked the surface of the front lawn in order to cover the seed. He did not find time to do this in the back yard before a period of wet weather set in, and this prevented further working of the soil.

In a week he noticed that many seeds in the back yard had germinated; in two weeks it was green with a good stand of seedlings. In the front yard, however, he found that only a small percentage of the seeds had grown; after six weeks a very poor stand of grass was still there. What was responsible for the difference in growth of the grass in the two areas?

PARALLEL READING: Some Concepts of Light Energy

The germination of the seeds of some varieties of lettuce, like that of bluegrass seeds, has been observed to be quite erratic. Light has been shown to be an important factor in the germination of these seeds. In order to explore the role of light in germination, the investigator must have a working knowledge of the nature of light. In preparation for the investigations which follow, a brief discussion of this subject is appropriate. It takes the place of the literature search and specialized study which the researcher in science properly makes before beginning work on a problem.

Present-day study of the nature of light is based largely on a combination of two theories, each of which explains some of the complex phenomena. According to one theory, light travels in waves somewhat similar in nature to water or sound waves; therefore, we speak of the frequency and amplitude of light waves. (Frequency measures the number of wave crests which pass a given point in a given time; amplitude measures the height of the waves.) We recognize differences in frequencies of light waves as colors and differences in amplitudes as light intensities.

The corpuscular theory considers light to be composed of small particles traveling at extremely high speeds. Each particle, or *photon*, possesses a definite amount of energy, a *quantum*. According to this theory, light intensity depends on the amount of energy delivered to a given area in a given amount of time by all the photons striking it. Light quality (color) is determined by the energy of the individual photons.

Attempts to unify these two theories have led to useful mathematical descriptions of the energy relationships of light, although a completely unified theory of light has not yet been formulated. However, the blending of the two theories permits us to deal with different aspects of radiation. The properties of reflection and refraction are best explained by the wave theory. Photochemical effects, however, are most easily explained by the corpuscular (quantum) theory.

Light, considered as a form of wave motion, is only one part of a wide span of electromagnetic radiation consisting of waves ranging in length from extremely short distances to thousands of miles. The waves that cause visible light are not the shortest of these waves, but they are too short to measure by using ordinary units of length. To deal with them, the millimicron (mμ) and the Angstrom unit (Å) are used. A millimicron is $1/1,000,000,000$ of a meter (1×10^{-9} meter). An Angstrom unit is $1/10,000,000,000$ of a meter (1×10^{-10} meter).

Thus we speak of certain radiant energy as having a wavelength of 5000 Å, or 500 mμ. As a matter of fact, 445 mμ is one way of describing a particular shade of blue, and we can describe a particular shade of red as being 6540 Å (654.0 mμ).

Figure 39, a diagram of the spectrum of the electromagnetic radiations known to man, gives an idea of the position occupied by visible light. The visible part of the spectrum is the range of wavelengths of energy which our eyes can sense. It is bounded on the one side by violet, the shortest visible wavelength, and on the other side by red, the longest visible wavelength. These two limits encompass the series of colors familiar to us in the rainbow because water droplets in the path of light rays separate the light into its several component wavelengths. Certain prisms will also divide white light into its various parts.

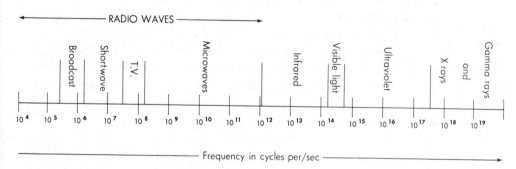

FIGURE 39. Spectrum of wavelengths.

Radiation of wavelengths too short to be seen is called ultraviolet, and that of wavelengths too long to be seen is called infrared. A considerable range of wavelengths much shorter than those ordinarily referred to as ultraviolet may be produced by X-ray machines or by events within the nuclei of atoms. These are called X rays and gamma rays, respectively. The distinction between these two is based on the sources and not on the wavelength.

It should be noted that the frequency varies inversely with the wavelength. Radiation of short wavelength has high frequency. The quantum (the energy possessed by a single photon) is proportional to the frequency. Thus, a quantum at 350 Å has twice as much energy as one at 700 Å.

When an object in sunlight reflects all the visible wavelengths, we say that the object is white. If the object reflects none of the visible wavelengths, our eyes receive no radiation, and we say the color is black. Those wavelengths which are not reflected are either absorbed by the object or transmitted. A piece of clear glass transmits most of the light falling on it, and a piece of colored glass transmits some, absorbs some, and reflects some. We say that an object is green if it transmits or reflects green and absorbs a high percentage of the other colors present in the white light.

Light energy which is absorbed may be converted to another form of energy—heat. However, the energy of a photon may be transferred to an electron, causing the electron to leave its original place. An example is the photoelectric effect, the basis for such instruments as the photoelectric light meter. Again, the energy of photons may be absorbed by certain molecules in such a way that an electron in a compound is raised to a higher energy level. The structure of the molecule may be changed by this, or the molecule may react more readily. The chemical change which results is called a photochemical reaction. The conversion of ergosterol to vitamin D and the light reaction in photosynthesis are examples of photochemical reactions.

In order to study the effects of light on plants in the laboratory, a source of light is necessary. If our results are to be compared with those of other workers, we must know the intensity and quality of the radiation. (Remember that intensity refers to the amount of energy, and quality refers to the wavelength.)

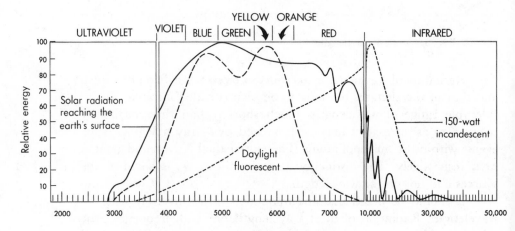

FIGURE 40. Distribution of wavelengths in Ångstrom units.

Figure 40 shows the distribution of wavelengths given as percentages of the total light emitted by different sources. Notice that the frequencies present in sunlight are represented by the solid line. Of the many types of artificial lamps, the common incandescent and fluorescent lamps come closest to producing light similar in quality to that of the sun. Notice that incandescent lamps have relatively less energy in the lower wavelengths (below 5500 Å) and that fluorescent lamps have relatively less energy in the higher wavelengths (above 6500 Å). In attempting to reproduce sunlight, it is best to use a combination of incandescent and fluorescent light.

On the basis of this brief summary of the nature of light, we can now proceed to design a series of experiments on the effects of light on germination. The following Investigation is designed to explore the question, "What is the effect of light on germination?"

INVESTIGATION 22
Effects of Light on Germination of Seeds

There are three possible answers to the question.

1. Light promotes germination.
2. Light inhibits germination.
3. Light has no effect on germination.

First, we may ask, "Does light promote, inhibit, or have no effect on the germination of two varieties of lettuce?" To find the answer, we can set up germination test conditions with

1. Grand Rapids lettuce seeds in light and in darkness
2. Great Lakes lettuce seeds in light and in darkness

PROCEDURE

Develop your own procedures after class discussion of the problem. Your teacher may wish to suggest some modifications. When you have completed the experiment, first combine the data obtained by your team with those obtained by other teams using the same procedure and then evaluate the results.

QUESTIONS FOR DISCUSSION

1. Is it easy to make a decision about the effect of light on the germination of the two varieties of lettuce?
2. On the basis of the data obtained by all the teams in your class, use an appropriate test of significance to determine whether the difference between the

treatments in each variety is significant. Do the results for each variety represent a normal distribution within one population, or do they represent two populations?

If we can determine that light has affected the germination of one of the varieties of lettuce, the next question might be, "Do all light bands (wavelengths) equally affect germination of the seeds?" If, on the basis of the work just completed, no decision can be made as to the difference in this respect between the two varieties, then both varieties should be tested.

To answer our question, an experiment can be designed in which the wavelengths reaching the seeds are controlled. This can be done by producing the desired wavelengths with a special lamp or by removing with filters the wavelengths not desired in natural or artificial light. An example of a lamp producing a narrow wave band is the germicidal bulb found in many clothes driers and refrigerators (2300–2800 Å).

By proper selection of filters and light sources, we can obtain fairly narrow wave bands of light which will be satisfactory for our purposes. The relative energy emitted in the different wavelengths by a daylight fluorescent lamp and by a 150-watt incandescent lamp is shown in Figure 40, page 180. Dark blue cellophane transmits little light between 5575 Å and 6575 Å. It does, however, transmit a high percentage of light which is beyond 7000 Å. Green cellophane transmits relatively little light below 5800 Å. By combining these filters with the fluorescent or the incandescent lamps, it is possible to provide four different bands of light to which the seeds can be exposed. The following table shows some of the approximate light bands which may be obtained in this way.

TABLE 15. LIGHT BANDS OBTAINED USING DIFFERENT FILTERS

Light source	Filter		Approximate light bands
Daylight Fluorescent	+ Blue filter	=	3900–5500 Å
Daylight Fluorescent	+ Green filter	=	4700–5800 Å
Daylight Fluorescent	+ Red filter	=	5800–6800 Å
150-Watt Incandescent	+ Red and blue filter	=	7000 Å up

The design of the following experiment may be stated according to the "If . . ., then . . ." logic as follows: If the various wavelengths of light directly affect germination, then a difference in the percent of germination should be observable in seeds exposed to varying wavelengths.

INVESTIGATION 23
The Effects of Different Wavelengths of Light on Germination of Seeds

PROCEDURE

After discussion with the class and with your teacher, work out procedures which may be expected to test the above hypothesis.

QUESTIONS FOR DISCUSSION

1. What conclusions can you draw from this experiment?
2. Can you think of any other factor in this experiment which might have influenced your results?

INVESTIGATIONS FOR FURTHER STUDY

1. What would be the result if you were to alternate exposures to the different light bands which influenced the germination of lettuce seeds?
2. The modern botanist is still seeking answers to the exact mechanism that "triggers" growth in seeds which show dormancy. The seeds of many plants in your area have their own unique triggering mechanisms. It would be exceedingly difficult to give specific examples, since plants vary in different areas; however, several types of plants are listed below, one or more of which probably grow in your locality. Remember that any scientific research begins with a careful perusal of all library material that is available. In your local library, it may be possible to locate papers that have been published about dormancy of plants in your area. Don't overlook the texts and reference materials in your school or local library. *Scientific American* reprints that will give you a great deal of assistance are "Germination" by Dov Koller (April 1959) and "Light and Plant Development" by Butler and Downs (December 1960).

    Collect as many different types of seeds as you can. You can then design your own investigation of germination, using the materials you have available. You will need several seeds of each type so that you can vary the types of treatments used. You might try: various methods of scarification; differing concentrations of gases, salts, or acids; alternate freezing and thawing of moist seeds; alternation of wet and dry periods; or combinations of these treatments.

    You may find it useful to experiment with the seeds of one or more of the following species: hawthorn, cocklebur, maple, tobacco, wild sunflower

(*Helianthus annuus*), cactus (these must be taken directly from the fruit), any other desert plant, or any members of the genera *Pinus*, *Betula*, *Fraxinus*, *Nasturtium*, *Bidens*, or *Triflolium*. The seeds of many of these will exhibit some kind of dormancy.

These Investigations with germinating seeds may illustrate how some research develops. The tests for germination are rather simple, but the explanation for some types of dormancy may be very difficult to obtain. The effect of light on the germination of lettuce seeds is probably a result of the effects of light on the pigment phytochrome which is also involved in the response of plants to different day lengths (photoperiodism). There is still, however, a great deal to be learned about many of these phenomena. Most certainly it will require extended work by both botanists (plant physiologists) and biochemists before complete understanding of such reactions is possible.

The same can be said for nearly all aspects of plant growth and development. You may be familiar with Van Helmont's experiments from which he inferred that a plant takes everything which it needs for growth from air and water. He neglected the few ounces of weight which were lost from the soil during the growth of the plant, probably regarding them as an error of measurement. In this case, the neglect of some of the data led to an erroneous concept of the source of raw materials for plant growth.

Since Van Helmont's day many advances have been made in our understanding of mineral nutrition of plants. The following Investigation illustrates one way in which you might study the requirements of plants for certain chemical elements.

INVESTIGATION 24
Mineral Requirements of Sorghum Plants

MATERIALS (per team)

1. 16 test tubes, 22 mm × 175 mm
2. Forty seeds of *RS 610* sorghum
3. A cardboard box large enough to hold the prepared test tubes of each team and a styrofoam plastic top with holes drilled to accommodate the test tubes. (See Figure 51, page 211.)

MATERIALS (per class)

1. Stock solutions as shown in Table 16

2. Minus N, Minus P, Minus Fe, and complete mineral solutions for culture of plants as shown in Table 17, page 186

3. Cotton for stoppering test tubes

4. A supply of distilled or deionized water

PROCEDURE

1. The following stock solutions may be available, or students may be asked to prepare them. Be sure that all glassware is chemically clean and that only analytical grade chemicals are dissolved in distilled water. The amounts specified should be ample for the class.

TABLE 16. PREPARATION OF STOCK SOLUTIONS FOR GROWTH OF SORGHUM PLANTS

Chemical	Amount	Amount of distilled or deionized water
1. $Ca(NO_3)_2 \cdot 4 H_2O$	23.6 grams	200 ml
2. KNO_3	10.0 grams	200 ml
3. $MgSO_4 \cdot 7 H_2O$	4.2 grams	300 ml
4. KH_2PO_4	4.2 grams	300 ml
5. $CaCl_2$	5.6 grams	100 ml
6. KCl	1.6 grams	200 ml
7. Iron chelate (iron ethylenediaminetetra acetate)	1.0 gram	100 ml
8. Trace elements (A stock solution should be made up to contain the following salts in the concentrations shown.)		
a. $MnCl_2 \cdot 4 H_2O$	1.8 grams	
b. H_3BO_3	2.8 grams	
c. $ZnSO_4 \cdot 7 H_2O$	0.22 grams*	
d. $CuSO_4 \cdot 5 H_2O$	0.08 grams*	
e. $Na_2MoO_4 \cdot 2 H_2O$	0.025 grams*	
f. Distilled or deionized water to make 1,000 ml.		

* The difficulty in weighing these small amounts of material directly can be avoided by using aliquots of more concentrated solutions of each. For example,

Weight out:	Dissolve in:	Use:
2.2 grams of $ZnSO_4 \cdot 7 H_2O$	100 ml of water	10 ml per liter
0.8 grams of $CuSO_4 \cdot 5 H_2O$	100 ml of water	10 ml per liter
2.5 grams of $Na_2MoO_4 \cdot 2 H_2O$	100 ml of water	1 ml per liter

2. Prepare the four mineral-culture solutions in which you will grow the plants. To do this, add each of the indicated stock solutions to about 500 ml of distilled or deionized water and then add enough water to make a total of one liter. (See Table 17). If you mix the stock solutions without diluting them, precipitates are likely to form; these may be difficult to re-dissolve.

TABLE 17. PREPARATION OF GROWTH SOLUTIONS FOR MINERAL-NUTRITION INVESTIGATION

Stock solution	Number of milliliters to be added for 1 liter of culture solution			
	Complete	Minus N	Minus P	Minus Fe
$Ca(NO_3)_2 \cdot 4 H_2O$	10	—	10	10
KNO_3	10	—	10	10
$MgSO_4 \cdot 7 H_2O$	10	10	10	10
KH_2PO_4	10	10	—	10
$CaCl_2$	—	10	—	—
KCl	—	10	10	—
Iron chelate	10	10	10	—
Trace element stock	10	10	10	10
Water (distilled or deionized) enough to make:	1 liter	1 liter	1 liter	1 liter

3. Each team should place about 40 seeds of RS610 Sorghum in a roll of moist paper towel and leave for 3 to 4 days. After the seeds are planted, stand the roll on end so that the roots will grow straight down and the shoots straight up.

4. Each team should prepare 16 test tubes as follows: wash with a detergent; rinse thoroughly with tap water; then rinse thoroughly with deionized water. It is most important that the test tubes used in the experiment be chemically clean.

5. Label each test tube by team, class, experiment, and date. Also label 4 of the tubes "complete," 4 "minus N," 4 "minus P," and 4 "minus Fe."

6. If you did not make them up, carefully study the formulas and procedures for making up the stock solutions and the

culture solutions. Place enough of the culture solutions in the test tubes to fill them to within about 5 or 6 centimeters from the top. Each team should have 4 test tubes each of "minus N," "minus P," "minus Fe," and "complete" solutions.

7. Select 16 uniform sorghum seedlings from the germination roll. Using a razor blade, carefully remove the cotyledon of each. Handle the seedlings with care and keep the roots moist at all times.

8. Select pieces of cotton large enough to serve as stoppers for the test tubes. Fold the cotton around the seedling as shown in Figure 41 and insert the roots of the seedling into the culture solution. The point where the cotyledon was attached should be near the bottom of the cotton. Be careful to prevent the cotton from coming into contact with the solution. If the cotton gets wet, molds are likely to develop in it. Also be sure that the roots of each seedling dip into the solution. It is not necessary, however, that all the root system be submerged. If the plug does not fit snugly, pack some additional cotton around it.

9. Place each labeled test tube containing a seedling in the cardboard box as illustrated in Figure 41. Then place the box in the best light available.

FIGURE 41. Technique for setting up materials to study mineral nutrition.

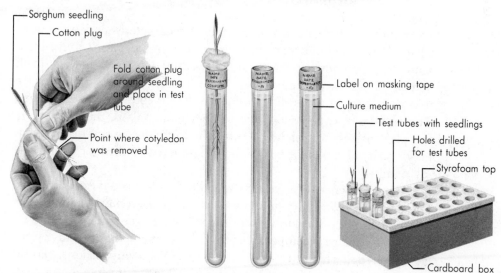

Sorghum seedling
Cotton plug
Fold cotton plug around seedling and place in test tube
Point where cotyledon was removed
Label on masking tape
Culture medium
Test tubes with seedlings
Holes drilled for test tubes
Styrofoam top
Cardboard box

10. Observe the plants three times a week for at least two weeks. Keep the level of the solution high enough so that the roots always reach into it. If necessary, add distilled or deionized water. A better procedure, however, is to discard the remaining solution and refill the tube with fresh culture medium. Study the plants carefully at each observation time. Record any changes either in the character of the root system or in color or appearance of the leaves.

QUESTIONS FOR DISCUSSION

1. Why were the cotyledons removed at the beginning of the Investigation?
2. Why must the glassware be chemically clean for this Investigation?
3. If you found a plant that exhibited chlorosis, how would you decide whether the deficiency was due to an inadequate supply of iron or of nitrogen?
4. On the basis of this Investigation and any other information you may find, what conclusions can you draw as to the role of various elements—especially nitrogen, phosphorus, and iron—in plant growth and development?

SECTION **11** GROWTH AND DEVELOPMENT IN ANIMALS

Since there is no characteristic or group of characteristics which enable us to draw a definite distinction between simple plants and simple animals, we cannot say that there are essential differences in the growth and development of these forms. In the higher animals, however, we find that there is more differentiation of cells into specialized tissues. The embryological development of higher animals is, therefore, more complex than that of higher plants. The formation of sperms and eggs may also involve more specialization than was described for the flowering plants.

Gametogenesis. As we have noted before, meiosis is essentially the same in both plants and animals. In the higher animals, meiosis leads almost directly to the formation of sperms and eggs since there are no intervening stages comparable to microgametophytes and megagametophytes, which are found in plants. Following meiosis, certain changes occur in the formation of sperms which result in a long-tailed, motile spermatozoan, or sperm cell. In the development of the egg, meiotic divisions of the nucleus are accompanied by unequal divisions of the cytoplasm, leading to the formation of one ovum (or egg) and three polar bodies which do not function.

Development of sperms and eggs in higher animals is shown schematically in Figure 42.

FIGURE 42. A comparison of spermatogenesis and oögenesis in higher animals.

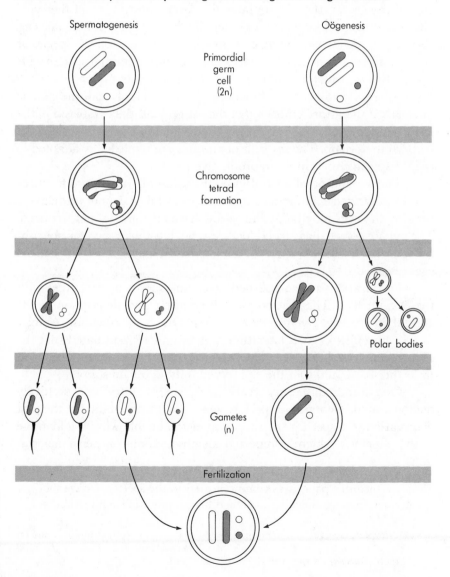

Various sperms may look very different, but they all contain three basic parts. One part is the head, which consists of a nucleus of tightly packed chromosomes. Another basic part is the middlepiece with its centrosome, which will function in fertilization, and its mitochondria, in which presumably the processes which provide the energy for movement occur. The third part is the tail, which consists mainly of a central fiber and which is the locomotor organelle.

An egg is a giant cell compared to a sperm. In addition to a nucleus, which contains the genetic material and is similar in size to that of the sperm, the egg contains the usual constituents of the cytoplasm and the yolk or food supply for the developing embryo. The egg also has a protective covering of some sort, depending on the species of animal. There are great differences among eggs of animal species with respect to size, which depends on the amount of yolk present.

All eggs perform three functions: (1) they supply half the genetic material of the future embryo; (2) they supply all the cytoplasm of the fertilized egg; and (3) they supply food reserves that will enable the embryo to develop. The sex cells of animals are remarkably adapted to their respective roles in the reproductive process.

The Beginning of an Individual: Fusion of Gametes. If fertilization in most species of animals is to be successful, it cannot be delayed, because the eggs may die within a few hours if they are not fertilized. The egg, of course, has a good food reserve, but this is used for nourishment of the developing embryo, not for maintenance of the unfertilized egg.

Fertilization is accomplished when the sperm nucleus fuses with the egg nucleus. The first correct observation and description of the animal sperm was given in 1677 by the Dutch microscopist, Antony van Leeuwenhoek. At a later time, Leeuwenhoek postulated that the sperm penetrated the egg. He believed that the embryo developed from the sperm and that the egg served only as a food supply.

One hundred seventy years later, sperm entry was positively demonstrated. George Newport is generally given credit for the first observation of a sperm penetrating an egg. The following is a footnote to the paper[1] in which he reported his observations of sperm penetration:

> Since this paper was communicated to the society, I have succeeded, through the adoption of a different mode of examination, in

[1] Newport, George. "On the impregnation of the ovum in the Amphibia and on the direct agency of the spermatozoan." *Philosophical Transactions of the Royal Society of London* for the year 1853—pp. 233–290.

detecting spermatozoa within the vitelline cavity in direct communication with, and penetrating into the yelk [yolk]. They were first seen by myself, in company with a friend, on the 25th of March of the present year [1853] within the clear chamber above the yelk, at about forty minutes after fecundation, when the chamber begins to be formed. I have since repeatedly observed them within the chamber, and in some instances still in motion, in which state I have had opportunities of showing them to my friend Professor Ellis of University College, and to two other medical friends, so that the presence of active spermatozoa within the vitelline cavity in the fecundated egg of the Frog may now be regarded as indisputable. The details of my investigation I reserve for a future communication, and will merely now add, that the spermatozoa do not reach the yelk of the Frog's egg by any special orifice or canal in the envelopes, but actually pierce the substance of the envelopes at any part with which they may happen to come into contact; as I have constantly observed while watching their entrance: sometime after they have entered the yelk chamber they become disintegrated and are resolved into elementary granules.

For purposes of discussion, it is convenient to consider fertilization as a sequence of phases. The first phase, or release of gametes, must occur at about the same time in both male and female because the gametes are short-lived cells. Once the gametes are released, the sperm must approach the egg. Contact between the sperm and egg is usually the result of random movement; however, some chemical attraction may occur. Although the sperms may be unguided missiles, a hit should be fairly easy when we consider the size of the egg in relation to the number of sperms. When the sperm contacts the egg, it must then enter into it; this entry is called *penetration*. The sperm head, or nucleus, may fuse with the egg nucleus. As a result of penetration followed by nuclear fusion, the egg is activated and embryo development begins. The activation of the egg by the sperm cannot as yet be completely explained. We can, however, observe these early events of development in the laboratory by using the frog as our experimental animal.

PREPARATORY TECHNIQUE 1
Obtaining Frog Pituitaries

To obtain frog eggs for observation and experimentation, the experimenter can induce ovulation in female frogs by injecting pituitary glands into their abdominal cavities. The number of pituitaries needed varies with the season of the year as follows:

October: 8 to 10
November and December: 5 to 6
January and February: 3 to 5
March and April: 1 to 3

To obtain the pituitaries, anesthetize as many frogs as will be required by placing them in a large container containing a wad of cotton saturated with ether or chloroform. Seal the container securely. Make certain that the frogs are under deep anesthesia before removing them from the jar.

PROCEDURE

1. The top of the frog's head must be removed by three cuts, to avoid cutting through the pituitary. Remove the top of the head by a cut extending from the angle of the jaw past the

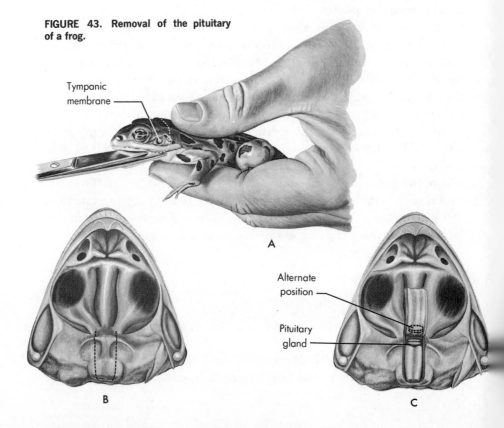

FIGURE 43. Removal of the pituitary of a frog.

Tympanic membrane

Alternate position

Pituitary gland

A

B

C

tympanic membrane on each side of the head. Then make a transverse cut as indicated by the dotted line in A of Figure 43. Lay the top of the head mouthside upward and remove the skin from the roof of the mouth to expose the cross-shaped bone.

2. Locate the foramen magnum and carefully insert the tip of the scissors to one side. Cut through the brain case on each side as shown by the dashed lines in B of Figure 43.

3. Lift the flap of bone with forceps and locate the pituitary gland, a pinkish oval body. (It will be found either on the anterior end of the brain or on the flap of bone. See C of Figure 43.)

4. Remove the pituitary with forceps and strip off any adhering white tissue. Cut up the glands into small pieces and place them in a dish containing 10% Holtfreter's solution.

PREPARATORY TECHNIQUE 2
Injecting Pituitaries

PROCEDURE

1. Draw the pieces of the required number of pituitary glands, along with about 1 ml of Holtfreter's solution, into the barrel of a hypodermic syringe before attaching the needle. Place a size 18 needle on the syringe and remove all the air from the syringe.

2. Hold a female frog on its back and insert the needle through the belly skin and muscle, being careful not to injure the ventral blood vessels and the intestines. Inject the pituitaries; then pinch the skin around the needle as you remove it. Continue holding the puncture closed for a few seconds to prevent the possible loss of pituitary material.

3. Draw additional solution into the syringe to check whether or not any pituitary fragments lodged in the needle. If they did, remove most of the solution from the syringe; then inject these glands as before.

4. Place the injected frog in about one-half inch of water in a covered container and put it in a cool place (12 to 21°C).

5. Check the frog for ovulation after 48 hours by gently squeezing the abdomen toward the cloaca. Ovulation will usually occur

after a period of between 48 and 72 hours. You can expect about 1000 eggs from each successful injection. When you are sure the female is gravid, keep it in the refrigerator until the eggs are needed.

CAUTION: *To be relatively sure of success, you should inject two females at a time, since there is a chance of death or failure to ovulate. The hypodermic syringe and needle must be sterilized before and after the injection. (70% alcohol may be used for this.)*

PREPARATORY TECHNIQUE 3
Fertilization in Vitro

Before removing the testes from the males and preparing the sperm suspension, the females should be test-stripped to determine whether or not ovulation has been accomplished. The eggs which are removed in the test stripping should be discarded.

CAUTION: *The eggs must be stripped directly into the suspension of active sperms.* Immediately after ovulation the jelly surrounding the egg begins to swell, and those eggs not fertilized in a matter of minutes after they are stripped will not be fertilized.

Best results are usually obtained with the sperm from freshly removed testes. The testes may be dissected out a day or two before they are used, and the sperm within such testes will remain viable provided the testes are kept refrigerated.

Note: The testes must not be put in water. If they are to be stored, place them between layers of slightly moist cotton in a covered petri dish in the refrigerator.

MATERIALS

1. A mature male frog
2. A mature female frog which has been previously injected with pituitary glands
3. Sharp, pointed scissors
4. Sharp forceps
5. Cotton
6. Covered, glass dishes
7. Medicine dropper
8. 10% Holtfreter's solution

PROCEDURE

1. To remove the testes, use a male frog that has been double pithed—that is, the central nervous system has been destroyed so that it can no longer feel anything. Open the body cavity of the frog and move the digestive organs aside to expose the testes which are whitish-yellow, ovoid bodies. Use scissors and forceps to remove the testes from the body cavity. Rinse the testes by dipping them in saline solution; then blot them gently on paper toweling.

2. To prepare the sperm suspension, place the pair of testes in a petri dish moistened with a thin film of 10% Holtfreter's solution. Use scissors and forceps to cut and tear the testes into the tiniest pieces possible. If any sizable pieces persist, mash them with the end of a solid glass rod. Then add 20 ml of 10% Holtfreter's solution to the dish and allow 10 minutes for the sperm to begin swimming actively.

3. To strip and to fertilize the eggs, hold the injected female frog back downward in the palm of one hand. (Do not squeeze the frog until you are ready for the eggs.) Grasp the hind legs with the other hand; then squeeze gently over paper toweling until several eggs have been stripped. Wipe the cloaca dry; then proceed to strip 100 or more eggs into the sperm suspension. When stripping the eggs, squeeze gently and move the female around over the dish so that the eggs form a ribbon and do not pile up. Draw some of the sperm suspension into a medicine dropper and use it to bathe the eggs. Allow the eggs to remain in the sperm suspension for 10 to 15 minutes. During this time, examine the orientation of the eggs and compare the number of eggs that show the black area up with the number that show the white area up. If any eggs that were fertilized earlier in the day are available, examine them to see if their orientation has changed. After 15 minutes, pour the sperm suspension out of the petri dish and replace it with pond water. Let the eggs stand until near the end of the class period, while the jelly that surrounds the eggs swells slowly.

Once the eggs have been fertilized, allow as much time as possible for the jelly to swell before transferring them. If the eggs are moved too soon, damage may result. When the newly fertilized eggs are lifted from the glass, they must be handled with great care; they should not be sucked into a pipette or medicine dropper, or subjected to any other stress that would distort them and thus destroy their organization.

There should be enough eggs fertilized to provide for both establishment of cultures at room temperature and development in the temperature-gradient box. In addition, enough embryos might be permitted to develop at room temperature to provide tadpoles for future investigations.

As soon as the eggs have been fertilized and the jelly has swelled to the extent that it can be recognized as a separate mass surrounding each egg, the large mass of eggs should be divided, with the aid of scissors, into smaller masses of from 5 to 10 eggs. These smaller masses can then be transferred to the appropriate dishes. Either 100 ml of pond water or 10% Holtfreter's solution in each dish is adequate for 50 eggs to develop through Stage 25. (See Figure 45, page 199.) As soon as the majority of eggs in each dish hatches, it is advisable to place either a small mass of growing filamentous algae or a sprig of *Elodea* in the dish. This will provide both food and oxygen for the embryos as they develop.

Frog embryos have the ability to develop over a wide range of temperatures, although the rate of development varies. By maintaining small batches of embryos at different temperatures and examining them every day, you will see more stages of development than if you maintained the embryos at only one temperature. Using different temperatures will also enable you to make several important observations about the influence of temperature on the rate of development. To provide conditions of varying temperature, the temperature-gradient box is used. As shown in the diagram, this is a specially constructed box to be used inside a refrigerator. Note the following points:

(a) The temperature adjustment: determine which way you must turn the screw to raise (or lower) the temperature.

(b) The light bulb: a new, long-life, 100-watt incandescent bulb is used.

(c) The 110-volt, plug-in cord: this must be plugged into an outlet or extension cord outside the refrigerator. (When the door of a refrigerator is shut, the power supply to the outlet inside the refrigerator is automatically disconnected.)

(d) The box itself: when the door is closed, the box should be as nearly airtight as possible above the air space at the bottom. This measure prevents convection currents inside the box which will destroy the gradient of temperatures.

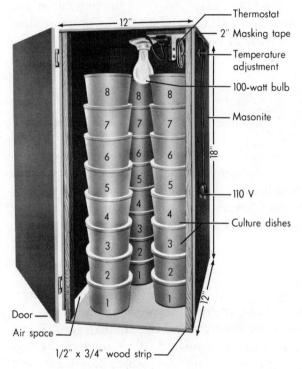

Thermostat
2" Masking tape
Temperature adjustment
100-watt bulb
Masonite
110 V
Culture dishes

Door
Air space
1/2" x 3/4" wood strip

FIGURE 44. A temperature-gradient box.

(e) The stacks of numbered dishes in the box: these are 8-ounce, polyethylene, refrigerator dishes with snap-on tops. They are numbered for identification so that the progression of numbers agrees with the progression of temperatures.

The box should be placed in the refrigerator with each of the dishes containing, as mentioned before, about 100 ml of either natural pond water or 10% Holtfreter's solution. The thermostat should be adjusted to obtain the desired temperatures; the temperature of the water should be checked and recorded daily for at least a week before the frog eggs are placed in the dishes. If the box is placed in a refrigerator that will maintain a temperature of 6°C and the thermostat on the gradient box is adjusted so that the temperature of the eighth dish is 28°C, you should be able to maintain a fairly even gradient of temperatures from approximately 8°C to 28°C. The space at the bottom of the box aids in temperature regulation. Partial closure of this space can be used to increase the gradient of temperature inside the box.

If the refrigerator will not maintain temperatures as low as 8°C, you can still achieve satisfactory results. Some of the earlier stages in development may be missed (see page 199); but, in general, at any temperature between 6°C and 14°C, there should be no more than one stage of development each 24 hours.

INVESTIGATION 25
Development of the Frog Embryo

Dishes containing fertilized eggs should be placed in the temperature-gradient box and in suitable locations at room temperature. They should be allowed to remain until the following day, when observation of development may begin. Observations should be compared with the stages shown on Shumway's Chart, Figure 45, on the opposite page.

PROCEDURE

1. First examine the eggs that have been left at room temperature. Have they changed since the previous day? Grasp an egg mass with forceps and flip it over, so that you can see the side which had faced downward. Place some of the eggs under a hand lens or dissecting microscope, illuminate them brightly, and determine as fully as you can what has happened to them in the past 24 hours. Refer to Shumway's Chart. Can you tell what stage they are in?

2. In observing the dishes from the temperature-gradient box in the refrigerator, notice that the dishes in each stack are numbered consecutively with No. 1 on the bottom and the largest number at the top. Every dish must be returned to the same place in the box after each class has observed it. Each stack of dishes will serve two teams—one team observing the odd-numbered dishes, the other observing the even-numbered ones. One member from each of the two teams using a stack should remove the dishes for his team and distribute them. As soon as you get the dishes, carefully check and record the temperature of the water.

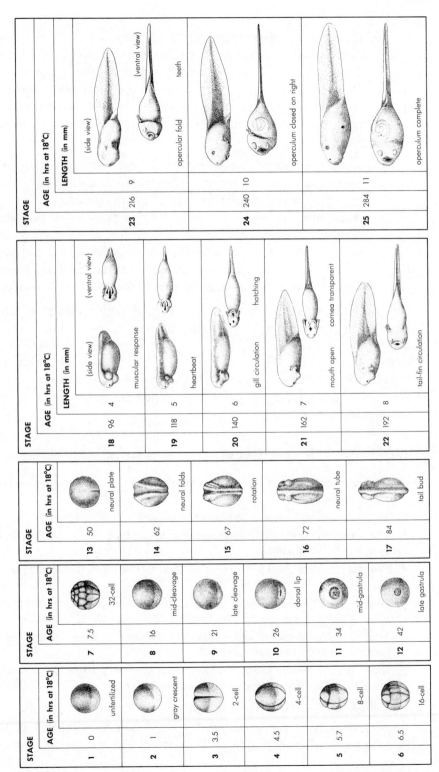

FIGURE 45. Shumway's Chart and drawings of the stages of development of *Rana pipiens*. (Courtesy of THE ANATOMICAL RECORD, vol. 78, no. 2, Oct., 1940.)

3. Examine the developing embryos, using bright illumination and a hand lens or dissecting microscope. Refer to Shumway's Chart. Can you count the cells in any of the younger embryos (that is, those developing at lower temperatures)? Are the cells of the light yolk area the same size as those in the dark pigmented area? Is the yolk completely covered in any of the eggs?

4. Record the stage of the embryos in each dish and note any special features that you see. All embryos in a single dish (that is, at the same temperature) will be in approximately the same stage; any that appear to be in a different stage are probably dead or damaged and should be removed. How much do the embryos developing at different temperatures differ from each other? Does this experiment determine the highest and lowest temperature at which frog embryos of this species will develop?

Between Stages 17 and 18, the nervous and muscular systems have developed sufficiently for some movement to be possible. As embryos reach Stage 17, notice that some do not move of their own accord. Using a fine bristle mounted on a handle, poke the embryos. Carefully note and record the extent and type of movement of embryos at various stages until you have determined the sequence of movements leading to the ability to swim.

QUESTIONS FOR DISCUSSION

1. What changes did you observe during the first 20 to 30 minutes after the eggs were fertilized?
2. Why were the egg masses cut into small groups?
3. What questions that you have answered in this section can be answered best by using statistical tests?
4. Why do you think that the zygote undergoes cleavages before developing further?
5. At what stage of development does the embryo begin to resemble an animal form? What features lend "animal form" to the embryo?
6. Do larvae raised at lower temperatures look the same as those raised at higher temperatures? Do they swim as well?
7. What does the answer which you have found to Question 6 tell you about the natural development of frog embryos out-of-doors?

Summary of Development of the Frog Embryo. Your observations of frogs have prepared you for the following summary of the various stages of embryonic development. When the amphibian egg cleaves, the actual cleavage plane can be seen to cross the surface of the pigmented animal hemisphere and slowly make its way downward into the yolky area until the entire egg is divided into two equal halves. (See Figure 46.) Before the first cleavage plane has completely extended to the light vegetal hemisphere, a second cleavage plane begins at right angles to the first and slightly above the equator. Subsequently, the egg splits into smaller and smaller cells, even though the cells of the animal

FIGURE 46. Stages in the early development of an amphibian: A, single-cell stage; B and C, early cleavage stages; D, blastula, external view; E, blastula with cells removed to show cleavage cavity; F, early gastrula, external view; G, section through early gastrula; H, late gastrula, external view; I, section through late gastrula just prior to organ formation.

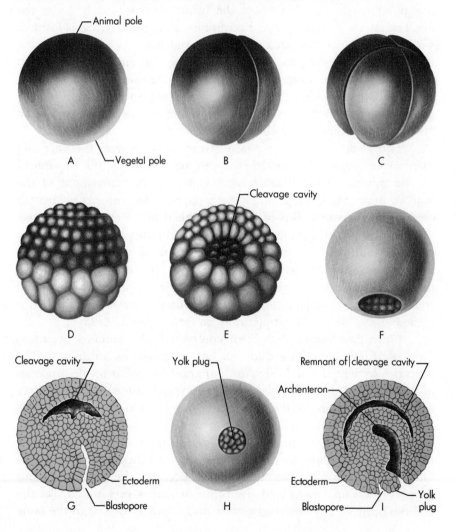

region are, for a long time, smaller than those of the vegetal region. When the cells have become quite small, they all tend to move close to the surface. As a result of this movement, a fluid-filled space is formed inside the embryo. This stage of the embryo is called a blastula. (See Figure 46.)

Since you have examined most stages of development in the laboratory, only points of special interest will be described here. At the end of the blastula stage, the embryo undergoes gastrulation to become a gastrula. In this process, some of the dark surface material folds inward and forms a new inner layer of tissue. The first sign of this event is the appearance of a short, faint, black line just below the edge of the pigmented area. Experimental embryologists have shown that this point marks the midline of the future animal's back. Later the line becomes a definite curve, and finally a complete circle is formed. The line represents the area where material from the surface is being folded under to form the new layer inside the embryo. As it folds under, the surface material also stretches downward. You may see some embryos in which the yolk region is completely, or almost completely, enveloped by the pigmented material, leaving only a little "polka-dot" blastopore on the under surface.

The surface material actually stretches down over the yolk and then moves inside. Its movements can be traced by placing marks on the outside of the blastula and watching the movement of the marked areas. An embryo of a species having light-colored pigment is used for such a study. Bits of agar are soaked in a vital dye (a colored substance that can stain living tissue without injuring it) and are then pressed against the embryo's surface so that small marks (blue or red) are printed on it. As gastrulation proceeds, the small, round marks may be seen to grow longer, move toward the region of infolding, and then turn and disappear inside. If the outer layer of the embryo is peeled away, the marks can actually be seen in the inner layer.

Once gastrulation has been completed, the embryo begins to form itself into something that can be recognized as an animal. The processes by which it does this are remarkably similar to the processes of development of the blastoderm of any vertebrate embryo and serve as a good illustration of the fact that young embryos tend to carry out their early developmental stages in a similar way. After gastrulation, the frog embryo develops a flattened neural plate. Folds rise up along the edges of this plate, then grow together until they meet in the midline. (See Figure 45.) The tube formed becomes the central nervous system—the spinal cord and the brain, which is very much like the embryo brain of any other vertebrate and which is formed in the same way. As in other vertebrates, the future eyes bulge out of the brain at its anterior end.

Soon a series of block-shaped somites appears along the sides of the spinal cord; these will produce the backbone and muscles as well as part of the skin. Later, a heart forms and begins to beat. The basic pattern of the circulatory system, at the time when the blood begins to flow, is identical in bird and frog embryos.

There is a brief period of time when the body plan of all vertebrate embryos is basically the same. Later, each embryo progresses toward the precise form it is to attain. The developing frog changes gradually from the short, compact larva to the active, swimming tadpole. Several months or more pass before the tadpole undergoes metamorphosis and becomes a mature frog.

Amphibians are very much limited by the fact that they must reproduce in water. The vertebrates took a big step forward in evolution when some of the ancient reptiles began to produce eggs that could be laid on land. Such eggs were necessarily large because they had to contain sufficient water to supply the young animal through its developmental period. This step was important because it enabled the reptiles, and subsequently their bird and mammalian descendents, to move away from the edges of ponds and streams to populate the continents.

Frogs and salamanders, like all amphibians, are often called "cold-blooded" animals. This term is really a misnomer, since the body temperature of these animals is variable rather than cold. It rises and falls with the temperature of the environment. Animals which cannot maintain their own body temperatures at a constant level cannot provide heat for incubating their developing young, as birds and mammals do. The development of the frog embryo is not limited to a certain definite temperature, but rather it slows down at low temperatures and speeds up at high temperatures, just as plants grow faster in warm weather than they do in cold.

INVESTIGATION FOR FURTHER STUDY

Effects of Ultraviolet Irradiation on Frog Development

In the previous Investigation, you were able to observe the normal development of frog embryos. Use the information derived from that work as a basis of comparison with the results of this Investigation.

The effects of radiation upon living organisms is a subject of increasing interest both to scientists and nonscientists. One of the most harmful effects of radiation is the alteration or mutation of the gene content of a cell's chromosomes. Contrary to popular notion, there is no minimum amount of radiation which must be exceeded before mutations occur; any amount, however small, which reaches tissue can cause mutations.

A suntan is produced by ultraviolet radiation from the sun. Ultraviolet radiation *is* high energy radiation, *but* the penetrating power of U.V. is so slight that it cannot pass through skin or, for that matter, through water. Thus, the only harmful effect we usually observe is a case of sunburn. (However, some recently published work has implicated excessive exposure to the sun in the production of skin cancer.) Nevertheless, if a *very thin* layer of sperm suspension is exposed to U.V. irradiation, it is possible to "knock out" sperm chromatin. When eggs are fertilized by sufficiently irradiated sperm, haploids (progeny with only one set of chromosomes) are produced.

The following experiment enables one to test the effects of ultraviolet irradiation on spermatozoa and thereby assay the ability of irradiated sperm to (a) fertilize and (b) take part in subsequent developmental processes. This experiment also enables one to recognize "normal" and "abnormal" developmental processes.

MATERIALS (per class)

Ultraviolet lamp (germicidal)

10% Holtfreter's solution

1 male frog/group

1 female frog (previously injected with pituitary)

9 fingerbowls/group

Dissecting instruments

1 petri dish

1 2.0-ml pipette

Note: A data sheet similar to the one illustrated on page 206 will be needed.

PROCEDURE

1. 9 fingerbowls will be used. Two of them will be labeled "controls," one will be used to cover other bowls, and the remaining six will be labeled "experimental." Both control and experimental vessels should bear a time label: 5 seconds; 15 seconds; 30 seconds; 1 minute; 5 minutes; or 10 minutes.

2. 8.0 ml of 10% Holtfreter's solution is added to each vessel.

3. Sacrifice a male frog. Dissect out the testes and macerate them in 10% Holtfreter's (10 ml per pair of testes).

4. Pipette 2 ml of the resulting sperm suspension into each of the two control vessels.

5. Place the remainder of the sperm suspension in a large petri dish. Thus a thin layer of suspension is obtained. (Remember, U.V. is not very penetrating.)

6. The U.V. lamp should be positioned 15 inches above the tabletop.

7. Turn on the lamp and allow it to "warm up." CAUTION: Do not look at the U.V. lamp when it is on and avoid unnecessary exposure of your skin to its rays.

8. Place the petri dish beneath the lamp; RECORD THE TIME.

9. Swirl the sperm suspension gently during exposure to the U.V. lamp. This insures equal exposure to all sperm.

10. After 5 seconds of irradiation, remove the petri dish, pipette 2.0 ml of the suspension and place it in the finger bowl labeled EXPERIMENTAL 5 SECONDS. (Thus, there will be a total of 10 ml in each vessel—that is, 8 ml of 10% Holtfreter's plus the 2 ml aliquots of sperm.)

11. Continue to irradiate the sperm suspension and remove 2 ml aliquots at the prescribed intervals and place them in their respective bowls.

12. When all the irradiated sperm suspensions have been collected, the eggs are then stripped from the female directly into the experimental and control vessels. (THIS TIME REPRESENTS THE TIME OF FERTILIZATION—"0-TIME" ON YOUR DATA SHEET. RECORD THIS TIME.)

13. From this point the material is handled as outlined in Investigation 25, except that it should be maintained at room temperature.

    NOTE:

    a. Each group of students will be given two controls of 50 eggs each and 50 experimental eggs of each irradiation time. The members of each group will then observe the development of all eggs at the times specified in the data sheet. The information requested in the data sheet should be carefully recorded.

EFFECTS OF UV IRRADIATION ON FROG DEVELOPMENT

Time		Hours									Days				
Hours of Development	Approximate	1	2	4	8	16	28	48	72	96	5	6	7	8	9
	Actual														
Shumway Stage	Stage														
1 unfertilized (or DEAD)															
2 grey crescent (rotated)															
3 2-cell															
4 4-cell															
5 8-cell															
6 16-cell															
7 32-cell															
8 mid-cleavage															
9 late cleavage															
10 dorsal lip															
11 mid-gastrula															
12 late-gastrula															
13 neural plate															
14 neural folds															
15 rotation															
16 neural tube															
17 tail bud															
18 muscular response															
19 heartbeat															
20 hatching															
23 opercular fold															
25 opercular complete															
TOTAL LIVING															
TOTAL DEAD (discard)	from previous rdg.														
	from original No.														
% LIVING	from previous rdg.														
	from original No.														
% HAPLOID	from previous rdg.														
	from original No.														
TOTAL NUMBER OF HAPLOIDS															

TREATMENT _____ GROUP _____ STUDENT(S) _____

b. If the "2-CELL STAGE" has not been reached by 4 hours after fertilization, record these eggs as DEAD (UN-FERTILIZED) AND *DISCARD THEM* IMMEDI-ATELY.

c. All dead eggs should be removed from the bowls and discarded after EACH observation period.

d. The % LIVING should be recorded after EACH observation period, and this number should be based on the number living at the PREVIOUS observation.

e. As soon as HAPLOID CHARACTERISTICS are evident, they should be recorded in the appropriate space.

PARALLEL READING: The Genetic Control of Differentiation

As outlined a single fertilized egg undergoes a series of mitotic divisions, and, in the process of dividing and multiplying, assumes a shape and gradually differentiates into cells of markedly different appearance, size, and function. We have previously learned to associate differences between organisms with differences in genetic makeup. How is it possible for genetically identical cells to become divergent?

One answer was suggested in 1893 by the well-known German biologist, August Weissmann, who suggested that not all daughter cells received identical sets of genetic determinants. Instead, cells in different parts of the embryo would receive different sets of "instructions," corresponding to the specializations in tissues and cells that are found in the developed organism. In modern terms this could correspond to an unlike distribution of chromosomes in cleavage.

The experiments of Hans Spemann, in the early 1900's, lent scant support to Weissmann's notion. Spemann separated the blastomeres of developing amphibian eggs at the 8-cell stage and found that he could produce two similar embryos. Evidently, at this stage, the amphibian blastomeres are still genetically identical, and this nuclear equivalence of early blastomeres has been demonstrated for the sea urchin and chick as well. In plants, F. C. Steward, of Cornell University, has isolated a single, adult carrot-root cell; from this cell, a complete plant developed.

If we cannot look to differences in gene composition between cells of a developing embyro to account for cell differentiation, what other possibilities arise? An early experiment of Spemann again points a direction. Spemann demonstrated that a portion of the cytoplasm of the fertilized amphibian egg, the grey crescent, played a vital part in determining the location of the blastopore in the early embryo. When Spemann transplanted a portion of grey crescent to the dorsal side of a blastula, the blastopore formed there instead of at the usual site.

Spemann's experiment suggests that conditions in the cytoplasm of the cell are capable of influencing the pattern of development. Clearly, differences in the environment of different cells of the embryo must arise as gradients of oxygen concentration, CO_2 tension, and concentrations of chemical substances are established. That environmental conditions can, in fact, influence cellular activity is shown in the following examples.

Euglena, a single-celled, free-swimming Protist, contains about 10 chloroplasts per cell, and these chloroplasts function in photosynthesis. When grown in the dark, the chloroplasts disappear and another mode of nutrition is used. When exposed to light, *Euglena* again makes chloroplasts. Lymphoid cells capable of antibody production will form antibodies only in the presence of an antigen; for example, a bacterial toxin. Without such exposure, the cell's potential ability to produce antibodies is not expressed.

During the past quarter of a century, unraveling the manner in which chromosomes affect and direct cellular activities has been one of the most exciting challenges in biology. Countless experiments have shown that the large nucleic acid molecules exert primary control over the life processes of all cells. Furthermore, nucleic acids are the vital hereditary links between generations. Much has been learned about the structure and function of these molecules.

The term, nucleic acid, was first applied in the 1800's by a Swiss biochemist, Friederich Miescher, who discovered acidic substances in the cell nucleus. Since nucleic acids are now known to exist also in the cytoplasm of cells, the term is somewhat inaccurate but nevertheless universally accepted.

In spite of their large size, the basic chemical structure of nucleic acids is relatively simple. They are composed of only a few kinds of smaller molecules.

The basic unit of the nucleic acid is a nucleotide; ordinarily, there are only four kinds of nucleotides in a molecule of nucleic acid. These are arranged in a long chain. Just as a chain with four different colors of beads can be arranged in many ways, so can the sequence of four nucleotides be altered to many different patterns. Species of organisms differ in the number and sequence—but not normally in the kinds—of nucleotides in their nucleic acids.

Nucleotides consist of three parts. One part is a 5-carbon sugar. In one kind of nucleic acid, called ribonucleic acid, or RNA, this sugar is ribose. In the other kind of nucleic acid, called deoxyribonucleic acid, or DNA, the sugar is deoxyribose. The chemical structures of these sugars are shown in the following diagram. Note the only difference: the shaded hydroxy (OH) group of ribose is replaced with a hydrogen atom in deoxyribose. Study but do not memorize these structures. We will use the symbols indicated for demonstrating the structure and function of nucleic acids.

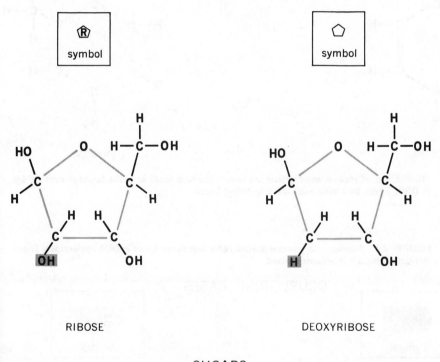

RIBOSE   DEOXYRIBOSE

SUGARS

FIGURE 47. The molecules RNA and DNA are named after the sugars ribose or deoxyribose in their nucleic acids. Note that the only difference between the sugar molecules is the shaded group.

In addition to their sugar segment, nucleotides contain phosphoric acid (H_3PO_4) and one of four compounds composed of carbon, hydrogen, and nitrogen, called nitrogen bases.

There are four kinds of nucleotides in DNA. Each contains one deoxyribose, one phosphoric acid, and one nitrogen base. Thus, the kind of nitrogen base distinguishes the four nucleotides of DNA from one another. The four nitrogen bases of the DNA nucleotides—cytosine, thymine, adenine, and guanine—are shown in Figures 48 and 49.

SINGLE-RING BASES

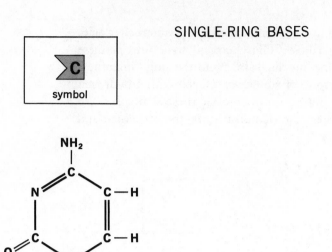

CYTOSINE

THYMINE

FIGURE 48. Cytosine and thymine are two of the four kinds of bases found in nucleotides in DNA. These two molecules are single-ring bases.

FIGURE 49. Adenine and guanine are the other two bases found in DNA nucleotides. These two molecules are double-ring bases.

DOUBLE-RING BASES

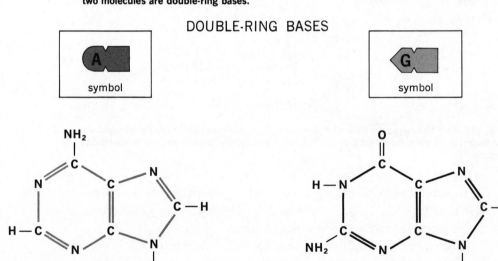

ADENINE

GUANINE

The nucleotides of RNA contain ribose (instead of the deoxyribose of DNA), phosphoric acid, and one of the four bases, also. These bases are adenine, guanine, and cytosine as in DNA. However, the thymine of DNA is replaced by uracil in RNA. The structure of uracil is seen in Figure 50.

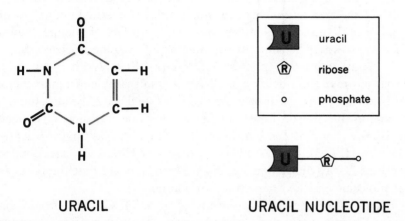

FIGURE 50. RNA has a uracil nucleotide made from the base uracil, instead of the thymine nucleotide that DNA has.

The sugar, phosphoric acid, and base parts of the nucleotide molecule join together through dehydration synthesis. One of the bases links with sugar, and the sugar in turn links with the phosphoric acid to form one of the four nucleotides commonly found in DNA. Symbolic representations of these nucleotides are shown below in Figure 51.

FIGURE 51. Symbolic representations of the nucleotides commonly found in DNA.

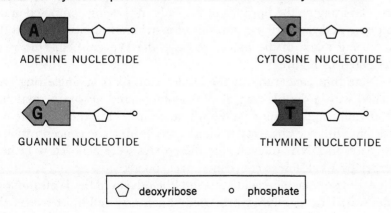

Once scientists understood the chemical composition of nucleotides, a great search was enjoined to determine how nucleotides were related to one another structurally within the nucleic acid molecule. If this mystery could be solved, then one might better determine how nucleic acids are able to direct the vital activities of the cell.

A clue to the structure of DNA was provided when it was noted that (1) the number of adenine nucleotides was always equal to the number of thymine nucleotides and (2) the number of guanine nucleotides was always the same as the number of cytosine nucleotides.

X-ray diffraction photos of DNA provided another clue as to the structure of this nucleic acid. When a small sample of a pure chemical is placed in the path of X rays, the rays will pass through the substance; in the process, however, they are bent or "diffracted," depending on the chemical composition of the material. The English scientist, Maurice H. F. Wilkins and his co-workers at King's College, London, obtained X-ray diffractions of DNA which indicate that the bases are flat molecules stacked one on top of another.

In 1953, James D. Watson and Francis H. C. Crick from Cambridge University, England, pieced together these clues about the structure of DNA and proposed a molecular model for DNA structure. In 1962 Watson, Crick, and Wilkins shared a Nobel Prize for this work which has established the chemical foundation for the study of modern genetics.

Watson and Crick decided that DNA consists of two strands of nucleotides joined together and twisted about one another in a form called a double helix. A double helix is a double coil and can be understood by imagining a ladder with flexible uprights as shown in Figure 52 on the opposite page.

As noted in the figure, if the bottom of the ladder is held in place and the ladder is twisted from the top, a double helix is formed. Now, if one can imagine the uprights of this ladder as being composed of the phosphate and deoxyribose portions of a string of nucleotides and the rungs being made of the bases, we have the Watson-Crick model as represented in Figure 53.

Note that each rung in the ladder consists of a single-ring base matched with a double-ring base. Adenine would always be matched with thymine and guanine paired with cytosine. Thus the structure was consistent with the chemists' evidence concerning the concentration of these bases in DNA. In each case these bases were believed to be held together by weak forces called hydrogen bonds.

A problem facing scientists before the advent of the Watson-Crick model was: How is DNA synthesized? Each new cell must receive the

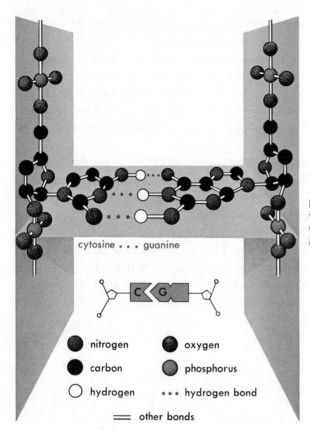

cytosine ... guanine

FIGURE 52. A diagram of one rung of the DNA "ladder," showing how cytosine connects to guanine by means of hydrogen bonds.

- ⬤ nitrogen
- ⬤ carbon
- ◯ hydrogen
- ⬤ oxygen
- ⬤ phosphorus
- ••• hydrogen bond
- ═ other bonds

FIGURE 53. In Watson and Crick's model of a DNA molecule, each upright is made of phosphate groups (P) connected to deoxyribose sugar groups (S). The ladder rungs are made of single-ring and double-ring bases.

same genetic instructions of DNA as the original cell. The DNA molecules, therefore, must be duplicated accurately.

The nucleotides of a single strand of DNA may be arranged in many ways with each nucleotide being repeated as often as necessary. As seen in Figure 54, however, once the order of the nucleotides in one strand is determined, the order in the other strand is likewise determined since thymine always pairs with adenine and cytosine with guanine.

Now to comprehend the duplication of DNA, imagine the hydrogen bonds that hold the double helix together to work somewhat like a zipper. By starting at one end of the molecule, each base can be "unzipped" from its partner—one at a time. This will leave an unpaired base on each strand. New nucleotides from the cell's store of raw materials now are free to fall in place. Thus the two single strands of DNA have been duplicated to form two new chains of double-stranded DNA, each identical in base sequence to the original strand. This process is shown, using our symbols in the diagrams of Figures 54–56.

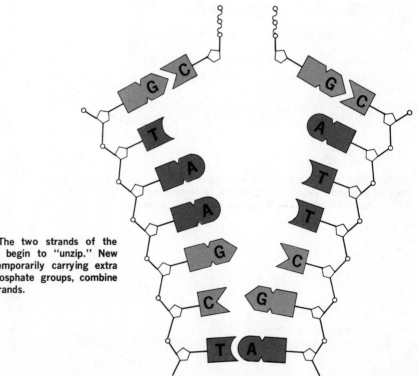

FIGURE 54. The two strands of the DNA molecule begin to "unzip." New nucleotides, temporarily carrying extra energy-rich phosphate groups, combine with the old strands.

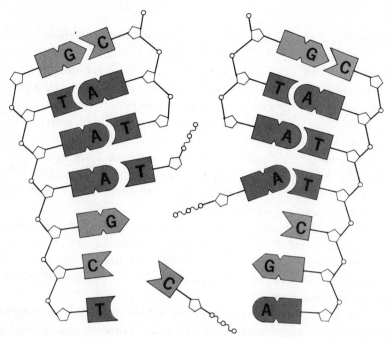

FIGURE 55. The two original strands of DNA are completely separated, and new nucleotides are being added to each strand to make two new molecules.

FIGURE 56. Each of the two strands of DNA has new partners to match its old ones. The DNA molecule has been exactly duplicated.

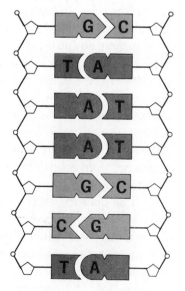

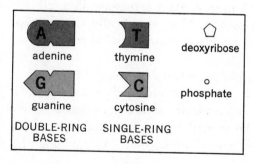

When living organisms or cells within an organism reproduce, how do they transmit their particular characteristics to their descendants? Biologists now believe that one group of molecules, the nucleic acids, contains the information about the making of all the other kinds of molecules in the cell. It is now known that nucleic acids control the cell activities and, as we have seen, can make exact copies of themselves. Furthermore, they are found in all reproductive cells and thus provide the chemistry linking one generation to another.

How does DNA achieve control over cellular activities? The Watson-Crick model provided a basis for the intensive investigations in the nineteen-fifties and nineteen-sixties which have done much to clarify this mystery. Today we have good reason to think that DNA is so constructed as to contain coded instructions for the synthesis of the various molecules required by the cell.

A code is a system of symbols used to transfer information from one form into another. Written language is in itself a code. For instance, the English alphabet uses just 26 letter symbols and 10 number symbols to transmit the experiences and thoughts of men into a nearly unlimited number of documents.

But what are the symbols of the DNA code? We have already learned that the huge molecules of DNA are composed of just four kinds of nucleotides. What kind of message might a cell receive from the code in its DNA? As you have already learned, each species of plant and animal contains many proteins. These proteins are of a particular kind for each species and even individuals within a species may have their distinct proteins. As you know, too, these proteins are made mostly of 20 common amino acids arranged in a seemingly endless variety of chains to form specific proteins. The specific proteins in turn are responsible for the structure and function of the cell through their activities as enzymes. So when the investigations into this phenomenon of DNA coding asked what kind of a message a cell might receive, the answer seemed to be that DNA must contain a message that tells the cell to make a certain kind of protein. More specifically, this message might be "Add a certain amino acid to a chain of amino acids to form a certain protein."

The question then facing biologists was how many DNA nucleotides would be needed to code a message giving directions about a single amino acid. Since 20 amino acids are involved, at least 20 separate messages would have to be formed by the four nucleotides with their distinctive "nitrogen base letters" of adenine (A), thymine (T), cytosine (C), and guanine (G).

It is easily discernible that the four letters are, when taken individually, sixteen letters short of being enough to code for 20 amino acids. What if one combines two nucleotides to make a code letter instructing the cell to add a particular amino acid? In this case, 4^2 or 16, combinations of nucleotides would be possible. These are: AA, AT, AC, AG, TA, TT, TC, TG, CA, CT, CC, CG, GA, GT, GC, and GG. This is still an insufficient number of code words for 20 amino acids. However, when 3 nucleotides are combined to form a code word, the four letters can be made into 4^3 or 64 possible combinations. This would be more than adequate to code the 20 common amino acids and others that are less common.

A relationship exists between the DNA and RNA codes. You will recall that three of the nitrogen bases are found in the nucelotides of both RNA and DNA. These are those of adenine, cytosine, and guanine. A base in the fourth nucleotide of RNA, uracil, is only slightly different from that of the fourth nucleotide in DNA, thymine. Thus, the two "alphabets" are quite similar. Both contain the letters A, C, and G; in addition, RNA contains a U and DNA a T.

Nirenberg and his co-workers at the U.S. National Institutes of Health synthesized RNA molecules with a known sequence of base(s). These RNA molecules were able to code for chains of amino acids in a test tube if the free amino acids and several enzymes were present. These chains were not complete enough to be called proteins. Such fragments of proteins are called peptides. It was found that each kind of RNA with its specific sequence of nucleotides would introduce a particular amino acid into the peptide chain. By determining what kind of amino-acid chain each kind of RNA made, the investigators were able to translate the RNA code.

For example, the Nirenberg group made an RNA strand containing only one kind of nucleotide, uracil (U). When put into a test tube containing all 20 amino acids and the necessary enzymes, this strand of uracil nucleotides, which may be written (-U-U-U-U-U-), caused a chain of amino acids to be formed that contained only a single kind of amino acid—namely, phenylalanine. From these results they concluded that the trinucleotide code word of RNA responsible for incorporating phenylalanine into protein was a tri-uracil nucleotide which might be represented by the word (U-U-U). Figure 57 shows diagrammatically how a long strand of polyuracil nucleotides would cause the synthesis of a chain of phenylalanine molecules.

It is believed that the coded messages of the DNA alphabet may be translated into the RNA alphabet which in turn conducts the actual synthesis of protein. Excellent evidence indicates that the sequence of basic code units of DNA are duplicated as mirror images of RNA in much the same manner that one DNA strand serves as the template for the synthesis of another DNA strand. Actually this is carried out with the DNA double strand intact, and the synthesis of bases in RNA is

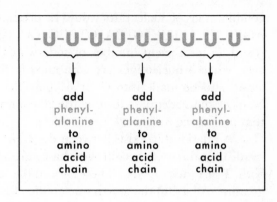

FIGURE 57. Each group of three uracil nucleotides specifies the addition of one molecule of phenylalanine to a protein molecule.

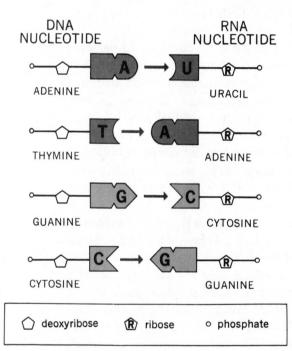

FIGURE 58. The letters of the DNA alphabet are translated into RNA letters.

thus dependent on the paired bases in DNA. This is shown symbolically in Figure 58.

Thus one could imagine that the polyuracil RNA nucleotide chain, as synthesized by the Nirenberg group, would have to be synthesized in the cell by a double strand of polyadenine DNA nucleotide. In other words, the RNA code word UUU is coded from the DNA code word ATATAT. The table on the next page shows the DNA and RNA code words for some other amino acids.

TABLE 15. DNA AND RNA CODE WORDS FOR SOME
OF THE AMINO ACIDS

DNA Code Word (nucleus)	RNA Code Word (nucleus to cytoplasm)	AMINO ACID Built into Protein (cytoplasm)
CCA	GGU	Glycine
AGA	UCU	Serine
CGA	GCU	Alanine
CAA	GUU	Valine

This code of life (DNA) seems to be the same substance in virtually all forms of life studied, lending strong evidence that all life is related and evolved from a common source.

It seems, therefore, that DNA makes a kind of RNA called "messenger" RNA, and this RNA somehow makes proteins. In most cells, however, DNA is located in the nucleus, and most cell activities, including protein synthesis, go on in the cytoplasm. Hence most of the RNA of cells is also found in the cytoplasm. The question is thus posed, "How do the coded instructions of nuclear DNA eventually reach the protein 'factories' of the cytoplasm?" These factories, rich in another kind of RNA (ribosomal RNA), are called ribosomes. Ribosomal RNA comprises 80% of the RNA of the cell.

Recent studies indicate that "messenger" RNA acts in transmitting the coded DNA information to the ribosomes. The messenger RNA then moves out of the nucleus into the cytoplasm and attaches to a ribosome. A much smaller molecule of RNA, known as "transfer" RNA, is already in the cytoplasm, picking up the individual amino acids and bringing them to the messenger RNA of the ribosome. The ribosome then responds to the instructions carried by RNA by arranging the amino acids in a characteristic fashion to form a specific protein. The transfer RNA's attach to the messenger RNA when both are on the ribosomes because they have a nucleotide sequence on the critical part of their molecule which is complementary to this message from RNA. That is, adenine must pair with uracil, cytosine with guanine, and so forth. After the protein molecule is formed, it breaks free from the ribosome, leaving the messenger RNA free to act again as a template for another molecule of protein; or, in many instances, messenger RNA is broken down after making a few protein molecules. The overall process is shown diagrammatically in Figure 59.

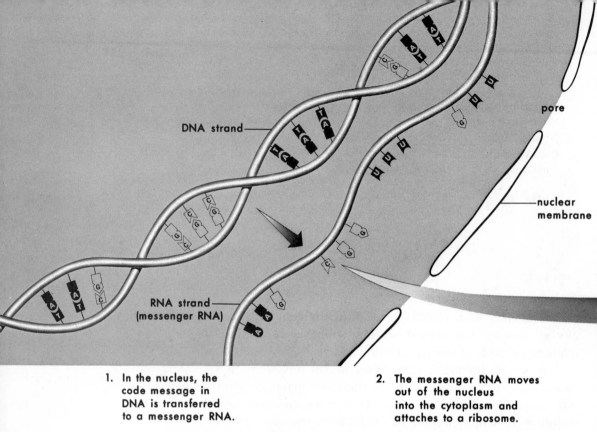

1. In the nucleus, the code message in DNA is transferred to a messenger RNA.

2. The messenger RNA moves out of the nucleus into the cytoplasm and attaches to a ribosome.

FIGURE 59. DNA directs protein formation by way of RNA. (Modified from the Japanese language adaptation of the BSCS BLUE VERSION.)

How genetic instructions, coded identically into each cell of the early embryo, are gradually changed as cells differentiate, remains one of the most active problems on the frontier of modern biology.

Here is a central, unanswered question: How do chromosomes, cytoplasm, and the environment regulate genetic expression differently in genetically identical cells? Furthermore, development seems to require a programmed regulation of cellular differentiation that is programmed in time as well as in space. How is a DNA "gene" turned on to produce one type of messenger RNA in some cells while the same gene in another cell is not so affected? Much research is currently going on to answer this most fundamental question of development and differentiation.

For example, in certain organisms swellings have been found at specific sites along the DNA-containing chromosomes. These swellings, called Balbiani rings or puffs, are sites of strong synthesis of messenger RNA. The position of the chromosomal puff is characteristic for the age and type of cell in which they are found. Furthermore, when the nuclei of salivary glands of fly larvae are transplanted into a preparation containing cytoplasm of developing eggs, it was found

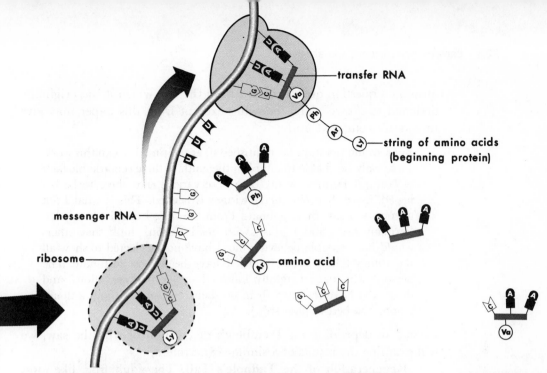

transfer RNA

string of amino acids
(beginning protein)

messenger RNA

ribosome

amino acid

3. Each triplet on the messenger RNA selects a transfer RNA with a specific amino acid attached.

4. The ribosome moves along the messenger RNA as it "reads" the code. The amino acids are joined to each other in the order coded, forming a protein molecule.

5. After delivering its amino acid, transfer RNA can pick up another amino acid molecule.

that the pattern of chromosomal puffs in the nuclei changed to resemble that of earlier stages of development. Hence, it seems that just as the activities of the cytoplasm are dependent on the nucleus, so is nuclear development dependent on changes in the cytoplasm. Certainly the exciting advances of modern-day biology have only begun with the development of the DNA model and the cracking of its code.

Regeneration

Embryologists have long been interested in the process of regeneration because it apparently represents a "return" to embryonic conditions in a fully developed animal. The phenomenon of regeneration was first discovered in 1740 by Abraham Trembley, a tutor at the French royal court, while he was attempting to answer the question, "Are polyps plants or animals?" Trembley published the results of a remarkable series of experiments on the little animal, *Hydra,* in a paper entitled *Memoirs Concerning the Natural History of Polyps.*[1] The paper is a good example of an event that occurs frequently in science: an experiment

[1]Translated and reprinted from Trembley's *Histoire des Polypes;* Verbeek, Leiden, Holland; 1744.

answers a question quite different from the one which it was originally designed to answer. The following excerpt from this paper may give direction to our next study:

> But I would not have been satisfied if, at the time I began this work, I had only been able to establish the truth of these remarkable facts of Natural History by my own observations. For these to be believed, more than one direct witness is needed. This is what I felt as soon as I saw these animals. From the first, I could scarcely believe my eyes, and I must, with good reason, think that others would have trouble believing me. I have not neglected to show all the things I have seen. In this I have been most fortunate. The persons who have carefully followed my own observations, and those who have repeated them on their own, are, without a doubt among the best of judges. . . .

Instead of depending on Trembley's description of what he saw, we will examine the results of a similar experiment.

Regeneration of the Tadpole's Tail. The adult frog, like most higher vertebrates, has no capacity for replacing appendages that have been lost. Interestingly enough, however, in its tadpole stage the frog does show the capacity for regeneration which its salamander cousins retain into adult life. When its tail is cut off, the tadpole will frequently grow a new one. Because the development of the new tail really represents the return to embryonic conditions, embryologists have often studied tail regeneration in an effort to learn more about the factors which cause a mass of undifferentiated cells to develop into a perfect organ neatly fitted to its place in the whole animal.

INVESTIGATION 26
Observation of Regenerating Tissues

In the following experiment you will amputate a portion of the tail of some anesthetized tadpoles in order to observe the phenomenon of regeneration. The objective is to determine, through your own observation, the major events in this process and to note whether the regenerated portion of the tail is of a size and shape which will be in proportion to the tadpole's body.

MATERIALS (per team)

1. Separate containers for tadpoles
2. Insect pins

3. Tadpoles

4. 3 petri dishes

5. Anesthetic

6. Viewing chamber

7. Microscope

8. New razor blade

9. 10% Holtfreter's solution

10. Lettuce or *Elodea*

PROCEDURE

1. Prepare the anesthetic, MS-222, by dissolving enough of the compound in 10% Holtfreter's solution to make a concentration of 1:10,000, and a second dilution of 1:20,000. Chloretone in water in 1:3,000 and 1:6,000 dilutions is a second choice.

2. Place tadpoles in a dish of the stronger dilution until they become immobilized.

3. Remove tadpoles immediately and rinse in a container of 10% Holtfreter's solution.

4. With a clean, sharp (preferably new) razor blade, amputate a portion of the tail. Do not cut off more than the final $\frac{1}{4}$, as this probably would sever some of the large blood vessels and cause the animal to bleed to death. Amputate with a firm downward pressure of the blade against the tail while the tadpole is resting on moist filter paper or a paper towel. Do not saw back and forth.

5. As quickly as possible, place the tadpole in the viewing chamber, which is filled with the 1:20,000 dilution of MS-222, and wedge the tail against the cover glass with the insect pins, as demonstrated by your teacher.

6. Though tadpoles tolerate the MS-222 well, you should be alert to detect signs of overexposure, especially if you are using chloretone instead. If the blood which you can see moving through the vessels in the tail begins to pulsate markedly, rather than flow smoothly, this is an indication that the heart action is becoming impaired. When this happens, you must terminate your observations immediately and remove the tadpole from the chamber. Rinse it in Holtfreter's solution (or pond water) to remove all the anesthetic from its body surface.

Return it to its holding container with fresh Holtfreter's or pond water.

7. Feed the tadpole daily but do not let the water in the holding container become contaminated with uneaten food.

8. Through daily observations during the period of regeneration, determine the events of the process.

QUESTIONS FOR DISCUSSION

1. Define regeneration.
2. Based on your observations, describe the stages of the regenerative process in the tadpole as you visualize them.
3. In what way do the regenerations you have studied resemble embryonic development?
4. When a part of the tadpole's tail is cut off and discarded, the rest of the body contributes the raw material from which a new tail is built. Does it contribute anything besides the raw material—that is, does it influence the form, size, or orientation of the new tail? On what observations are your answers based?
5. Referring to any experimental design that you have previously applied to embryos, do you think any of these techniques would help you to learn more about regeneration? Which techniques would be worthwhile applying?

PARALLEL READING: Regeneration in Other Multicellular Organisms

By "regeneration" we mean the ability of an organism to repair or replace parts of its body that have been lost or damaged. Fortunately, all organisms have some capacity for regeneration. In the course of your childhood, you probably suffered numerous cuts and abrasions that were smoothed over by the regenerative powers of your skin. Most parts of the body can repair wounds; broken bones will knit together, and even a cut or torn nerve may regenerate if the severed ends are brought together. Actually, the skin and the lining of the intestine are continuously regenerating, for the surfaces of both organs are steadily being worn away and replaced by new cells that move out to the surface.

However, while a man cannot replace an eye or a tooth or even a finger tip that has been lost, a few vertebrates are able to regenerate large and complex structures. Perhaps you have heard that certain lizards, when caught by the tail, quickly snap it off and subsequently grow a new one. As you have seen, frog tadpoles can regenerate a tail that has been cut off; adult salamanders can do so, also. The salamander, in addition, has the surprising ability to replace an entire limb that has been amputated. If a salamander's leg is cut off, even at the shoulder, a mass of undifferentiated cells appears at the wound surface. Over the course of several months, these cells gradually build up a perfect limb, with all the bones, muscles, nerves, and blood vessels correctly formed and in their normal positions.

One problem that has attracted a great deal of attention is the origin of the mass of cells (the blastema) from which the new limb or tail is formed. Does new bone come only from old, cut bone and new muscle only from old muscle? Or is damaged bone able to give rise to muscle, and skin to bone? Or does the animal contain some kind of "reserve cells" that swarm to the amputation site and organize a new limb or tail? This problem is of great theoretical significance, because an understanding of these aspects of regeneration would lead to an understanding of what happens as embryonic cells are converted into mature tissue.

It is of practical importance, too, because it might help us to learn whether it ever will be possible to induce regeneration of lost body parts in man. During the Second World War, one biologist succeeded in bringing about limb regeneration in adult frogs, which ordinarily do not replace lost structures. The technique used was to keep the wound open for a long time by interfering with the healing process. But no one has ever been able to bring about regeneration of a limb in a mammal.

There is practical interest in the possibility that human beings might some day be able to regrow all tissues and organs, but apart from this possibility, regeneration holds interest for the biologist because it demonstrates developmental processes in an adult organism. Experiments with frogs and salamanders indicate the possibility that every organ has the power of regrowth, needing only the appropriate environmental situation to bring it about. The proper environmental situation may include the presence of certain substances such as enzymes and hormones.

SECTION 12 HORMONAL REGULATION

The coordination of the cells and groups of cells which make up any multicellular organism is another problem in biology which is not yet completely understood. Our incomplete knowledge of the causes of cell differentiation has been pointed out. It is also true that, while much is known, much remains to be learned about the mechanisms by which the differentiated cells and tissues are controlled so that they act in harmony with others in the organism.

In animals, much of the coordination is controlled by the nervous system. Plants have no structures which correspond to the nervous system of animals. And even in higher animals, it is probable that the regulation of growth, development, and behavior is controlled as much by chemical regulators called hormones as by the nervous system.

Hormones are substances which are produced in one part of an organism but have their effect in some other part. Hormones were first recognized in animals, but several substances which fit this definition have since been found in plants.

Hormonal Regulation in Animals. Most of the hormones which have been identified in animals are produced in *endocrine* glands. The endocrines are glands which have no ducts. The names of some of these glands—such as the thyroid, the pituitary, the pancreas, the adrenals, and the gonads—should be familiar to you.

Each endocrine gland produces either one or several specific hormones. These are released from the gland into the blood stream and thus circulate through the body. Taken up by various tissues, they control many kinds of chemical processes. Insulin, a hormone produced by the pancreas, is essential for the release of energy from carbohydrates in food. Thyroxin, a product of the thyroid gland, allows the body to get the maximum amount of energy and body heat from food. Testosterone, produced by the testes, brings about the biochemical changes that cause secondary sex characteristics in the male, such as growth of a beard or change of voice. In fact, almost everything that goes on in our bodies is influenced by hormones. To remain in vigorous health, an animal must have normally functioning endocrine glands. Hormones are not only essential for the general day-to-day welfare of the body; they are also needed for many special functions. The production of eggs and sperm, and, in some organisms, the processes of courtship, mating, and care of the young are all controlled by hormones from the sex glands and the pituitary gland.

Hormones in Frogs. The influence of hormones on the function of reproduction in animals is strikingly illustrated in frogs, in which reproductive activity occurs only once a year. In the summertime, frogs spend their time hunting food; meanwhile, sperm develop in the testes of the males, and large masses of eggs are formed in the ovaries of the females. During the winter, the animals hibernate in sheltered spots under stones and logs. When spring comes, the frogs emerge and make their way to the nearest pond. There the males and females pair in a long-continued embrace called *amplexus*. After this embrace has lasted several days, the eggs from the female and sperms from the male are released simultaneously. The sperms fertilize the eggs just at the moment that the eggs are emerging from the female's body.

The fact that both males and females move to the water at the same time and are then able to release viable gametes that have been stored for several months poses some important problems. Evidently the frogs are responding to the change of season. Through the senses, the nervous system is informed of changes in light, temperature, and moisture. But how is this information relayed to the gonads? We now know that influences coming from the nervous system stimulate the pituitary gland, which is attached to the brain. The pituitary then releases the hormones called *gonadotrophins,* the hormones that tend to nourish and stimulate the gonads. When the time for egg-laying approaches in the spring, gonadotrophins are poured into the blood stream. Upon reaching the ovaries, these gonadotrophic hormones cause the eggs to be liberated into the oviducts. In the males, the testes are stimulated by the same hormones. The stimulated gonads in turn secrete hormones of their own, and these sex hormones, in some way not yet well understood, influence the nervous system, causing the frogs to leave their winter hiding places to migrate to the edges of the ponds.

Since frogs that lay their eggs in the spring build up a new crop of eggs in the summer when food is abundant, the eggs are fully formed when the females go into hibernation at the approach of cold weather. The ovaries need only the stimulus of gonadotrophins to release these eggs. As shown in Preparatory Technique 3, page 194, ripe eggs can be obtained out of season (in the fall or winter) by injecting gonadotrophic hormone into the mature female. Since the gonadotrophin content rises during the year, a large number of pituitaries are needed in the fall, but only one or two are needed just before the normal laying season.

Hormonal Control in the Frog Embryo. The tadpole contains millions of cells, all derived from the single fertilized egg. As long as these cells remain a part of the developing animal, they are marvelously coordinated—each committed to a precise role in the adult organism. It is obvious that during the period of development each cell comes under influences that restrict its growth and multiplication; if it did not, the final result would not be a tadpole.

The influences at work in an embryo, whatever they are, regulate the growth of the cells and assign to each a specific role in the total organism. These regulations are at work in all organisms and must have been perfected early in the evolution of living things. Thus we can study such organisms as oat seedlings, tadpoles, or chicks and learn about the basic problems of growth and differentiation in all organisms, including the human species.

The cycle of development in individual organisms involves a number of simultaneous changes. A process involving a number of simultaneous changes is called a *synchrony*. The problem is this: what causes a synchrony?

In the laboratory, it is possible to remove or to injure the thyroid gland in young or even unborn animals and to study the consequent effects on growth and development. Or, conversely, one can administer thyroxin to young animals and study the effects produced. Experiments like these have provided a great deal of information about the part which the thyroid gland plays in normal development.

In 1912, Frederick Gudernatsch tried feeding various dried glands (purified hormones were not available in those days) to frog tadpoles. In our laboratory experiments, we could try to repeat Gudernatsch's original experiment by feeding dried thyroid to some tadpoles. Since pure thyroxin is now available, however, we shall try to do the experiment with this thyroid hormone itself.

Thyroxin has been referred to as "the" hormone produced by the thyroid gland. Actually, thyroxin was one of the first hormones to be understood chemically. Its structural formula was determined in 1926, and, in England the following year, Sir Charles Harrington succeeded in making the hormone in the laboratory. For the next quarter of a century, it was believed that the problem of the chemical nature of thyroxin was solved.

In 1952, two British physiologists examined mammalian blood plasma by the then new technique of chromatography. In the plasma they found a previously unknown substance that chromatographed differently from thyroxin. They showed that the new substance could be found in extracts of the thyroid gland. Chemical analysis of the new substance showed that it differed from thyroxin in one respect: it contained three iodine atoms per molecule, whereas thyroxin contains four. Thus, the new substance was named triiodothyronine. (Thyroxin can be properly called "tetraiodothyronine.")

The big surprise came when triiodothyronine was tested and compared with thyroxin. Quite unexpectedly, the new substance turned out to be four or five times more effective than thyroxin itself. When we say that triiodothyronine was more effective than thyroxin, we are referring to tests of thyroid activity made in *mammals*. One might ask whether this triiodothyronine is also more effective in other animals. This question can be investigated by placing tadpoles in tri-iodothyronine solutions, one-half or one-quarter as concentrated as our thyroxin solutions. If triiodothyronine proves to be more effective in amphibians, we should expect that a given amount of it should affect metamorphosis to the same degree as does a larger amount of thyroxin.

There is also the possibility that triiodothyronine will have effects quite different from those produced by thyroxin. In biological research, it is essential to realize that results obtained with one species cannot be applied to another without first being tested. Even animals as much alike as the mouse and rat sometimes differ strikingly in their response to a certain drug or hormone. The fact that triiodothyronine is more effective than thyroxin in certain animals does not necessarily apply to relative effectiveness of the two substances in other animals. For example, in chickens it has been shown that triiodothyronine is *not* more effective than thyroxin.

Both thyroxin and triiodothyronine contain iodine. Actually, the essential function of the thyroid gland is the collection of iodine from the blood stream (we take it in with our food and drinking water) and the synthesis of thyroid hormones. Iodine is apparently indispensable if the body is to carry on its normal chemical processes. If there is not enough iodine in the diet, the thyroid may become very much enlarged and thus may trap more effectively the small amount of the element that is available; such greatly enlarged thyroids are called goiters and may grow to 50 times the normal size of the gland. If the thyroid fails to get sufficient iodine, the supply of thyroid hormones diminishes, and the affected individual becomes dull and sluggish.

One might ask what chemical is really responsible for the effects of the thyroid gland on bodily function. Is it the iodine, the thyroxin, or the triiodothyronine? Perhaps we can answer this question by comparing the effects of each of these substances on test animals.

INVESTIGATION 27
Hormonal Control of the Development of Frog Embryos

MATERIALS (per team)
1. Seven large, flat, culture dishes or pans. (Each dish should be able to hold 1 liter of culture solution.)
2. Seven stones or bricks (large enough to extend above the 1-liter water level, yet not so large as to prevent the free movement of tadpoles)

3. One 1-liter graduated cylinder

4. Thirty-five tadpoles just preceding, or in, the hindlimb-bud stage

5. Stock solutions of thyroxin (1 mg per liter) triiodothyronine (1 mg per liter), and iodine (1 mg per liter)

6. Three 10-ml graduated pipettes

PROCEDURE

1. Label each of the culture dishes (team, solution).

2. Add to each dish the proper solution as follows:

   a. Twenty-five micrograms of thyroxin per liter. (Prepare by adding 25 ml of the thyroxin stock solution to 975 ml of pond water.)

   b. Ten micrograms of thyroxin per liter. (Prepare by adding 10 ml of the thyroxin stock solution to 990 ml of pond water.)

   c. Ten micrograms of triiodothyronine per liter. (Prepare by adding 10 ml of the triiodothyronine stock solution to 990 ml of pond water.)

   d. Two and one-half micrograms of triiodothyronine per liter. (Add 2.5 ml of stock solution to 997.5 ml of pond water.)

   e. Two micrograms of iodine per liter. (Add 2 ml of iodine stock solution to 998 ml of pond water.)

   f. One microgram of iodine per liter. (Add 1 ml of stock solution to 999 ml of pond water.)

   g. One liter of pond water for pond-water control.

3. Divide the 35 tadpoles into 7 groups of 5 each, selecting them so that each group will include the same range of developmental stages. Place one group in each of the seven containers. Keep the containers at room temperature.

4. Observe and record your observations every second or third day. At the same time, feed the tadpoles, and replace the culture solutions with fresh ones. If any tadpoles die, remove them promptly (do not replace them).

5. As the tadpole hindlimbs become well developed, place a rock in each container for the tadpoles to crawl onto. Without this precaution, the tadpoles will drown.

In concluding this work, carefully compare the results obtained in each of the solutions. Then write a summary of the evidence which you have obtained in this experiment.

Metamorphosis involves a complete remaking of the body of the tadpole so that it becomes quite a different organism. Some of the changes that occur are internal and will not be visible. For example, the intestine becomes much shorter, and the gills are replaced by lungs. Other changes, however, are plainly visible.

Record the time at which each of the following external changes is observed in the individual animals cultured in the various solutions. The times at which these visible changes occur in the various culture solutions should provide you with data which you may use to evaluate the effectiveness of each solution in causing these changes. Note carefully when:

1. The tail shortens and finally disappears.
2. The hindlimbs, which begin as tiny buds, grow and develop joints and feet.
3. Eardrums appear on the surface of the head.
4. The forelimbs erupt through a window of skin onto the surface of the body. Note whether both of the forelimbs break through the skin at the same time.
5. The small mouth, which is toothless but has a horny beak, is replaced by a wide, gaping mouth which extends from the tip of the head back past the level of the eyes.
6. The body loses its ovoid shape and begins to resemble that of an adult frog.

QUESTIONS FOR DISCUSSION

1. What is the purpose of matching the development of controls with that of the thyroxin-treated tadpoles at the beginning of the experiment?
2. When a tadpole dies, why should it not be replaced with another from the stock?
3. Was the first developmental event to occur in one group also the first to occur in all the others?
4. Did some concentrations of thyroxin appear to overstimulate development? Why?
5. Did metamorphic changes occur at different rates in the various concentrations of thyroxin? If so, how much difference was there?
6. Did the untreated controls undergo any metamorphic change during the period of observations? Why were controls included in this experiment?
7. Does triiodothyronine increase or decrease the rate of metamorphosis more than thyroxin? On what evidence is your answer based?

8. Do your results suggest what may regulate the changes which occur in normal tadpoles when metamorphosis takes place spontaneously? What experiments could be used to test the correctness of your hypothesis?

9. If a very young larva were deprived of its thyroid gland, do you think it could undergo metamorphosis in response to thyroxin? How could this point be tested?

INVESTIGATION 28
Behavior and Development in Chicks

The influence of sex hormones on sex differentiation is a subject that has been extensively studied in laboratory animals. Under controlled conditions, one can inject male or female sex hormones into immature animals and learn something about the factors that control the animals' ability to respond to the hormones. To set up a simple study of this sort, it is a good idea to use animals in which there are clear-cut differences in the appearance of the two sexes, since such differences make it easy to detect what the hormone is doing. External differences are not always easily discernible; a human observer can, for example, distinguish between male and female flickers only by their behavior, since their outward appearances are nearly identical. Male and female chickens, on the other hand, differ strikingly from one another in appearance as well as in behavior.

To investigate the control of development in laboratory animals, we shall inject newly hatched cockerels and pullets with the sex hormone testosterone propionate, and the trophic hormone, chorionic gonadotrophin. Because these hormones are soluble in different carriers, it will be necessary to inject two sets of control birds, each with a separate, pure solvent. Uninjected chicks should also be used as a third control group against the action of the solvents.

MATERIALS (per team)

Five 1-day-old male and five 1-day-old female chicks

MATERIALS (per class)

1. Brooder for chicks
2. Four hypodermic syringes, graduated to 0.1 ml, and equipped with No. 21 or 22 needles
3. Testosterone propionate

4. Sesame oil

5. Gonadotrophin

6. Saline solution

7. A beaker of 70% ethyl alcohol

8. Chick food

9. Plastic tape—5 colors

Unless otherwise directed by the teacher, the class will work as five teams. Each of four teams will use a different hormone, or solvent for a hormone. The fifth team will maintain the noninjected control group of chicks.

The sex of each chick may be readily distinguished by placing one or two bands of colored plastic tape around one of its legs; two bands can be used to designate a male and one band to designate a female.

For example, the color code for plastic tape might be: testosterone, blue; sesame oil carrier, yellow; gonadotrophin, green; saline carrier, white; untreated controls, black.

PROCEDURE

1. Carefully apply a band of plastic tape of the proper color around one leg of each female chick. Do not make it too tight, but be sure it is tight enough so that it will not fall off. Place two bands on a leg of each male. It will add some interest and some value if you can number the individuals and keep separate records for each. As the chicks grow, it will be necessary to replace these bands once or twice in order to avoid constriction of the legs.

2. Determine each chick's weight and "comb factor" (see Figure 60) and record these in your notes. The comb factor is calculated from the following equation:

$$CF = \frac{CL \times CH}{2}$$

where CF = comb factor
CL = comb length
and CH = comb height

Comb length is measured as the distance between the points indicated as A and B in Figure 60. Comb height is measured perpendicular to the skull at the point where the comb is tallest.

FIGURE 60. Comb development in chicks.

Newly hatched chick showing comb

Normal cockerel about 3 weeks old, showing comb

3. Prepare to inject your chicks as follows. Keep your sterile hypodermic syringe, with needle attached, partly submerged in a beaker of 70% ethyl alcohol. When you are ready to inject, take the vial containing the solution you are to use and swab off the rubber cap with cotton soaked in alcohol. Insert the needle into the rubber top of the vial and draw some of the solution into the barrel of the syringe. Holding the syringe with

the point of the needle up and the vial above, push back all but 0.1 ml of the solution, along with any air that may have entered the syringe. Withdraw the needle and inject the chick as described in Step 4. Chicks should receive a daily dose of 0.1 ml of either the hormone solution or carrier every day of the first week unless this includes a weekend. Then a single double dose should be administered Friday. After the first week, the dosage should be doubled to 0.2 ml daily.

4. The cockerels and pullets should be less than three days old when you start injecting them. Because it is difficult for one individual to inject a young chick, two people should cooperate in the operation. One person grasps the chick, belly up, so that its back is in the palm of the hand and its head is held between the thumb and forefinger. The other then pinches up the loose skin under the chick's wing or leg and barely inserts the needle through it. The needle, attached to the syringe with the proper volume of solution, should be held parallel to the chick during insertion. A No. 21 or 22 needle should be used because the sesame oil carrier is fairly viscous.

5. Place the chick back in the brooder, making sure food and water are available. Rinse out the syringe in a beaker of alcohol by drawing it full of alcohol and then emptying it several times. Clean your equipment and return it to its proper place.

During the course of the experiment, watch for and record any change in the color, shape, and turgidity (swollen appearance) of the combs. Weigh the chicks every other day, and keep a record of their weights. Plot the weights and comb factors on appropriate graphs. Graphing the results of the experiment will bring out most clearly any differences that might be produced by the various treatments. Indicate individual points on the graph; you do not properly represent your data by showing a plotted line with no points to indicate the values from which it was derived.

As you observe the chicks which have been injected with the various solutions, you should note the behavioral changes resulting from sex-hormone treatment. These observations will be of as much value as the comb factor and weight in determining the effectiveness of the treatments. To make these observations, you must approach the chicks quietly, without disturbing them. What you want to see is how the chicks behave among themselves when they are not frightened. Therefore, you should make these observations each day *before* you pick up the chicks to examine or inject them.

Keep an accurate record of the time at which various behavioral changes are first observed in each chick. Watch for the following behavior traits in each set of chicks and record their presence or absence and the time of their onset.

1. Scratching, with and without food

2. Pecking, with and without food

3. Preening

4. Vocalizations, which begin as peeps and trills, and later become crowing in some groups

5. Huddling together, and the reaction to being isolated at increasing ages. (Huddling may be a response to being too cold.)

6. Strutting, posturing, and "neck-stretching"

7. Fighting

QUESTIONS FOR DISCUSSION

1. Do your data indicate that the hormones affect the rate at which the chicks gain weight? If so, what are the effects?

2. Do the data indicate that the injected hormones affect the rate of development of the combs on the males? on the females? If so, what are the effects?

3. Are the changes in weight and comb growth correlated with any behavioral effects you have observed?

4. In what ways does testosterone propionate influence the development of young chicks?

5. What aspect of experimental design enables you to feel sure that the effects of testosterone which you have observed are due to testosterone specifically?

6. Would you conclude that testosterone (which is called "the" male sex hormone) affects only sexual characteristics? Does it seem to have any other function?

7. How can you explain the observed effects of gonadotrophin?

PATTERN OF INQUIRY 8
Pituitary-gonad mechanism

Organ A, a ductless gland, was removed from a number of adult, female rats. By comparison with control animals, it was found that organ B (also a ductless gland) and organ C ceased functioning in the animals deprived of organ A. Clearly, A is a necessary factor in the functioning of organs B and C. Describe and diagram (using an arrow to mean "controls" or "stimulates") three quite-different routes by which the control of B and C by A may take place.

QUESTIONS FOR DISCUSSION

1. Using what you now know about the interaction between hormones and tissue, what do you think may have taken place in both the male and female chicks with respect to hormonal interaction and gonad size? If you were to dissect out and weigh the testes and ovaries of the control and testosterone-injected chicks, what might you find? Why?

2. If you were to dissect out and weigh the testes and ovaries of the control and gonadotrophin-injected chicks, what might you find? Why?

3. Does a young chick have to be a genetic male in order to respond to testosterone?

4. Using your knowledge of hormones and genes, explain what determines whether the infant animal will mature to look and act like a male or female.

PARALLEL READING: Regulation in Plants

At first thought it would seem that the problem of regulation in plants should be much better understood than is the problem of regulation in animals. There is less specialization of cells in plants than in animals. Plants exhibit little motility or motion; motility is characteristic of most animals. So far as we know, plants have nothing which compares with the nervous system of higher animals as a coordinating system.

Cells of plants do differentiate as the organism develops, and the several parts of the plant operate in patterns serving the whole. Young plants mature, produce flowers and seeds, and eventually die. Annual or biennial plants complete this cycle within a fairly definite time schedule.

Our understanding of the control of these processes is still quite incomplete. As with animals, we have learned much, but no doubt even more remains to be learned.

We will study only a few examples of regulation in plants.

The factors which affect the growth of plants are similar to those affecting growth in animals. Some of these are internal and may be determined chiefly by heredity. Others are environmental.

INVESTIGATION 29
The Effect of Light on the Growth of Seedlings

You know that light is necessary for the formation of foods by photosynthesis. However, what is the effect of light on the process of growth in plants? Three possibilities can be considered: (1) light accelerates the growth of the plant, (2) light inhibits the growth of the plant, or (3) light has no effect on the growth of the plant. The following investigations should yield data which will answer these questions.

MATERIALS (per team)

1. Germination trays   2. Seeds of Alaska peas

PROCEDURE

1. Each team should place about 40 Alaska peas in each of two germination trays containing sand or vermiculite.
2. Water the sand or vermiculite well but drain off excess water. Place one tray in the dark and leave the other tray in the room exposed to light.
3. After 7 to 9 days, examine all the plants and make detailed observations and measurements which might provide information to help you to evaluate your hypothesis. Record these data in your notebook.

QUESTIONS FOR DISCUSSION

1. What effect does light have on the growth of stems?
2. What effect does light have on the expansion of leaves?
3. What effect does light have on the formation of chlorophyll?
4. Is there any need for the use of statistics to determine that the effects of the two treatments are different?
5. Do you have any information which will help explain how light affected the growth of the stems?

PARALLEL READING: A Brief History of Our Knowledge of Auxins

At least a part of the effects of light on the growth of plants can be explained by the action of growth hormones called auxins. Our present knowledge of these substances has developed quite slowly.

Man has long observed the effect of light on stem growth, including the bending of plants towards light. In 1880 Charles Darwin and his son Francis investigated the role of light in plant-growth movements. They are given credit for laying the foundation for the modern study of plant-regulating mechanisms. A portion of their pioneering study is reprinted here.

LOCALISED SENSITIVENESS TO LIGHT, AND ITS TRANSMITTED EFFECTS[1]

by CHARLES and FRANCIS DARWIN

Phalaris canariensis. Whilst observing the accuracy with which the cotyledons of this plant became bent towards the light of a small lamp, we were impressed with the idea that the uppermost part determined the direction of the curvature of the lower part. When the cotyledons are exposed to a lateral light, the upper part bends first, and afterwards the bending gradually extends down to the base, and as we shall presently see, even a little beneath the ground. This holds good with cotyledons from less than .1 inch (one was observed to act in this manner which was only .03 inches in height) to about .5 of an inch in height; but when they have grown to nearly an inch in height, the basal part, for a length of .15 to .2 of an inch above the ground, ceases to bend. As with young cotyledons the lower part goes on bending, after the upper part has become well arched towards a lateral light; the apex would ultimately point to the ground instead of to the light, did not the upper part reverse its curvature and straighten itself, as soon as the upper convex surface of the bowed-down portion received more light than the lower concave surface. The position ultimately assumed by young and upright cotyledons, exposed to light entering obliquely from above through a window, is shown

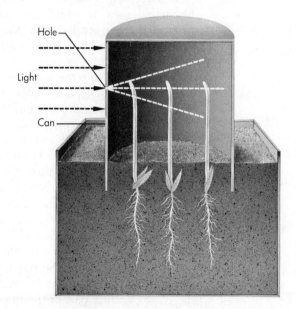

FIGURE 61. Bending of grass seedlings in unilateral light.

[1]Darwin, Charles and Francis. 1880. *The Power of Movement in Plants.* London.

in the acompanying figure and here it may be seen that the whole upper part has become very nearly straight.

When the cotyledons were exposed before a bright lamp, standing on the same level with them, the upper part, which was at first greatly arched towards the light, became straight and strictly parallel with the surface of the soil in the pots; the basal part being now rectangularly bent. All this great amount of curvature, together with the subsequent straightening of the upper part, was often effected in a few hours.

After the uppermost part has become bowed a little to the light, its overhanging weight must tend to increase the curvature of the lower part; but any such effect was shown in several ways to be quite insignificant. When little caps of tinfoil (hereafter to be described) were placed on the summits of the cotyledons, though this must have added considerably to their weight, the rate or amount of bending was not thus increased. But the best evidence was afforded by placing pots with seedlings of *Phalaris* before a lamp in such a position, that the cotyledons were horizontally extended and projected at right angles to the line of light. In the course of $5\frac{1}{2}$ h. [hours] they were directed towards the light with their bases bent at right angles; and this abrupt curvature could not have been aided in the least by the weight of the upper part, which acted at right angles to the plane of curvature.

It will be shown that when the upper halves of the cotyledons of *Phalaris* and *Avena* were enclosed in little pipes of tinfoil or of blackened glass, in which case the upper part was mechanically prevented from bending, the lower and unenclosed part did not bend when exposed to a lateral light; and it occurred to us that this fact might be due, not to the exclusion of the light from the upper part, but to some necessity of the bending gradually travelling down the cotyledons, so that unless the upper part first became bent, the lower could not bend, however much it might be stimulated. It was necessary for our purpose to ascertain whether this notion was true, and it was proved false, for the lower halves of several cotyledons became bowed to the light, although the upper halves were enclosed in little glass tubes (not blackened), which prevented, as far as we could judge, their bending. Nevertheless, as the part within the tube might possibly bend a very little, fine rigid rods or flat splinters of thin glass were cemented with shellac to one side of the upper part of 15 cotyledons; and in six cases they were in addition tied on with threads. They were thus forced to remain quite straight. The result was that the lower halves of all became bowed to the light, but generally not in so great a degree as the corresponding part of the free seedlings in the same pots; this may perhaps be accounted for by some slight degree of injury having been caused by a considerable surface having been smeared with shellac. It may be added, that when the cotyledons of *Phalaris* and *Avena* are acted on by the apogeotropism, it is the upper part first which begins to bend; and when this part was rendered rigid in the manner just described, the upward curvature of the basal part was not thus prevented.

To test our belief that the upper part of the cotyledons of *Phalaris,* when exposed to the lateral light, regulated the bending of the lower part, many experiments were tried; but most of our first attempts proved useless from various causes not worth specifying. Seven cotyledons had their tips cut off for lengths varying between .1 and .16 of an inch, and these, when left exposed all day to a lateral light, remained upright. In another set of 7 cotyledons, the tips were cut off for a length of only about .05 of an inch (1.27 mm) and these became bowed towards a lateral light, but not nearly so much as the many other seedlings in the same pot.

We next tried the effects of covering the upper part of the cotyledons of *Phalaris* with little caps which were impervious to light; the whole lower part being left fully exposed before a south-west window or a bright paraffin lamp. Some of the caps were made of extremely thin tinfoil blackened within; these had the disadvantage of occasionally, though rarely, being too heavy, especially when twice folded; the basal edges could be pressed into close contact with the cotyledons; though this again required care to prevent injury. Nevertheless, any injury thus caused could be detected by removing the caps, and trying whether the cotyledons were then sensitive to light. Other caps were then made of the thinnest glass,

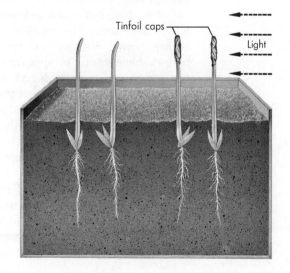

FIGURE 62. Light-proof caps over the coleoptile tips prevent unilateral light from bending the coleoptile, even if its lower portion is exposed to the light.

which when painted black served well, with the one great disadvantage that the ends could not be closed. But tubes were used which fitted the cotyledons almost closely and the black paper was placed on the soil round each, to check the upwards reflection of light from the soil. Such tubes were in one respect better than the caps of tinfoil, as it was possible at the same time to cover some cotyledons with transparent and others with opaque tubes; and thus our experiments could be controlled. It should be kept in mind that young cotyledons were selected for the trial, and that these when not interfered with bowed down to the ground towards the light. . . .

The summits of nine cotyledons, differing somewhat in height, were enclosed for rather less than half their lengths in uncolored or transparent tubes; and these were then exposed before a southwest window on a bright day for 8 h. All of them became strongly curved towards the light, in the same degree as the many other free seedlings in the same pots; so that the glass tubes did not prevent the cotyledons from bending toward the light. Nineteen other cotyledons were, at the same time, similarly enclosed in tubes thickly painted with Indian ink. On five of them, the paint, to our surprise, contracted after exposure to the sunlight, and very narrow cracks were formed, through which a little light entered; and these five cases were rejected. Of the remaining fourteen cotyledons, the lower halves of which had been fully exposed to light for the whole time, 7 continued quite straight and upright; 1 was considerably bowed to the light, and six were slightly bowed, but with the exposed bases of most of them almost or quite straight. It is possible that some light may have been reflected upwards from the soil and entered the bases of the seven tubes as the sun shone brightly, though bits of blackened paper had been placed on the soil around them. Nevertheless, the seven cotyledons which were slightly bowed, together with the 7 upright ones, presented a most remarkable contrast in appearance with many other seedlings in the pot to which nothing had been done. The blackened tubes were then removed from 10 of these seedlings, and they were now exposed to the lamp for 8h; 9 of them became greatly, and 1 moderately, curved toward the light, proving that the previous absence of any curvature in the basal part, or the presence of only a slight degree of curvature there, was due to the exclusion of light from the upper part. . . .

From these several sets of experiments, including those with the tips cut off, we may infer that the exclusion of light from the upper part of the cotyledons of *Phalaris* prevents the lower part, though fully exposed to the lateral light, from becoming curved. The summit for a length of .04 or .05 of an inch, though it is itself sensitive and curves toward the light, has only a slight power of causing the lower parts to bend. Nor has the exclusion of light from the summit for a length of .1 of an inch a strong influence on the curvature of the lower part. On the other hand, an exclusion for a length of between .15 and .2 of an inch, or of the whole upper half, plainly prevents the lower and fully illuminated part from becoming curved in the manner (see figure) which invariably occurs when a free cotyledon is exposed to a lateral light. With very young seedlings, the sensitive zone seems to extend rather lower down relatively to their height than in older seedlings. We must therefore, conclude that when seedlings are freely exposed to a lateral light some influence is transmitted from the upper to the lower part, causing the latter to bend. END OF ARTICLE

Growth movements in response to stimuli from one direction are called tropisms. The work began by the Darwins was continued by others. Notable among the early workers were Boysen-Jensen, Paal, and F. W. Went. These men showed that (1) the stimulus for the response studied by the Darwins could move downward from the tip of a cut coleoptile through a block of gelatin placed between the tip and the lower part of the coleoptile, (2) the stimulus could not pass through a substance such as mica. They proved that the stimulus (now suspected to be a substance) could be trapped in a gelatin block, and, if such a block were then replaced on a decapitated coleoptile, it would act much as does a tip in the intact plant. They showed that placing a block which contained the growth substance on one side of a decapitated coleoptile resulted in curvature away from the side on which the block was placed. They developed methods of quantitatively estimating the amount of the growth-regulating substance.

After this early research indicated the presence and action of plant hormones, specific identification of the chemical compounds involved soon followed. In 1934 the specific substance indoleacetic acid (IAA) was isolated and identified. Initially this substance was isolated from urine, but it was shown to have growth-promoting activity in plants. It has subsequently been found to be an important, naturally occurring, growth substance in plants. This and other substances which have similar effects, have been given the name *auxin*, which means "to increase." Many naturally occurring auxins have now been identified, and many more synthetic ones have been manufactured by chemists.

The amount of auxin in plants is so small that a bioassay is used to measure the quantity in a given plant part. One of the most sensitive structures for such measurement is the coleoptile of the oat seedling.

When an oat grain germinates, the first part of the plumule to emerge is the coleoptile. (The Darwins called this structure a cotyledon, but present knowledge of the structure of grass seeds is greater than in the Darwins' time. The structure which we now call the cotyledon remains within the seed during germination.) The coleoptile consists of a sheath which covers the remainder of the plumule until after emergence from the soil. Soon after exposure to light, the coleoptile stops growing, and the enclosed leaves split its tip and continue their growth upward. The very young coleoptile has a region of rapidly dividing cells near its tip, but by the time of emergence almost all its growth is the result of cell elongation, and very little cell division occurs. The elongation of the cells is strongly influenced by auxin. Under normal conditions of seedling growth, the auxin is produced in the tips, and it then moves downward. If the tip is removed, then the main source of auxin is removed, and the portion of the coleoptile behind the tip will grow in proportion to the amount of auxin that is supplied externally provided the supply of food is adequate. Therefore, we can supply known quantities of auxin and determine the amount of growth (increase in length) which will result. The growth resulting from different concentrations of auxin can be used as standards to determine the amount of auxin or auxin-like substances in an unknown sample.

In the following Investigation, you will determine first, the relationship between auxin concentrations and the growth in length of oat coleoptiles. Second, you will use this information to determine the concentration of auxin present in an unknown solution. These are the techniques used in one method for the bioassay of an auxin.

INVESTIGATION 30
A Biological Assay

MATERIALS

1. Several hundred grains of oats, *Avena sativa*, variety Victory
2. Germination box equipped with red cellophane and light shield. (See Figure 63.)
3. Indoleacetic acid
4. Sucrose
5. Cutter for cutting uniform sections of coleoptiles. (See Figure 64.)

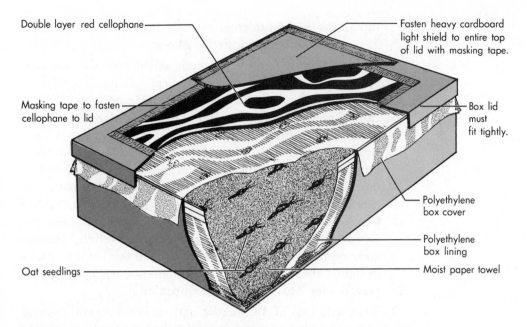

Double layer red cellophane

Fasten heavy cardboard light shield to entire top of lid with masking tape.

Masking tape to fasten cellophane to lid

Box lid must fit tightly.

Polyethylene box cover

Polyethylene box lining

Oat seedlings

Moist paper towel

FIGURE 63. Germination box with red filter and light-proof shield.

6. Six petri plates per team
7. Millimeter rulers
8. One 10-ml pipette per team

PROCEDURE

1. Each team should count out 50 grains of oats.
2. Divide the grains equally among the team members and remove the husks. Husking is a somewhat tedious operation, but it may be made easier if the grains are placed on a piece of paper towel and rolled with the palm of the hand. Label the grains (class, team, experiment, date) and store at room temperature until the next class day.
3. Before placing the husked grains in the germination box, soak them for *not less than 2 hours* and *not more than 4 hours.*
4. Place the husked grains in uniform rows on several layers of moist paper towels in a germination box. Cover the box with a sheet of clear polyethylene and then cover the box with the lid with the red filter and light shield as shown in Figure 63. Be sure the lid fits snugly and that the light shield over the

top allows no light to enter even through the red cellophane. Leave at room temperature for 24 hours.

5. After 24 hours, remove the light shield (not the box top). Allow light to pass through the red cellophane for a period of 30 minutes. Replace the light shield on the box cover.

6. Immediately after the 30-minute exposure, darken the room as much as possible. (Ideally you should work in a dark room with only a red light, but if you work rapidly with low-light intensity the experiment should be successful.) Working quickly, remove the box cover and the polyethylene cover and carefully add previously moistened vermiculite to a depth of $\frac{1}{2}$ inch over the germinating grains in the germination box. Gently pack the vermiculite to a uniform level and immediately replace the polyethylene cover and the box cover with its light-tight shield. Leave the germination box at room temperature for 2 to 3 days without opening it.

7. Stock solutions of 4% sucrose and various concentrations of IAA should be prepared. A stock solution containing 100 mg of indoleacetic acid per liter may be prepared as follows: Dissolve 100 mg of IAA in 1 to 2 ml of alcohol and add this to about 900 ml of distilled water. Warm this mixture gently on a hot plate or steam bath to drive off the alcohol. Then make up to 1 liter with distilled water.

Solutions containing 20.0, 2.0, 0.2, and 0.02 mg of IAA per liter can be prepared by serial diluting this stock solution.

These dilutions should be prepared shortly before using, because IAA is not very stable. The stock solution may be kept for one or two weeks if refrigerated in a dark bottle. Prepare enough of each dilution for all teams. (Each team will use 10 ml of each.) The instructor will provide a solution of IAA the concentration of which will be unknown to you.

8. On the day that the seedlings will be ready, each team will prepare 6 clean petri plates. After labeling each dish, add solutions to each as indicated in Table 16 at the top of the next page. Calculate the final concentration of IAA and sucrose in all dishes.

9. Approximately 2 to 3 days after covering the germinating grains with vermiculite, darken the room as much as possible and remove the germination box cover and polyethylene cover. Select 30 seedlings with straight coleoptiles from 1.5 to 2.5 cm in total length.

TABLE 16. PREPARATION OF PETRI PLATES FOR IAA ASSAY

Dish No.	Add
1	10 ml water + 10 ml 4% sucrose solution
2	10 ml 0.02 mg/1 IAA stock solution + 10 ml 4% sucrose solution
3	10 ml 0.2 mg/1 IAA stock solution + 10 ml 4% sucrose solution
4	10 ml 2.0 mg/1 IAA stock solution + 10 ml 4% sucrose solution
5	10 ml 20.0 mg/1 IAA stock solution + 10 ml 4% sucrose solution
6	10 ml unknown + 10 ml 4% sucrose solution

10. Cut off and discard the root and remaining part of the seed-ling and carefully line up the coleoptiles, two or three at a time, on a block of paraffin. Using the special cutter, as shown, cut from each coleoptile a 10-mm section back of a 3-mm tip and place five of the 10-mm sections in each of the petri plates previously prepared. Store them in the dark for 24 hours at a temperature as near 25°C as possible.

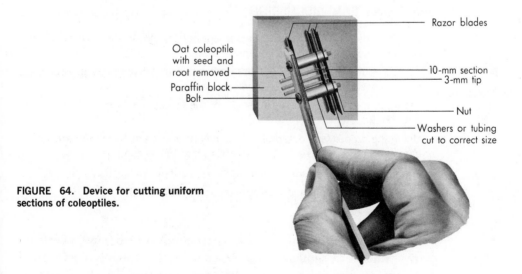

Razor blades

Oat coleoptile with seed and root removed

10-mm section
3-mm tip

Paraffin block
Bolt

Nut

Washers or tubing cut to correct size

FIGURE 64. Device for cutting uniform sections of coleoptiles.

11. At the end of the 24 hours, remove the coleoptile sections from the dishes and measure their lengths to the nearest 0.5 mm. Record the results in a chart similar to Chart D, page 248.

CHART D. **EFFECT OF IAA ON COLEOPTILE ELONGATION**

Coleoptile section	Dish No. 1 no IAA	Dish No. 2 0.01 mg/1 IAA	Dish No. 3 0.1 mg/1 IAA	Dish No. 4 1.0 mg/1 IAA	Dish No. 5 10.0 mg/1 IAA	Dish No. 6 unknown
1 2 3 4 5						
Average length						
Initial length						
Average change in length						

QUESTIONS FOR DISCUSSION

1. Plot your data on a graph with increments of growth on the vertical axis and concentration of IAA on the horizontal axis.
2. Compare the results of your team with those of other teams in your class and, if possible, with those of other classes.
3. Compute the mean increase in length of the coleoptiles in each dish. Are the data continuous or discrete?
4. Using the proper test, determine if the means in each dish with IAA are significantly different from each other and from that of the controls.
5. What is the concentration of your unknown solution?
6. Explain the mechanism of phototropism and geotropism.

Biologists have found many compounds which have effects on plants which are similar to indoleacetic acid. Probably very few of these are synthesized by plants. But one substance which appears to be as important in the regulation of growth in plants as is IAA was discovered in the following manner. A disease of rice plants results in overly rapid growth of the seedlings, which as a result become tall, weak, and finally fall over. (This was called "foolish seedling" disease, *bakanae,* by the Japanese farmers.) A scientist on Formosa found that the disease was caused by the fungus *Gibberella fujikuroi*. Japanese scientists were able to produce symptoms of the disease with cell-free extracts of the fungus and later isolated a substance in the extract which was the active agent. This substance was named gibberellin. Since that time, several related compounds have been found in the culture medium in which the fungus has been grown. One of these is gibberellic acid.

In the study which follows, you will investigate the effects of gibberellic acid on pea plants whose genetic constitution causes them to be dwarfs. Our problem is to determine whether genetic characteristics can be modified by the application of gibberellic acid.

INVESTIGATION 31
Effect of Gibberellic Acid

MATERIALS (per team)

1. Seeds of Alaska and Little Marvel peas
2. Four 4-inch flower pots (or similar containers)
3. Solution of gibberellic acid (100 mg per liter)
4. Germination trays

PROCEDURE

1. Place the two varieties of pea seeds in clearly separated halves of a germination tray between layers of moist paper towel. Store in the dark or keep covered for 3 to 4 days.

2. Mix equal parts of sand and soil, then moisten lightly. Add this mixture to four 4-inch flowerpots or similar containers. Each container should be about $\frac{2}{3}$ full. Transfer the 3- or 4-day-old germinated seedlings to separate pots as follows:

TABLE 17. PREPARATION OF PEA PLANTS

Pot	Variety of Peas	No. of Seedlings
1	Alaska	5
2	Alaska	5
3	Little Marvel	5
4	Little Marvel	5

Label each pot (team, class, pot number, date). Add about $\frac{1}{4}$ to $\frac{1}{2}$ inch of the sand-soil mixture to each pot. Keep soil moist *but not wet!*

3. When the seedlings are about 10 days old, or several centimeters high, measure the height in millimeters from the soil to the tip of the shoot apex of each seedling and record the measurements for each plant separately in a chart such as Chart E. Record these figures as the initial measurement.

4. Using a hand atomizer, spray one pot of Alaska peas and one pot of Little Marvel peas with a solution containing 100 mg of gibberellic acid per liter (experimental group). With a different atomizer, spray the plants of each variety in the other pots with pure water (control). Spray the plants until the leaves and shoot apex are wet enough to form droplets which will almost run off but do not permit *appreciable amounts* to drip into the soil. Since some of the spray for the experimental treatments may drift, it is advisable to place control plants and experimental plants in different parts of the room. However, be sure that they are both exposed to similar and maximum light conditions. Label each group as "control" or "experimental."

5. Measure the height of each plant in both experimental and control groups on each of the four days following the initial measurement, and then on the seventh day after the initial measurements. Record the measurements on your chart. As the plants in both experimental and control groups grow tall, it may be necessary to place stakes in the pots and tie the plants loosely to them.

QUESTIONS FOR DISCUSSION

1. Plot on millimeter graph paper the average measurements of each of the four groups in the chart you have made. Plot the days along the horizontal axis and the height of the plants along the vertical axis.

2. Compare the heights of the treated and control plants of each variety. Are the differences significant?

3. Compare the heights of Little Marvel peas sprayed with gibberellic acid with the heights of Alaska peas sprayed with water.

4. What hypotheses can you make with regard to the mechanism by which the genetic trait of dwarfism may operate in Little Marvel peas?

CHART E. EFFECT OF GIBBERELLIC ACID TREATMENT ON THE GROWTH OF TWO VARIETIES OF PEA PLANTS

Pea variety	Treatment	Individual	Length (mm) initial measurement	Length (in mm) on days following initial measurement				
				1st	2nd	3rd	4th	7th
Alaska	Sprayed with gibberellic acid	1						
		2						
		3						
		4						
	(Experimental)	5						
		Average						
	Sprayed with water	1						
		2						
		3						
		4						
	(Control)	5						
		Average						
Little Marvel	Sprayed with gibberellic acid	1						
		2						
		3						
		4						
	(Experimental)	5						
		Average						
	Sprayed with water	1						
		2						
		3						
		4						
	(Control)	5						
		Average						

A brief review of what we have learned up to this point about growth regulation in plants may be useful. A group of growth-regulating substances called auxins accelerate the elongation of plant cells. Unilateral stimuli such as light or gravity may result in unequal distribution of the auxin, IAA, on two sides of a stem or root and thus cause the curvatures which result in phototropism or geotropism. Light may destroy or inactivate IAA, and this probably accounts for the inhibiting effect of light on elongation of plants. Gibberellins also appear to be involved in the elongation process, at least in some plants. Research in these areas has revealed many other roles of plant-growth regulating substances. Some of these will be investigated in the laboratory, and others will be discussed.

One interesting problem in the behavior of plants is illustrated by the geotropic response of stems and roots. Stems tend to bend upward; roots tend to bend downward. How can gravity produce opposite responses in these two kinds of organs? Review your data from Investigation 31. Then study the relationships shown in Figure 65 and answer these questions:

1. Explain why roots grow down (positive geotropism) and shoots grow up (negative geotropism).

2. Explain how the tissues of two different types of plants might react to the same concentration of IAA.

3. Suppose research showed that narrow-leafed plants responded very weakly to a concentration of a growth regulator which

FIGURE 65. Approximate growth responses of roots and stems to indoleacetic acid. (Courtesy of K. V. Thimann, AMERICAN JOURNAL OF BOTANY, vol. 24, 1937.)

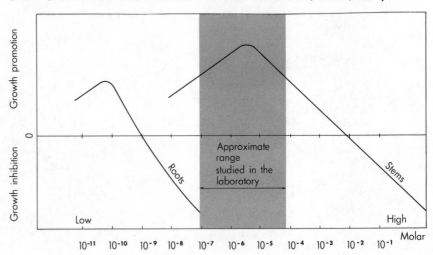

had a very striking effect on broad-leaf plants. Can you think of any practical way to use this information?

You have now studied the way in which two compounds, indoleacetic acid and gibberellic acid may act in the regulation of certain aspects of growth and development in plants. These and other compounds may have other effects in other circumstances. Such phenomena as the development of adventitious roots, control of axillary bud growth by the terminal bud, imitation of flowering, setting of fruits, fruit fall, and leaf fall are examples of other aspects of plant development which may be controlled by plant-growth regulating substances. (If such compounds are produced by the plant, we call them hormones.)

The purpose of the next Investigation is to explore the effects of some well-known growth substances on a common plant.

INVESTIGATION 32
Effects of Growth-Regulating Substances on Plants

We might well select different growth regulators from the ones mentioned here and plants other than beans for this Investigation. It is suggested that some teams may wish to modify their experiments. The following outline may serve as a useful guide.

MATERIALS (per team)

1. Sixty to eighty seeds of pinto beans. (Any other variety of *Phaseolus vulgaris* might be used, or you may wish to use plants from other genera or families.)

2. One germination tray.

3. Twenty-one bottles (baby-food jars work well) for holding cuttings, and if possible a holder for the bottles. (See Figure 66.)

4. Plant-growth regulators as follows:
   a. Indoleacetic acid, IAA; 100 ppm in 4% ethanol
   b. IAA, 0.1% and 1.0%, in lanolin paste
   c. Indolebutyric acid, IBA; 100 ppm in 4% ethanol
   d. IBA, 0.1% and 1.0%, lanolin paste
   e. Gibberellic acid, G.A.; 100 ppm in 4% ethanol
   f. G.A.; in 0.1% and 1.0% in lanolin paste

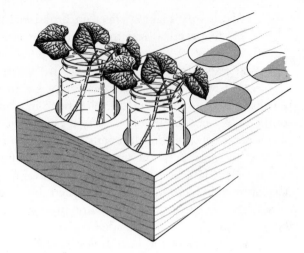

FIGURE 66. A convenient holder for bottles made by boring holes in a 2-inch by 8-inch board.

5. Control carriers; 4% ethanol and lanolin

6. Dropper pipettes, toothpicks, and razor blades

PROCEDURE

1. Each team should plant 60 to 80 bean seeds in a germination tray.

2. Select the 40 most uniform plants in 9–14 days—that is, when the seedlings are ready for cutting and the heart-shaped leaves should be fairly well expanded.

3. Label 21 bottles as shown in the following list. Fill the bottles with tap water. Place two cuttings in each bottle. The cuttings may remain in better health if you cut off the bottom 2 or 3 cm of each under water. This helps to prevent air plugs in the xylem vessels. Cut off the stems of each plant about 1 or 2 cm above the pair of heart-shaped leaves. Then treat each pair of cuttings as indicated. Use a different toothpick or medicine dropper for each different formulation of growth regulator.

No. 1: Lanolin Control: With a toothpick place a dab of plain lanolin on one of the remaining leaves and another dab on the cut end of the stem of each plant as illustrated in Figure 67.

No. 2: Ethanol control: With a small-tipped medicine dropper, place a drop of 4% ethanol on one leaf and another on the cut end of the stem of each plant.

No. 3: Control: No treatment.

No. 4: IAA, 100 ppm in 4% ethanol. Apply a drop of the solution to one leaf of each plant.

No. 5: IAA, 100 ppm in 4% ethanol. Apply a drop of the solution to the cut end of the stem of each plant.

No. 6: IAA, 0.1% in lanolin. Apply a dab of lanolin containing 0.1% IAA to one leaf of each plant.

No. 7: IAA, 0.1% in lanolin. Apply a dab of lanolin containing 0.1% IAA to the cut end of the stem of each plant.

No. 8: IAA, 1.0% in lanolin; apply lanolin containing 1.0% IAA to one leaf of each plant.

No. 9: IAA, 1.0% in lanolin; apply lanolin containing 1.0% IAA to the cut end of stem of each plant.

No. 10: IBA, 100 ppm in 4% ethanol, applied to leaf; same as No. 4 but use IBA in place of IAA.

No. 11: IBA, 100 ppm in 4% ethanol, applied to cut stem; same as No. 5 but use IBA.

No. 12: IBA, 0.1% in lanolin, applied to leaf.

No. 13: IBA, 0.1% in lanolin, applied to cut stem.

No. 14: IBA, 1.0% in lanolin, applied to leaf.

No. 15: IBA, 1.0% in lanolin, applied to cut stem.

No. 16: GA, 100 ppm in 4% ethanol, applied to leaf.

No. 17: GA, 100 ppm in 4% ethanol, applied to cut stem.

No. 18: GA, 0.1% in lanolin, applied to leaf.

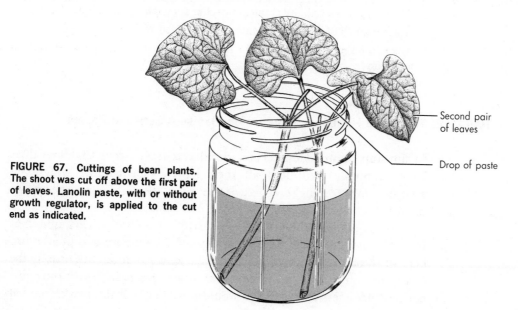

Second pair of leaves

Drop of paste

FIGURE 67. Cuttings of bean plants. The shoot was cut off above the first pair of leaves. Lanolin paste, with or without growth regulator, is applied to the cut end as indicated.

No. 19: GA, 0.1% in lanolin, applied to cut stem.
No. 20: GA, 1.0% in lanolin, applied to leaf.
No. 21: GA, 1.0% in lanolin, applied to cut stem.

4. Place all the cuttings in a well-lighted but cool place. Avoid direct sunlight if the room is above 25°C to 28°C. Observe them daily. Note the development of lateral shoots from the buds in the axils of the two heart-shaped leaves. Record these data each day in a chart similar to Chart F.

5. Measure the length of the axillary shoots on each plant each day. Carefully remove each cutting from the water, and, being careful not to injure the plant, measure the lengths of several representative roots on each. Count the number of adventitious roots (if they are not too numerous to count). Record these data each day in your chart. Make other charts for other growth regulators.

6. After 5 or 6 days, complete the experiment by making final measurements and noting any abnormalities or other effects which may have resulted from the treatment.

QUESTIONS FOR DISCUSSION

1. Which of the growth regulators has the greatest tendency to stimulate the formation of adventitious roots? Which has the greatest tendency to increase the growth in length of the roots?
2. Which substance had the greatest effect in stimulating elongation of stems?
3. Which method of application seemed most effective?
4. Which concentrations were most effective? Were the higher concentrations inhibitory?

PARALLEL READING: Hormonal Influence on Plants—A Series of Papers

As autumn approaches, certain deciduous trees (trees that lose their leaves annually) will lose all their leaves except those which are very near a street light. These leaves near the light may remain on the tree for weeks after the rest of the tree is bare.

By checking leaf fall systematically, scientists have found that leaves fall not only in autumn, but also fall from the trees in small numbers all through the growing season. The leaves that fall are usually the older ones on a branch. Some of the more observant early biologists noticed that even young leaves would fall quickly if the blade (the flat

CHART F. EFFECTS OF INDOLEACETIC ACID ON THE GROWTH OF ROOTS AND SHOOTS OF PINTO BEANS

Plant number	Average Length of Roots (r) and Shoots (s) by Days					Number of roots	Abnormalities
	Day 2	Day 3	Day 4	Day 5	Day 6		
1 r s							
2 r s							
3 r s							
4 r s							
5 r s							
6 r s							
7 r s							
8 r s							
9 r s							
10 r s							
11 r s							
12 r s							
13 r s							
14 r s							
15 r s							
16 r s							
17 r s							
18 r s							
19 r s							
20 r s							
21 r s							

part of the leaf) were injured or eaten away by insects. In 1916 a German scientific magazine carried a paper by a biologist, Küster, describing detailed experiments on leaf fall. Küster cut off the leaf blade and found that the remaining leafstalk fell quite soon; in fact, it fell many weeks before the intact leaf. But if only a tiny bit of the blade were left on, the fall was very much delayed. This disproportionate effect of a bit of the leaf blade suggested to Küster that the blade must be producing a chemical which, even in tiny amounts, could prevent leaf fall.

Küster's paper gives us a good example of why scientists work on a particular plant or animal. Trees are very awkward organisms to study. They are too big, grow too slowly, and take up too much space. Although their autumn leaf fall is spectacular, it is quite inconvenient to wait from one year to the next to study it. Küster made a survey of many plants, looking for a small, fast-growing plant that could be grown in the laboratory or greenhouse and which had a regular, fairly continuous leaf fall. He found such a plant in *Coleus blumei*, a member of the mint family from the tropics. This is the common house

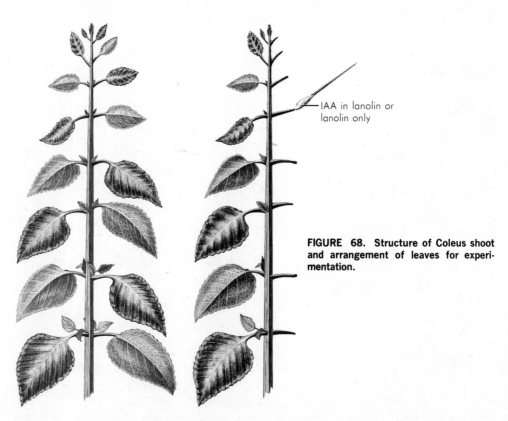

IAA in lanolin or lanolin only

FIGURE 68. Structure of Coleus shoot and arrangement of leaves for experimentation.

plant whose genus name has become so familiar that we use it as a common name. As often as once a week, this plant sheds its oldest (bottom) pair of leaves and makes a new pair of young leaves at the top, in the apical bud.

Ever since Küster reported the special advantages of *Coleus*, it has been the favorite plant for use in studies of what makes leaves grow old and fall.

The general structure of *Coleus* is shown in Figure 68. In this illustration the upper leaves have been removed. Like most plants, *Coleus* has a shoot portion (above the ground) and a root portion. The shoot portion, consisting of one main shoot (the upright stem) and one or more side shoots, grows at its tip by developing new leaves and stem tissue. The cluster of primordial leaves, within which much of this development goes on is called the apical bud. In the axil of each leaf is another bud, called an axillary bud. These may give rise to axillary, or lateral branches. Toward the bottom of the main shoot, the axillary branches grow larger and the leaves of the main shoot become progressively older. By the time a leaf of the main shoot is approximately 8 or 9 leaves down from the apical bud, it falls off. (The axillary branches, as you can see from Figure 68, remain on the main shoot and continue to grow.)

Several years ago, Dr. W. P. Jacobs became interested in the problem of why leaves fall. He has published several papers on this subject.

The following exercise is designed to show the aspects of experimental design used by Jacobs and others, and for our present purposes, precise knowledge of plant hormones and growth is considered of secondary importance.

It is often possible to follow all the steps of a previous experiment by reading about them and thus gain a great deal of understanding. If time permits, you might try to duplicate the steps of Dr. Jacobs and his co-workers to see if you obtain similar results.

In order to reduce the number of factors involved in their experiments on *Coleus*, Wetmore and Jacobs removed all axillary branches, and during each experiment they removed any new side shoots that developed. The plants which were used were all derived from a single original plant—not by collecting seeds and growing them but by taking repeated stem cuttings. The experimental design is outlined in the following steps which you can repeat, if time permits.

1. Select six plants and arrange into three sets of two plants each. The two plants of any one set should be similar in height, color,

and leaf size. Remove all side shoots. In selecting plants, match each pair by the lengths of their younger leaf blades. Use the youngest leaf pair which is from 50 to 100 mm long. Leaves on the plants of any matched pair should differ no more than 5 mm in length.

2. Cut off the leaf blades of one of the leaves at each node all the way up the stem to leaf pair 1. Leave the petioles intact.

3. Select at random one plant of each matched pair to be treated with indoleacetic acid (IAA).

4. One half of the IAA treated plants will receive 1.0% IAA in lanolin; the other half will receive 0.1% IAA in lanolin. Cover cut ends of each of the debladed leafstalks of those plants selected for treatment with a dab of lanolin containing the designated concentration of IAA.

5. Treat the debladed leafstalks of the second plant in each pair with plain lanolin, applied in exactly the same way.

QUESTIONS FOR DISCUSSION

1. Why is such extreme care in matching the plants necessary?
2. Why is the control treated with plain lanolin instead of being left untreated?
3. What are the advantages of using repeated stem cuttings rather than growing the plants from seeds?

The foregoing was a description of the way in which Jacobs designed an experiment. Several reprinted papers now follow, as illustrations of the way in which he and others reported their results.

WHAT SUBSTANCE NORMALLY CONTROLS A GIVEN BIOLOGICAL PROCESS?[1]

by WILLIAM P. JACOBS

Probably the question asked most often by developmental physiologists is the one posed in the title. In the current stage of our branch of science, we are culturally conditioned to ask this question, and to try to answer it. But there is no uniformity in the evidence provided in our answers. A search through the literature has uncovered no explicit formulation of what constitutes worth-while evidence. Yet it

[1]Reprinted by permission from *Developmental Biology,* Volume 1, No. 6, December, 1959. Copyright © 1959 by Academic Press Inc. Printed in U.S.A.

is obvious that there *are* rules, at least some of which are tacitly assumed by all researchers in the field. The purpose of this paper is to formalize the rules already in use, discuss their application to some already published research, and then in the companion paper to demonstrate their application to a well-known developmental problem.

But before formulating the rules, we should explain what assumptions lie behind the question in the title. (We are using the word "controls" in the title as a conventional, shorthand way of saying "is the factor which limits the rate of.") The most obvious assumption is that the process is being controlled by only one substance. It is theoretically possible that more than one substance is controlling the process. However, applying the suggested rules to each substance in turn will elucidate any such phenomenon. And rule 6 (see below) is specifically designed to ferret out this case, if it exists.

A second assumption, scarcely less obvious, is that the process is controlled by a substance rather than by a physical process. Although it is the current fashion to think in terms of chemical substances, we should keep in mind the possibility that any given biological process may be controlled by, for instance, a rate of diffusion.

With these qualifications in mind, then, what general types of evidence would be considered to verify the hypothesis that "the development of structure S is normally controlled in organisms by chemical C"? There are six basic types incorporated in the rules listed below. Each type of evidence confirms the hypothesis; and the more of them that are satisfied, the greater the likelihood that the hypothesis is correct.

1. Demonstrate that the chemical is normally present and that the amount of the structure varies in the *intact* organism in parallel fashion with the amount of the chemical.
   a. If a quantitative relation can be demonstrated in such *normal* parallel variation, this strengthens the case.
2. Remove the source of the chemical, and demonstrate subsequent absence of formation of the structure.
   a. If a quantitative relation can be demonstrated between the amount of this *artificial* decrease of the chemical and the amount of decrease of the structure, this strengthens the case.
   b. The chemical is sometimes removed by selecting genetic mutants.
   c. A less direct—and therefore less satisfactory—method of meeting this requirement is to add other chemicals pre-

sumed to more or less specifically block chemical *C,* and to show that the structure is blocked, too.

3. Substitute pure chemical *C* for the organ or tissue which had been shown to be the normal source of the chemical in the organism and demonstrate subsequent formation of the structure.

   a. Quantitative evidence, as usual, strengthens the case. In the form of *exact substitution*—i.e., adding exactly the amount of chemical which is normally produced by the excised organ —it is particularly important when considering natural inhibition effects attributed to hormones. (Adding surplus hormone will unspecifically inhibit a variety of processes.) The ideal application of this rule would be to add a number of concentrations of the chemical—with one of the number providing exact substitution—and to demonstrate quantitatively parallel variation in the amount of the structure. Exact substitution should, of course, give exactly the normal amount and type of structure.

   b. Since the isolation of a naturally occurring chemical (particularly when it is a hormone) is a task which has often thwarted developmental physiologists, they may be driven to the following progressively less direct modifications of the rule: substitute an *extract* of the organ; substitute chopped-up pieces of the organ; merely replace the excised organ with a diffusion gap between it and the region where the structure is to form.

4. Isolate as much of the reacting system as possible and demonstrate that the chemical has the same effect as in the more intact organism. This reduces the probability that the chemical is acting primarily on some other process, or part of the organism, and only secondarily on the structure in question.

   a. The part of biochemistry concerned with the *in vitro* reactions of extracted enzyme systems to added chemicals can be considered to represent the most extreme application of the isolation rule. Progressively less extreme isolations would be reactions of subcellular organelles; individual cells, tissues, pieces of organs, or whole organs (as in sterile cultures); leaf cuttings, etc.

5. Demonstrate the generality of the results by showing that the other five points hold for species from a number of different

families, as well as for the development of the structure in different kinds of organs.

6. Demonstrate that natural occurring chemicals other than C have no such effect on the structure.

(These six rules can be more easily remembered by using the mnemonic device "PESIGS"—representing Parallel variation, Excision, Substitution, Isolation, Generality, Specificity.)

If covering these six "rules" in their quantitative form represents a useful ideal, how many times has this ideal been realized? Very seldom, to my knowledge. Most of the thorough, quantitative work known to me does not fully follow the last two rules. Most papers in the field of plant development follow only rule 3 ("substituting the chemical") and that only qualitatively: exact substitution is very rare.

The action of auxin as the normal controlling factor for the growth of *Avena* coleoptiles is one of the few cases in plants which has been thoroughly investigated. Parallel variation in the normal plant is shown in Fig. 27 of Went and Thimann (1937). The decreased growth resulting from excision of the source of auxin is shown in their Fig. 7. Substitution of auxin for the excised natural source causes increased growth, the amount of extra growth being proportional to the amount of auxin added (their Fig. 24). By adding 50 parts per million of synthetic indolaeacetic acid to decapitated coleoptiles, Went (1942) was able to bring the growth rate up to that of the intact controls (his Table II). However, I have not been able to find a case where *exact* substitution of the auxin coming from the coleoptile tip has been attempted. Isolated sections of the coleoptile were found to give increased growth with increases in added auxin (Bonner, 1933), and these isolated sections have been used for many later papers, including some elegant recent work in which kinetic treatment has been applied to the growth data (Bonner and Foster, 1955). The general applicability of the results with *Avena* coleoptiles has been checked to some degree in many plants by many different investigators (cf. Went and Thimann, 1937; Söding, 1952). The chemical specificity has also been thoroughly checked.

Similarly, auxin has been shown to be the normal limiting factor for the differentiation of xylem cells in *Coleus* shoots, in quantitative studies of normal production, of excision effects, and of exact substitution (Jacobs, 1947, 1952, 1954, 1956; Jacobs and Morrow, 1957). Isolation has not yet been applied to *Coleus*, but in two other genera the addition of synthetic auxin speeded differentiation of xylem in tissue cultures (Torrey, 1953; Wetmore and Sorokin, 1955). The generality of the results with *Coleus* has been checked even less—one can only say that there are indications in the literature that auxin controls xylem differentiation in many different plants. The specificity of auxin in this effect has not yet been checked at all.

The only other area where the PESIGS rules have been followed in quantitative detail is the role of auxin in preventing the fall (abscission) of leaves. Parallel variation and the effect of excising the sources were demonstrated in *Coleus* (Wetmore and Jacobs, 1953). Synthetic auxin could completely substitute for the normal auxin sources; however, exact substitution for the normal auxin production was not attempted. Addicott and co-workers were the first to isolate the abscission area and to demonstrate that the isolated system of bean leaves functioned qualitatively the same as the intact one (particularly well shown in Addicott and Lynch, 1951). That auxin can inhibit abscission has been shown in many plants and in a variety of organs, and the chemical specificity of auxins for this effect has been looked into more than usual.

If there are so few cases of thorough application of the PESIGS rules even in the field of auxin physiology—a field which has been under intensive investigation for thirty years—one may well ask why more researchers do not routinely use these rules. One part of the answer may be that the rules have not previously been explicitly stated in connection with developmental research. It is also immediately obvious to any practicing researcher that there are many scientists who would be antipathetic to the chore of gathering such detailed evidence. Because of basic personality traits, they would not function as well as scientists if required to follow quantitative PESIGS rules as they do by checking, say, one of the rules in a qualitative way. This difference in the investigators is to some extent aesthetic—akin to the difference between Frans Hals and van Eyck. Both Hals and the qualitative investigators probably thought that their work adequately represented the real world. But, in the case of experimental biology this is not always the case; qualitative checking of only one of the PESIGS rules has often led to erroneous conclusions. Consider the following examples.

Elegant quantitative experiments showed that auxin moved with strict basipetal polarity in sections cut from a specialized organ of oat seedlings (Went and White, 1939). Quantitative work on other organs from other seedlings also showed strict basipetal polarity. Generalizing from these results, similar results from less thorough investigations of plants which were past the seedling stage were taken as evidence that strict basipetal polarity was a general phenomenon in the plant world. Phenomena which could, with some likelihood be ascribed to the action of auxin, would be considered to be not due to auxin "because auxin cannot move upwards." In essence, this was generalizing to all organs, including trees, on the basis of work with a few organs from seedlings. When auxin transport was quantitatively checked in stems of an independently growing plant, substantial upward transport was found. It could be demonstrated quantitatively, by parallel variation studies, that this upward transport was physiologically significant (Jacobs, 1952, 1954). One of the special virtues of quantitative work was shown in the cited work: although the experimenter, familiar by both reading and practice with the "fact" that auxin only moved down, did not believe his own results at first, their quantitatively reproducible nature finally forced him to change his preconceptions. If the work had been qualitative, he could easily have cast aside the evidence—as some of the earlier qualitative workers had done (cf. Jacobs, 1954, p. 334).

A well-publicized case in which parallel variation was, by itself, taken as sufficient evidence of a causal relation is that of dwarf corn. The type of corn called "Dwarf-1" results from a single gene mutation. Its mature height is about half that of the normal strain. Once van Overbeek (1938) showed that coleoptile tips from the dwarf seedlings produce about half as much auxin as the normal seedling, the small amount of growth hormone was taken as the cause of the decreased growth. After twenty years, this was reinvestigated by Phinney. He confirmed the parallel variation between growth and auxin production, but, in addition, tried substitution. Auxin added to the dwarf plants gave no increased growth (Phinney, 1956a), nor was there any increased growth from a variety of related compounds. However, when the more recently exploited substance, gibberellic acid, was added to the dwarf plants, their growth was brought up to that of the normal plants (Phinney, 1956b). (Normal plants were essentially unaffected by the same amount of added gibberellic acid.)

A third example, that of the role of auxin in apical dominance, we shall deal with in a subsequent paper. This is a case where exact matching, under the substitution rule, has not been tried except in the first, classic paper on the subject (Thimann and Skoog, 1934).

Summary

Six rules (the "PESIGS" rules) are stated for determining what naturally occurring substance normally controls a given biological process. The application of these rules, in quantitative form, is urged on developmental physiologists.

REFERENCES

Addicott, F. T., and Lynch, R. S. (1951). Acceleration and retardation of abscission by indoleacetic acid. *Science* 114, 688–689.

Bonner, J. (1933). The action of the plant growth hormone. *J. Gen. Physiol.* 17, 63–76.

Bonner, J., and Foster, R. J. (1955). The growth-time relationships of the auxin-induced growth in *Avena* coleoptile sections. *J. Exptl. Botany* 6, 293–302.

Jacobs, W. P. (1947). The effect of some growth hormones on the differentiation of xylem around a wound. *Am. J. Botany* 34, 600.

Jacobs, W. P. (1952). The role of auxin in the differentiation of xylem around a wound. *Am. J. Botany* 39, 301–309.

Jacobs, W. P. (1954). Acropetal auxin transport and xylem regeneration—a quantitative study. *Am. Naturalist* 88, 327–337.

Jacobs, W. P. (1956). Internal factors, controlling cell differentiation in the flowering plants. *Am. Naturalist* 90, 163–169.

Jacobs, W. P., and Morrow, I. B. (1957). A quantitative study of xylem development in the vegetative shoot apex of *Coleus*. *Am. J. Botany* 44, 823–842.

Phinney, B. O. (1956a). Biochemical mutants in maize: dwarfism and its reversal with gibberellins. *Plant Physiol.* 31, (Suppl.), XX.

Phinney, B. O. (1956b). Growth response of single-gene dwarf mutants in maize to gibberellic acid. *Proc. Natl. Acad. Sci. U. S.* 42, 185–189.

Söding, H. (1952). "Die Wuchsstofflehre." Georg Thieme, Stuttgart.

Thimann, K. V., and Skoog, F. (1934). On the inhibition of bud development and other functions of growth substance in *Vicia faba*. *Proc. Roy. Soc.* B114, 317–339.

Torrey, J. G. (1953). The effect of certain metabolic inhibitors on vascular tissue differentiation in isolated pea roots. *Am. J. Botany* 40, 525–533.

Van Overbeek, J. (1938). Auxin production in seedlings of dwarf maize. *Plant Physiol.* 13, 587–598.

Went, F. W. (1942). Growth, auxin, and tropisms in decapitated *Avena* coleoptiles. *Plant Physiol.* 17, 236–249.

Went, F. W., and Thimann, K. V. (1937). "Phytohormones." Macmillan, New York.

Went, F. W., and White, R. (1939). Experiments on the transport of auxin. *Botan. Gaz.* 100, 465–484.

Wetmore, R. H., and Jacobs, W. P. (1953). Studies on abscission: the inhibiting effect of auxin. *Am. J. Botany* 40, 272–276.

Wetmore, R. H., and Sorokin, S. (1955). On the differentiation of xylem. *J. Arnold Arboretum* (*Harvard Univ.*) 36, 305–317.

QUESTIONS FOR DISCUSSION

1. What does PESIGS stand for?
2. How could a scientist determine if his results were reasonably valid without using all six steps?
3. What two obvious assumptions does the author list as underlying the question in his title?

STUDIES ON ABSCISSION:
THE INHIBITING EFFECT OF AUXIN[1]

by R. H. WETMORE and WM. P. JACOBS

The first indication that abscission of leaves is controlled by auxin was given by the work of Laibach (1933). He showed that pollinia of orchids, applied to the distal cut end of a petiole, inhibited abscission of the petiole. Orchid pollinia were known to contain auxin, and a few years later La Rue (1936) reported an experiment showing that synthetic auxin in the form of indoleacetic acid (IAA) was also very effective in delaying abscission, if it were applied in the dark or in shade. The first thorough investigation as to whether auxin was the factor which typically in nature controlled petiolar abscission was the work of Myers (1940a). Using *Coleus* as his experimental material, he showed that those leaves which normally stayed longest on the plant were also those which gave the greatest amount of diffusible auxin; that removing the leaf blade greatly accelerated petiolar abscission; and that the presence of IAA on the debladed petiole inhibited abscission as well as did the presence of the leaf blade.

The experiments cited above give a picture of abscission in which the speed with which any given petiole abscises is directly controlled by the amount of diffusible auxin coming down from the leaf blade. That is, so far as abscission is concerned, each leaf seems to act as a physiological entity.

The experiments to be described below confirm and extend these earlier experiments, while the experiments in the second paper of this series (Rossetter and Jacobs, 1953) show that we must discard as being too simple the view that "with respect to abscission, each leaf is a physiological entity."

METHODS.—A clonal stock of *Coleus blumei* Benth. was grown in the greenhouse. The variety was the same as that used by Jacobs (1952),

[1] Reprinted by permission from the *American Journal of Botany*, Vol. 40, No. 4, 272–276, April, 1953. Printed in U.S.A.

Rossetter and Jacobs (1953), and Jacobs and Bullwinkel (1953). Supplementary lighting was used during winter months to obtain faster growth.

The plants were trained to a two-branched form by excising the main stem and allowing only two of the axillary branches to develop. All other axillary branches were removed. The terminology used in referring to parts of each shoot is the same as that used in Jacobs (1952) and is shown diagrammatically in Figure 1.

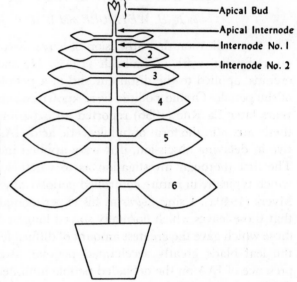

FIGURE 1. Diagram of Coleus plant showing the methods of designating leaves, internodes, and apical bud. (The leaves, the arrangement of which is actually decussate, are here represented as being all in one plane.)

A "spiral" pattern of deblading was used: that is, the leaf blade of one member of each leaf pair below the apical bud was excised, and the debladed petioles were arranged in a spiral on the stem (see Figure 1 of Rossetter and Jacobs, 1953). The intact leaf of each pair is listed as the control; the debladed petiole as the treated petiole.

IAA and naphthaleneacetic acid (NAA) were applied at 1 percent concentrations in lanolin; 8 shoots were used for each treatment. Determinations of diffusible auxin were run as described in Jacobs (1952).

The plants were checked every day for abscission by tapping each main stem rather vigorously before counting leaf fall.

Material for histological study was embedded in paraffin, sectioned at 10μ, and stained with haematoxylin and safranin.

Statistical methods and terminology follow Snedecor (1946).

RESULTS.—1. *Normal development of abscission layer.*—Figures 2–5 show the abscission region in leaves 2, 3, 5, and 6. A readily detectable abscission layer is first visible in leaf 3; the fully differentiated layer

FIGURE 2–5. Photomicrographs of the prospective abscission zone in petioles 2, 3, 4, and 5, respectively.

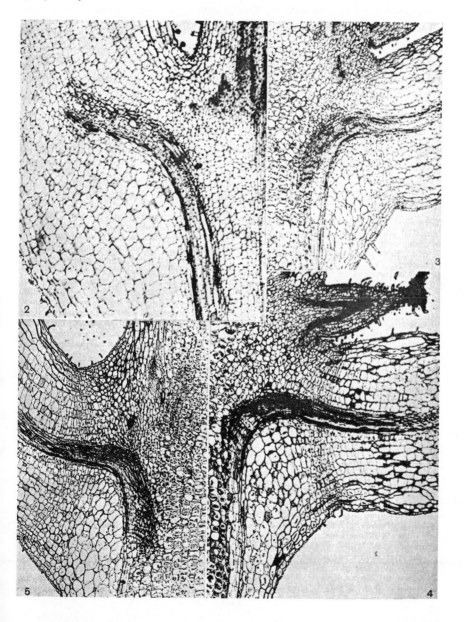

being visible is leaf 6. The normal order of abscission with the characteristic time intervals can be seen from the control leaves of Figure 7. The younger the leaf (i.e., the nearer the apical bud it is), the longer it stays on the plant. Another way of saying this is that *Coleus* leaves usually abscise when they are in leaf position 6–8.

2. *Diffusible auxin from the leaf blades shows a high correlation with the length of time petioles are normally retained.*—Figure 6 shows typical results of a diffusion experiment on *Coleus* leaves. In general agreement with the findings of predecessors, successively greater absolute amounts of auxin were obtained from successively younger leaves, but with leaf 1 showing a decrease relative to leaf 2. Calculation of the correlation coefficient for amount of diffusible auxin and length of petiolar retention, omitting the values for leaf 1, gave r = 0.9525, a value which is statistically significant.

3. *Removing leaf blades speeds abscission.*—Excising the leaf blade of one member of each leaf pair, and placing plain lanolin on the petiolar stump gave results as shown to the left in Figure 7. The controls show the expected gradient of petiolar abscission. The debladed petioles, on the other hand, abscise very much faster and show no gradient along the axis with respect to the time of abscission. (The lack of a gradient in the debladed petioles does not agree with Myers' (1940a) results.)

4. *Auxin replaces the leaf blade in slowing abscission of the proximal petiole.*—When plants are treated as above except that synthetic auxin is applied in lanolin to the petiolar stump, the treated petioles stay on as long or longer than do those which have the leaf blade still attached. Results with IAA are shown to the right of Figure 7. One per cent IAA

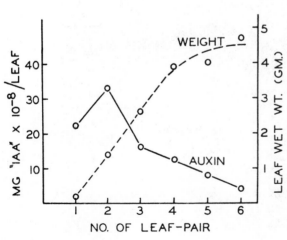

FIGURE 6. Amounts of diffusible auxin obtained from Coleus leaves from various positions on the stem.

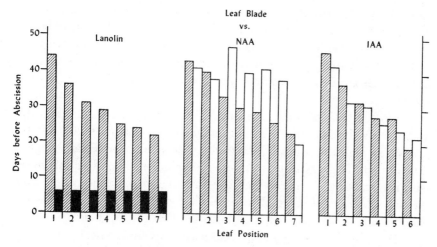

FIGURE 7. Pre-abscission intervals for intact leaves (shaded) and debladed petioles, in Coleus plants with spiral deblading, when the debladed petioles are treated with plain lanolin, NAA-lanolin, and IAA-lanolin, respectively.

almost exactly replaces the leaf blade in causing petiolar retention. NAA in lanolin acts even more effectively to prevent petiolar abscission (center of Figure 7.)

DISCUSSION.—Study of histological developments in the abscission zone of *Coleus* shows that this plant resembles cotton, pepper and poinsettia studied by Gawadi and Avery (1950) in that a distinct abscission layer is differentiated prior to abscission. In agreement with Myers' (1940a) report for *Coleus,* the layer is first recognizable in leaf pair 3 and is fully differentiated in leaf pair 6. As our Figures 2–5 show, the abscission layer is remarkably well developed even in leaf pair 3, leaves which are only two-thirds of their full size and which will not actually abscise for about 30 days (Figures 6, 7). These observations support the contention of Gawadi and Avery that the so-called abscission layer has no *causal* relationship to the actual abscission of leaves.

These experiments confirm those of Myers in showing the importance of diffusible auxin in controlling the normal pattern of leaf abscission. In agreement with the results of Myers (1940a), excision of the leaf blade speeds up markedly abscission of the debladed petiole. Myers found that debladed petioles showed a gradient like the intact leaves, in that petiole 1 abscised later than petiole 2, etc. No such gradient was detectable in our plants (cf. also Rossetter and Jacobs, 1953, Figures 2 and 3). IAA added to the cut surface of the debladed petiole was found, in agreement with Myers (1940a), to replace very closely the effect of the leaf blade in delaying abscission. NAA was for most of the petioles *more* effective than the intact leaf blade. Finally, the general agreement noted by Myers between the amount of diffusible auxin produced by the leaves and the length of time they would normally stay on the plants was confirmed, and it was further shown that there was a high and statistically significant correlation between the amount of diffusible auxin from leaves 2–6 and the pre-abscission period of those leaves.

On the basis of these experiments we conclude that the diffusible auxin from the leaf blade is the factor which normally controls the pattern of leaf abscission. Once the petiole reaches a certain stage of development (e.g., as in leaf 2), its pre-abscission interval is correlated with the amount of auxin which is being released into the petiole from the distal leaf blade. That this correlation is causal is indicated by the experiments in which auxin was substituted for the leaf blade.

It is important to note two uncertainties which have not yet been resolved. First, we have no indication as to *where* in the petiole the auxin exerts its effect. The most obivous guess is directly on the abscission layer. But, since auxin seems necessary for growth of the petiole (Mai, 1934; Myers, 1940a, b) and since the petioles normally continue to elongate after the blades have ceased growing (Myers, 1940a; and our observations), it may be that auxin prevents abscission only indirectly (i.e., through its effect on growth of the petiole). This possibility was apparently first suggested by Myers (1940b). Support for this interpretation comes from a detailed comparison of the growth of debladed petioles treated with IAA-lanolin and of normal petioles with distal intact leaves. In Figure 7 it can be seen that IAA-lanolin does not quite "replace" the leaf blade in inhibiting abscission of petioles 1 and 2, but "*more* than replaces" the leaf blade for petiole 6. The same result is shown in Figure 4 of Myers' paper (1940a), where data for petioles 7 and 8 show even longer retention when IAA is added. If one now inspects Figure 1 of Myers' other paper (1940b), where the effect of IAA-lanolin versus the leaf blade is shown for the *elongation* of petioles, one can see that there is an exact parallel with the abscission effects. This hypothesis, that auxin inhibits abscission indirectly through its stimulating effect on petiole growth, has two advantages. First, in contrast to the hypothesis that auxin inhibits abscission by a direct effect on the abscission layer, this hypothesis explains the apparently contradictory findings that A) the amount of auxin from the various leaf blades shows such a high correlation with the pre-abscission interval of the leaves, although B) auxin added to debladed petiole 6 gives relatively fast abscission even though the *same* concentration of auxin keeps debladed petiole 3 on for the same time as does the intact leaf blade. Secondly, the hypothesis of the "indirect effect" has the advantage of explaining the action on abscission of the growth-substance, auxin, by a growth effect instead of necessitating the assumption of still another type of effect to the already bewildering multiplicity of auxin effects.

The second uncertainty which has not been discussed is that although the experiments reported above seem unequivocal in showing the primary role of auxin from its own leaf blade in preventing abscission of a given petiole, no study has yet been reported which investigates the influence of the *rest* of the organism on the abscission of a leaf petiole. Such a study will be reported in Rossetter and Jacobs (1953).

Summary

The anatomy and physiology of the development of the abscission layer in *Coleus* leaves have been investigated. The abscission layer is first apparent, and is almost fully differentiated anatomically, in leaf pair 3. The pre-abscission interval of leaves at various positions along the stem (and thus of varying ages) shows a significant correlation with the amount of diffusible auxin produced by the leaves. Removing the leaf blades (and thus the auxin-sources) speeds up abscission very markedly. The leaf blades can be replaced in their inhibiting effect on abscission by the application of synthetic auxin. The experiments are interpreted as showing the dominant role of diffusible auxin from the leaf blades in controlling the normal order and intervals of leaf abscission.

Biology Department,
  Harvard University,
    Cambridge, Massachusetts

Biology Department,
  Princeton University,
    Princeton, New Jersey

LITERATURE CITED

Gawadi, A. G., and G. S. Avery, Jr. 1950. Leaf abscission and the so-called "abscission layer." Amer. Jour. Bot. 37: 172–180.

Jacobs, W. P. 1952. The role of auxin in differentiation of xylem around a wound. Amer. Jour. Bot. 39: 301ff.

———, and B. Bullwinkel. 1953. Compensatory growth in *Coleus* shoots. Amer. Jour. Bot. In press.

Laibach, F. 1933. Versuche mit Wuchsstoffpaste. Ber. Deutsch. Bot. Ges. 51: 386–392.

La Rue, C. D. 1936. The effect of auxin on the abscission of petioles. Proc. National Acad. Sci. (U.S.) 22: 254–259.

END OF ARTICLE

QUESTIONS FOR DISCUSSION

1. Compare this paper by Wetmore and Jacobs to the paper by the Darwins (page 239). Note the difference in style. Was the experimental design as obvious in both papers?

2. Which of the PESIGS rules do the authors follow?

3. While the author's use of the statistic called the correlation coefficient may be unfamiliar to you, his statement that "r = 0.9525, a value which is statistically significant," does have meaning. Tell briefly what this means, in relation to the rest of the paper.

STUDIES ON ABSCISSION:
THE STIMULATING ROLE OF NEARBY LEAVES[1]

by F. N. ROSSETTER and WM. P. JACOBS

Current interpretations of the physiological basis of petiole abscission, as described in Wetmore and Jacobs (1953), are slightly unsettling to a researcher accustomed to interpreting plant behaviour in terms of the organismic concept. The view that each leaf is a physiological entity as far as abscission is concerned is quite unexpected if one stops to think of the great mass of evidence indicating dependencies and interdependencies among various organs of higher plants.

Scepticism as to the completeness of the view that auxin from its own leaf blade was the controlling factor in preventing abscission of a given petiole, led one of us (WPJ) to begin some of the investigations reported in this paper. It will be seen that as far as abscission is concerned, leaves do not act like independent "colonies" fastened to a common stem, but are influenced by the rest of the organism.

MATERIALS AND METHODS.—A variety of *Coleus blumei* Benth. was used for these experiments. The plants were grown as described in Jacobs (1952) and belong to the same clone as the plants used by Jacobs (1952), Wetmore and Jacobs (1953), and Jacobs and Bullwinkel (1953). The method of designating leaf numbers, etc. is the same as that used by Jacobs (1952) and is illustrated in Figure 1 of Wetmore and Jacobs (1953). Only single stemmed upright plants were used. Axillary buds and branches were removed. In their experiments Wetmore and Jacobs (1953) used a 'spiral'' arrangement of debladed leaves. That is, one

[1] Received for publication on August 7, 1952.
Report of work supported in part by an American Cancer Society Grant recommended by the Committee on Growth of the National Research Council and in part by funds of the Eugene Higgins Trust allocated to Princeton University.
Reprinted by permission from the *American Journal of Botany*, Vol. 40, No. 4, 276–280, April, 1953. Printed in U.S.A.

blade of each leaf pair was excised, and the debladed petioles were in a spiral arrangement on the stem (Figure 1). Although it is not stated explicitly, Myers (1940) apparently used a spiral arrangement, too. Such a spiral arrangement means that each debladed petiole has an intact leaf just above, just below, and immediately opposite it. The main patterns of leaf deblading used in our experiments were as follows:

1) Spiral deblading with plain lanolin on the petiolar stumps (Figure 1).

2) "Two-sided" deblading with plain lanolin on the petiolar stumps (Figure 1).

3) "Four-sided" deblading with plain lanolin on the petiolar stumps. That is, *all* leaf blades below the apical bud were excised.

4) "Four-sided" deblading with indoleacetic acid in lanolin on the petiolar stumps.

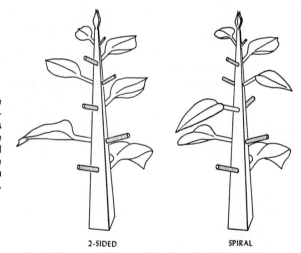

FIGURE 1. Diagrams to show the "spiral" and "two-sided" deblading patterns. One petiole of each leaf pair has been debladed. The debladed petioles are designated by stippling. The top debladed petiole (that is, the one nearest to the apical bud) is petiole no. 1; the bottom one (in these diagrams) is petiole no. 6.

2-SIDED   SPIRAL

In each pattern of leaf deblading, the leaves of the apical bud, which had not yet unfolded, were left untreated and untouched. In-doleacetic acid (IAA) was used in 1 per cent concentration in lanolin. The lanolin or IAA-lanolin was applied to the distal cut surface of the debladed petioles. The plants were checked every day for abscised petioles. A seemingly more uniform method than that used in Wetmore and Jacobs (1953) for testing leaves about to abscise was employed by applying a slight but relatively constant pressure to each petiole daily to ensure that the petiole was not being retained by only a few intact strands of vascular tissue, the cells of the abscission layer having already separated. The pressure was applied with a few strands of fine soft wire in a transfer-needle handle.

Except where noted under "Results," there were at least three replications of each experiment, with statistically significant results within each replication. Results were essentially the same in each replication, the variability in terms of absolute numbers being apparent from Figures 2–4. Five to fifteen plants were used for each treatment.

The "t" test was used as a measure of statistical significance. Statistical methods and terminology follow Snedecor (1946).

RESULTS.—To test the effect of nearby leaves on the abscission of debladed petioles, the four treatments described under "Materials and methods" were used. Results of a typical replication are shown in Figure 2 and Table 1.

In confirmation of earlier work, spiral deblading resulted in very fast shedding of the debladed petioles, no matter what their position on the stem.

The "two-sided" deblading, however, resulted in significantly longer retention of the debladed petioles. In the particular replication shown in Figure 2, there was a significant difference between the overall averages for petioles 1–7 in each treatment; but when differences between petioles at any *one* position were considered, both in

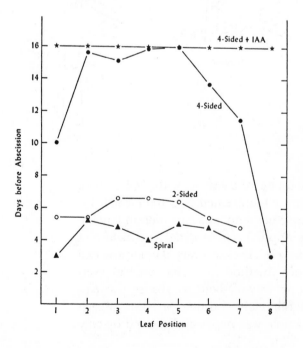

FIGURE 2. Influence of deblading pattern and of IAA-lanolin versus plain lanolin on the retention of debladed Coleus petioles at different positions along the stem. The horizontal line at 16 days for IAA-treated petioles shows that these petioles had not abscised when the experiment was ended after 16 days. Casual observation indicated that they were retained for periods approximating those in figure 6 of Wetmore and Jacobs (1953).

TABLE 1. INFLUENCE OF DEBLADING PATTERN AND IAA
ON RETENTION OF COLEUS PETIOLES

Petiole position	Spiral	Two-sided	Four-sided	Four-sided plus IAA[b]
1	3.0 ± 0.3 (5)	5.4 ± 1.9 (5)	10.0[a] (10)	16+ (10)
2	5.2 ± 0.6 (5)	5.4 ± 0.4 (5)	15.6[a] (10)	16+ (10)
3	4.8 ± 0.6 (5)	6.6 ± 0.2 (5)	15.1[a] (10)	16+ (10)
4	4.0 ± 0.3 (5)	6.6 ± 0.6 (5)	15.9[a] (10)	16+ (10)
5	5.0 ± 0.4 (5)	6.4 ± 0.2 (5)	16+ (10)	16+ (10)
6	4.8 ± 0.9 (4)	5.4 ± 0.2 (5)	13.7[a] (10)	16+ (8)
7	3.8 ± 0.5 (4)	4.8 ± 0.2 (5)	11.5[a] (8)	16+ (4)

[a] Denotes averages which include petioles which stayed on for 16 days or longer (records ceased after 16 days). These are minimum averages, since 16 days was taken as the true value in such cases.

[b] Every petiole treated with IAA stayed on for 16 days or longer.

The data in the table are expressed as "Average number of days before abscission ± standard error." The numbers in parentheses represent the number of petioles included in the averages.

the cited replication and in the other replications, significant differences were found only at positions 3, 4 and 5.

As might be expected from previous reports, "four-sided" deblading coupled with treatment of the petiolar stumps with auxin gave retention of all petioles, no matter what their position, for 16 days or longer.

Quite unexpected were the results obtained from "four-sided" deblading when plain lanolin was added to the cut surface. All petioles were retained much longer than with plain lanolin treatment of any other deblading pattern used. Even those petioles which already have abscission layers fully differentiated (i.e., No. 6 and 7) are retained longer than in the "spiral" or "two-sided" groups. This result was obtained with each of 5 replications.

These results demonstrate that the presence of other leaf blades speeds up the abscission of debladed petioles. The presence of an intact leaf just above, just below, and just opposite each debladed petiole (as in "spiral" deblading) gives faster abscission than having no intact expanded leaves above or below the debladed petiole in the same orthostichy, even though intact leaves are opposite as well as above and below at right angles (as in "two-sided" deblading). Slowest abscission of all results from having no intact leaves above, below *or* opposite the debladed petioles (e.g., "four-sided" deblading).

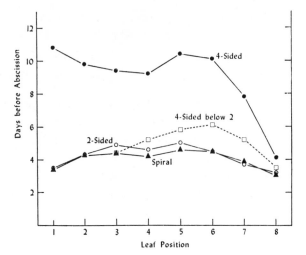

FIGURE 3. Influence of deblading pattern on the retention of debladed petioles which had been treated with plain lanolin on their distal cut surface.

In both the "two-sided" and "spiral" groups, intact leaves were opposite each debladed petiole as well as above and below it at the next nodes (i.e., on the adjacent orthostichies). In an attempt to separate effects of opposite leaves from leaves above and below, there was added to the three plain lanolin treatments used above a new treatment in which all leaves below the No. 2 position were debladed ("four-sided below 2"). Results are shown in Figure 3 and Table 2. It can be seen that the presence above the debladed petioles of intact leaf pairs 1 and 2, in addition to the intact leaf pair of the apical bud, markedly speeds up abscission of the subjacent petioles, although the petioles are retained longer than with "spiral" or "two-sided" deblading. Results from a similar experiment, but with all leaves debladed below leaf pair 3 in one of the treatments is shown in Figure 4.

TABLE 2.

Petiole position	Four-sided deblading	Four-sided below Leaf 2	Two-sided deblading
3	9.4 ± 0.4 (20)	4.4 ± 0.2 (20)	4.9 ± 0.2 (15)
4	9.2 ± 0.4 (20)	5.2 ± 0.2 (20)	4.6 ± 0.2 (15)
5	10.4 ± 0.4 (20)	5.8 ± 0.2 (20)	5.0 ± 0.3 (15)
6	10.1 ± 0.5 (20)	6.1 ± 0.2 (20)	4.5 ± 0.3 (15)
7	7.8 ± 0.7 (20)	5.2 ± 0.3 (20)	3.7 ± 0.2 (15)
8	4.1 ± 0.2 (18)	3.5 ± 0.3 (18)	3.2 ± 0.2 (12)

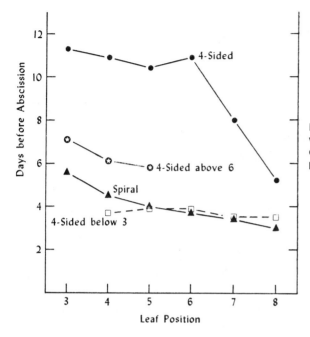

FIGURE 4. Influence of distal leaves versus proximal leaves on the retention of debladed Coleus petioles treated with plain lanolin.

It can be seen that the presence of all leaves at and above the leaf 3 position speeds up abscission in petioles 4–8 to the same degree as does the presence of one member of each leaf pair all the way down the stem.

To determine the effect, if any, of older proximal leaves, an experiment was tried with "four-sided" deblading below the apical bud as far down as leaf pair 6 (Figure 4, Table 3). With leaf pairs 6, 7, and 8 left intact, debladed petioles 1–5 abscised faster than their counterparts in the "four-sided" debladed controls. The difference was statistically highly significant for each petiole position.

(The special treatments described in the above two paragraphs were each tested only once.)

TABLE 3.

Petiole position	Four-sided deblading	Four-sided above Leaf 6	Spiral deblading
3	11.3 ± 0.3 (10)	7.1 ± 0.2 (10)	5.6 ± 0.3 (10)
4	10.9 ± 0.3 (10)	6.1 ± 0.3 (10)	4.5 ± 0.3 (10)
5	10.4 ± 0.4 (10)	5.8 ± 0.5 (8)	4.0 ± 0.2 (9)

DISCUSSION.—The experiments described above show that, contrary to the impression left by the experiments of Myers (1940), abscission of a petiole is not independent of nearby leaves and thus dependent solely on amounts of diffusible auxin coming from the leaf blade of that petiole. When the pre-abscission period of debladed petioles was determined, it was found that the more leaves there were around the debladed petiole and the closer they were, the faster the petiole abscised. Maximum retention of petioles treated with plain lanolin was obtained when all the leaf blades below the apical bud were excised.

The abscission-stimulating effect of intact leaves is not strictly polar in its movement, judging by the treatment in which leaves 6–8 were left intact (Figure 4). This treatment resulted in petioles 4 and 5 (just above the intact leaves) abscising significantly faster than in the "four-sided" controls, although they abscised significantly slower than did corresponding petioles which were just below intact leaves 1–3.

We have no evidence so far as to whether this difference is due to a polar difference in *movement* of the abscission-stimulating effect or to a difference in the *strength* of the effect as produced by old versus young leaves.

The general conclusions are confirmed by a more thorough investigation of the individual petioles. For instance, if the opposite and distal leaves are more important than the proximal (and older) ones in determining the speed of abscission, we could predict that debladed petioles 1 and 2 of the "spiral" and "two-sided" groups would fall off at nearly the same time, since a glance at Figure 1 shows that these petioles have exactly the same pattern of intact leaves opposite and distal to them. Corresponding to expectation, the pre-abscission period of "spiral" petiole 1 was not significantly different from that of "two-sided" petiole 1 in any of the three replications in which such data were recorded. The same was true of petiole 2 in the two treatments.

It is only in relation to debladed petioles 3, 4, etc., that there are differences between "spiral" and "two-sided" plants in the position and number of distal intact leaves; and it is between these petioles that significant differences in pre-abscission times first appear.

What is the physiological basis of this abscission-stimulating effect? The two most obvious hypotheses are: 1) that intact leaves produce some substance which speeds abscission of nearby petioles, 2) that removing leaf blades induces compensatory growth of the remaining debladed petioles, with a consequent inhibition of their abscission. Such compensatory growth in the petioles would scarcely be surprising in view of the complex compensatory relations described by Jacobs and

Bullwinkel (1953) for this same clone of *Coleus*. However, experiments bearing, perhaps, on the first suggestion are already in the literature. In view of the interesting finding of Addicott and Lynch (1951), who studied the abscission zone between a bean leaflet and its petiole, that IAA applied *proximally* to this abscission zone *speeded* abscission one might suspect that the intact leaves of *Coleus* speeded abscission of nearby debladed petioles by acting as sources of proximal auxin. Even the abscission-stimulating effect of leaves *below* the debladed petioles would not be unexpected, according to this auxin hypothesis, since it has been shown by direct transport tests that auxin can move acropetally in *Coleus* stems as well as basipetally (Jacobs, 1952; confirmed by Leopold, private communication). However, evidence that the leaves are *not* speeding abscission by means of their production of auxin can be deduced from Figure 5[2] and Figure 6 of Myers' (1940) paper. After "four-sided" deblading, one member of each petiole pair was treated with IAA-lanolin, the other member with plain lanolin. When results are compared with "four-sided" deblading in which all the petioles were treated with plain lanolin, one finds that treating half the petioles with auxin has no effect on the speed of abscission of the petioles treated with plain lanolin. We have obtained the same result with our clone of *Coleus* in a recent experiment. Thus, although IAA can substitute for the leaf blade in retarding the abscission of that leaf blade's petiole (Wetmore and Jacobs, 1953), it apparently can not substitute for a leaf blade in speeding the abscission of nearby petioles in *Coleus*.

First results from a series of experiments now in progress indicate that the active agent is a gas. Ethylene, a gas long known to speed abscission, is suspected to be the active agent. Results of these experiments will be reported in a later paper.

The following interpretation of the normal abscission pattern seems indicated by our experiments. Each leaf has acting on it an abscission-stimulating effect from the nearby leaves. Whether due to polar movement, or to a gradient among the leaves in the strength of this effect, the younger expanded leaves (leaf pairs 1–3) have a greater influence than the older leaves. This abscission-stimulating effect is opposed, for any one petiole, by the diffusible auxin moving into it from its attached leaf blade. The younger leaves normally stay on longest both because they are the greatest producers of diffusible auxin and also because they have a smaller number of abscission-stimulating leaves above them. The older leaves normally fall off very quickly

[2] The legends for Fig. 4 and Fig. 5 (Myers, 1940) should be exchanged.

both because they have a much larger number of abscission-stimulating leaves above them and because they produce the smallest amounts of diffusible auxin. The high correlation found by Wetmore and Jacobs (1953) between the production of diffusible auxin and the pre-abscission period of various leaves indicates that normally the auxin effect over-rides the other effect.

This physiological mechanism (elucidated here for the control of abscission) whereby a stimulator is balanced by an inhibitor, is probably of widespread occurrence in regulatory mechanisms for biological organisms. Such a "balancing" mechanism would be expected to have marked selective value. It ensures that the younger leaves not only stay on longer under normal conditions but that they also hasten the shedding of the older, less actively metabolizing leaves. Furthermore, it ensures that accidental loss of some of the leaves will be compensated for by the longer retention of the remaining leaves—a result which will be made possible by the very loss of the other leaves (and of their abscission-inducing effect).

Summary

Evidence that intact leaves speed up abscission of nearby debladed petioles has been obtained by deblading in various patterns a clonal stock of *Coleus* plants. The younger, still growing leaves have a greater abscission-stimulating effect than the older, fully-grown leaves, but whether this is due to polar movement of the effect or to stronger stimulus from the younger leaves has not yet been established. Reasons for thinking that the abscission stimulator is not auxin but may be ethylene are briefly discussed.

Department of Biology,
Princeton University,
Princeton, New Jersey

LITERATURE CITED

Addicott, F. T., and R. S. Lynch. 1951. Acceleration and retardation of abscission by indoleacetic acid. Science 114: 688–689.

Jacobs, W. P. 1952. The role of auxin in differentiation of xylem around a wound. Amer. Jour. Bot. 39: 301–309.

——, and B. Bullwinkel. 1953. Compensatory growth in *Coleus* shoots. Amer. Jour. Bot. In press.

Myers, R. M. 1940. Effect of growth substances on the absciss layer in leaves of *Coleus*. Bot. Gaz. 102: 323–338.

Snedecor, G. W. 1946. Statistical methods. Iowa State College Press. Ames, Iowa.

Wetmore, R. H., and W. P. Jacobs. 1953. Studies on abscission: the inhibiting effect of auxin. Amer. Jour. Bot. 40: 272–276.   END OF ARTICLE

QUESTIONS FOR DISCUSSION

1. Why was the study reported in this article undertaken?

2. This paper reports some experimental refinements over the techniques used in the last report. What were they?

3. The authors report they used the *t* test in this study. Can you use the *t* test on the data reported in Table 1 or in Table 2?

At the present time, much research is being done in attempts to further explain regulation in plant growth and development. Although there is very good evidence that flowering is controlled in at least some plants by hormones, no one has been able to isolate the hormone. Many plants will flower only when exposed to proper periods of light and darkness. The hormone in this case appears to be produced in leaves and is then translocated to buds where it causes the initiation of flower parts.

We know that some substance which is produced in apical buds of many plants inhibits the growth of axillary buds. If the apical bud is damaged or removed, the axillary buds develop. The hormone in this case appears to be auxin (IAA).

Some of the effects of auxins are shown in Figure 69.

SUMMARY

We have employed the basic processes of science as we explored some of the ways in which different parts of an organism are coordinated. We have seen that coordination is effected by means of nervous systems and by hormones. In plants, nervous tissue is not known. In most animals, regulation is maintained by interlocking reactions of the nervous system with the endocrine system, although elements of either system may be predominant in the control of many individual sets of reactions.

Control of the reactions of most animals is very complex, and as yet it is not well understood. The next section will consider in some detail a few examples of animal behavior.

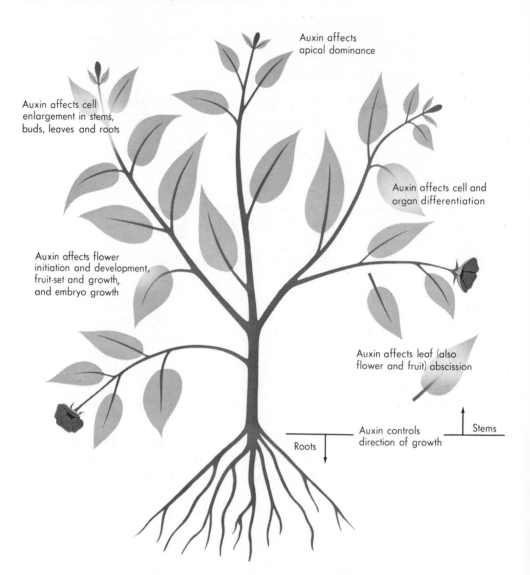

Auxin affects
apical dominance

Auxin affects cell
enlargement in stems,
buds, leaves and roots

Auxin affects cell and
organ differentiation

Auxin affects flower
initiation and development,
fruit-set and growth,
and embryo growth

Auxin affects leaf (also
flower and fruit) abscission

Auxin controls
direction of growth

Roots

Stems

FIGURE 69. Effects of auxins on plant growth and regulation.

SECTION 13 ANIMAL BEHAVIOR

At dawn of the last day of the third quarter of the moon and the first
day of the fourth quarter, during October and November, the Palolo
worm, *Palolo viridis,* swarms in the ocean near Samoa. At that time
only, the elongated posterior segments break away from the body of

the worm and spawn near the surface in an eerie luminescent display in which sperm and egg unite in fertilization. In Africa, the migratory locusts periodically transform from a scattered population of harmless, solitary individuals to a dense population of gregarious forms which assemble in vast numbers. These huge swarms of locusts travel over the continent of Africa, eating vegetation in their path, and destroying in excess of 90 million dollars worth of vegetation per year during locust outbreaks. The western newt *Taricha* breeds in streams of northern California but spends most of the year in the wooded hillsides above the streams. Each year the newts return to the same part of their home stream in order to breed, Newts will return to their stream site after being displaced as much as five miles from it. In Mexico and Central America, the orange-fronted parakeet (Figure 71) builds its nests and rears its young in active nests of a species of tree-dwelling termites. The bird hollows out a cavity in the termite nest, and the termites then seal off the inner exposed portions of the nest, separating the termites from the intruders.

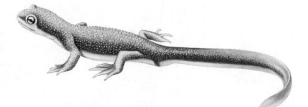

FIGURE 70. The western newt, Taricha, performs overland migrations of several miles to return to its home stream site to spawn.

How do the Palolo worms time their annual display so precisely? What causes the transformation of locust populations from solitary to gregarious phases, and what stimulates these insects to aggregate and migrate in such large numbers? By what means does the western newt locate the same spawning site each year? How does the orange-fronted parakeet recognize the termite's species in whose nest it builds? Is there some advantage to the parakeet in nesting in termite nests? How did this behavior evolve?

Answers to many of these questions are not yet available, and some of these behavior patterns, such as those of the migratory locusts, are currently under investigation. However, these are the kinds of phenomena, together with the questions which they raise, that form the basis for our study of animal behavior. The behavior of an animal cannot be defined precisely, but it includes activities carried on by the entire animal such as movement, sleeping, feeding, reproducing, and finding its way.

In trying to understand behavior, biologists want to know:

a. What patterns of behavior does a particular species possess?
b. How do the behavior patterns of an animal contribute to the survival of its species?

In attempting to understand specific behavior patterns, biologists ask:

a. Under what conditions does the behavior appear?
b. What anatomical, physiological, and biochemical features of the animal are involved in the expression of the behavior?
c. How does the expression of the behavior change with time and with the experiences of the animal?
d. How, if at all, is the pattern inherited?
e. How did the behavior pattern evolve?

FIGURE 71. In Central America, the orange-fronted parakeet hollows its nest out of active termite nests built on the branches of trees.

Satisfactory answers to these kinds of questions permit biologists to make generalizations about the nature of behavior in animals and to provide meaningful hypotheses for the investigation of human behavior.

Extensive studies of the behavior of a wide range of animal types among the invertebrates and vertebrates have already led to important generalizations. It is now clear, for instance, that the behavior exhibited by an animal at a specific time depends on the animal's developmental history, on its anatomical, physiological, and biochemical characteristics, and on the environment in which the animal finds itself. Biologists have gathered evidence which demonstrates that a specific animal's perception of its environment differs from our own or from that of other kinds of animals. This difference can be explained in the light of the organization of the nervous system and in the limitations and special capabilities of sense organs. Bees, for example, can detect infrared light; male gypsy moths can detect the female moth's scent in concentrations below that of the most sensitive instruments; rattlesnakes can detect temperature changes of $0.001\,°C$; Mormyrid fish can detect alterations in a surrounding electric field; and bats can hear ultrasonic sounds.

Furthermore, it is characteristic of many behaviors that the stimuli which elicit one particular behavior are highly specific. For example, during the breeding season, male yellow-shafted flickers will attack other males within their territories but not females. The specific stimulus by which a male flicker recognizes another flicker as a male is the presence of a black patch at the corner of the mouth, the so-called moustache. When a female flicker is captured and painted with a black patch at the corner of the mouth, she is attacked by her mate. When the patch is removed she is again accepted. Evidently, among all the characteristics by which male flickers differ from females, only the black patch is significant in eliciting attack.

FIGURE 72. The black patch at the base of the bill (left) serves as a signal for the recognition of male flickers. Females (right) lack this patch.

Young ducklings show escape responses to a short-necked, long-tailed model moving overhead. This model resembles the outline of

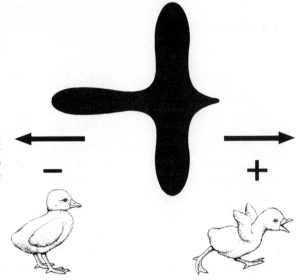

FIGURE 73. When the silhouette in the illustration is moved to the left over a group of ducklings, the birds show no response. When moved to the right, the birds attempt to escape.

a predator, such as a hawk. When the same model is moved in the opposite direction, now resembling the outline of a duck or goose, the ducklings show no response. We see that the animal detects certain specific features from the total environment; these features then are capable of acting on the animal as stimuli which elicit responses.

Leopard frogs feed exclusively on moving prey, especially insects, but they do not perceive insects as we do. Microelectrodes implanted into single fibers within the optic nerve indicate that impulses are sent to the brain by any small form with a convex edge moving

FIGURE 74. Frog vision differs greatly from that of man. A small form, with convex edges, moving toward the center of the frog's vision is recognized as an insect.

toward the center of the field of vision for that fiber. In this way, the frog recognizes an abstract of features common to a majority of insects and responds by leaping to catch its prey.

Kinds of Behavior

An important task of the biologist is to recognize and describe entities which exist in biological systems. Once the gene was identified as a biological unit whose duplication and transmission determined the pattern of inheritance of traits, rapid progress was made in our understanding of the mechanisms of genetics.

In similar fashion, it is desirable to recognize as an entity, or unit of behavior, the single behavioral action and to group as a category similar-appearing patterns or sequences of behavioral actions. Unfortunately, the internal mechanisms underlying a behavior pattern are usually not known. This places the grouping of similar-appearing behaviors on uncertain footing. Nevertheless, it appears desirable to classify behaviors in order to generalize about them, and several approaches have been developed.

Studies of the relationships of animal groups demonstrate that major groups (phyla) of animals differ in a graded fashion in the complexity of their nervous system. Correspondingly these groups differ in their ability to carry out elaborate and difficult tasks, to solve complicated kinds of problems, and to cope with changes in the environment in a flexible and reasoned way.

The following classification of behavior patterns appears to be based upon an increasing complexity of the underlying neural mechanisms. The study of the neurological bases underlying behavior is still so new that a more definite statement cannot be made. However, the broad distribution of these behaviors among animal groups reflects this assumption, since the more complex behaviors are found among those groups with the most highly developed and elaborate nervous systems, such as those found in birds and mammals.

Kinesis (plural, *kineses*) is an unlearned and undirected response of an animal to a stimulus. The response may be an increase or a decrease in the rate of turning, of movement, or of locomotion. For example, planarian worms are usually found in wet, dark habitats, such as the underside of stones, leaves, or twigs. When brought into the laboratory and placed in uniform conditions under dim, diffuse light, the planarian, *Dendrocoelum lacteum,* turns occasionally; the number of degrees turned, right or left, in one minute can be measured. When the light intensity is increased, the rate of turning increases. In this

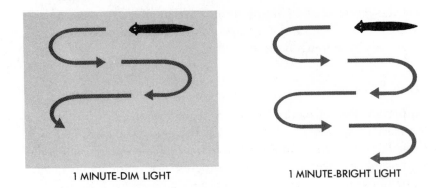

1 MINUTE-DIM LIGHT   1 MINUTE-BRIGHT LIGHT

FIGURE 75. *Kinesis.* The rate of turning of the planarian worm increases in response to an increase in light intensity. This increase in locomotion will continue until the worm accommodates to the new conditions. In nature, this behavior will often bring the worm into contact with leaves or stones where it finds shelter.

manner, the chances of the planarian's finding a darkened habitat are increased. (See Figure 75.)

Woodlice, sometimes called sowbugs or pillbugs, are land isopods (Crustacea) which occur commonly in rotten logs, under stones, and in moist leaf litter. The ability of most species to resist desiccation is poor. In the laboratory, *Porcellio scaber* and *Oniscus asellus* (two species of woodlice) are active in dry air and show a marked increase both in movement and in the rate of turning. In a saturated atmosphere they are almost motionless. Without any directional movement, this kinetic response enables woodlice to encounter and remain in moist habitats in their vicinity.

Taxis (plural, *taxes*) is an unlearned, oriented response of an animal with respect to a stimulus such as light, heat, gravity, chemicals, or surfaces. The response may simply be a moving toward or away from the source of the stimulus, as a moth responds by moving toward light, or the response may maintain some constant relationship to the stimulus, such as an ant's holding a 45° angle to the sun when returning to its nest. Taxes function to help the animal avoid unfavorable environmental situations, find more optimal conditions (such as warmer or drier locations), locate places in which food is more likely to be found, and maintain its position while flying, walking, or swimming.

For example, the brine shrimp *Artemia,* which lives in the Great Salt Lake and other inland saline waters, normally swims with its ventral side up. If a light is placed underneath it, *Artemia* swims dorsal side up. Thus *Artemia,* as well as many fairy shrimp such as *Bran-*

chinecta, use the direction of light, normally the sun's rays, as a stimulus for maintaining their swimming position. (See Figure 76.)

When placed in an aquarium at room temperature in which there is a tube warmed from 33°C to 35°C by circulating water through it, the medicinal leech *Hirudo medicinalis* moves directly toward the tube. By this means, this parasite can locate its host and move to its vicinity. The clam mite *Unionicola ypsilophorus* is a parasite of the freshwater, bottom-dwelling clam, *Anodonta,* in which it lives and feeds in the mantle cavity and on the gills. In an aquarium, lighted from above and containing only clam mites, the mites are photopositive and move to the top of the aquarium. When an extract of *Anodonta* tissue is added to the water, the mites become photonegative and move to the bottom where they would have a greater chance of encountering their host.

Several different patterns of orientation to a directional stimulus are now known to occur, enabling taxes to be further subdivided on the basis of the mode of orientation used. It is important to remember (1) that an animal is capable of responding to more than one stimulus at a time and (2) that variations in the internal state of the animal, such as those which occur with hunger, alter responses to stimuli.

Reflexes involve simple, stereotyped movements, usually only by part of an organism in response to a stimulus. The stimulus in a reflex is generally localized, the response following shortly after the stimulus. Reflexes invariably involve the central nervous system. The underlying

FIGURE 76. *Taxis.* Swimming position of fairy shrimp is maintained by reference to the direction of light. As the direction of light changes, the swimming position changes accordingly. Under usual conditions, light comes from above and the shrimp swim upside down.

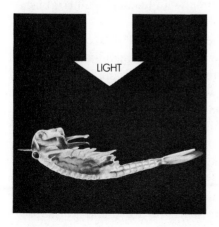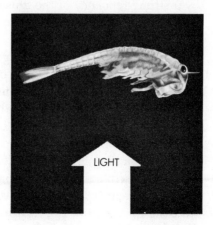

neural mechanism may be relatively simple, as in the *stretch reflex* affecting flexion and extension of muscles in vertebrates, or it may be complex, involving portions of the brain as in *righting reflexes*. Reflexes can be either innate or learned, as in the conditioned reflex. In man, for example, both the pupil reflex in response to increasing light intensity and the knee-jerk reflex are well known. In the frog and in the cockroach, a drop of dilute acetic acid placed on the dorsal region will elicit scratching movements by a hind leg, even in anmials which have been decapitated.

Instincts form the basis of another behavioral classification. The concept of instinctive behavior is very old, and its meaning has undergone modification over the years as a result of healthy criticism. As currently used by animal behaviorists, instincts refer to those stereotyped, sequentially patterned acts the expression of which is typically not influenced by the past experiences of the animal. Because of the emphasis on the *unlearned* nature of instinctive behavior, classifying a behavior as instinctive is essentially a statement about the development of that pattern. The best studied examples of instinctive behavior are to be found among the birds, fishes, and insects.

Because of their tendency to be specific for each species, instincts are ideal taxonomic characteristics and highly useful in the study of the evolution of behavior. The inflexibility of some instinctive patterns is nicely shown in the following example from the work of Dr. T. C. Schneirla.

The army ant of the American tropics conducts above-ground foraging raids. These raids are conducted by the workers and soldiers which form tight columns reaching over 900 feet in length and containing thousands of ants. In studying the organization of these columns, it has been found that the continuity of the column is maintained by a strong tendency of the ants to follow a chemical trail established by the ants ahead of them in the column.

One day, quite by chance, the leading end of a column under observation began to circle a large wooden post which had once served as a building support. The rest of the column automatically followed producing a complete ring of ants which continually circled the post for an entire day and terminated only after the exhaustion and death of the foragers. Under natural conditions, the substrate surrounding trees is so disrupted by roots and other irregularities that this circular column of milling behavior, or "suicide mill," seldom occurs.

The usefulness of instinctual behavior studies to taxonomy has become evident in the study of similar, closely related species. While studying the digger wasp, *Ammophila campestris,* Reverend Adriaanse of

FIGURE 77. When army ants are maneuvered into forming a circular trail, they continue to circle, each ant following the worker preceding it, until the assemblage dies of exhaustion. This lack of flexibility in behavior is a characteristic of many instinctive patterns in lower animals.

the Netherlands noticed some individuals in the population with strikingly different behavior traits involving care of the young. In typical *A. campestris* individuals, the female digs a hole, deposits an egg, collects paralyzed insects which it places with the egg, and then closes the hole. Careful study of the behavior of the aberrant individuals in the population led to the discovery of a new species of wasp, now called *Ammophila pubescens*.

Table 18. BEHAVIORAL DIFFERENCES BETWEEN THE TWO SPECIES

Behavior	Ammophila campestris	Ammophila pubescens
Nest hole filled with material from	nearby digging	flown in
Choice of food for developing young	sawflies	caterpillars
Sequence of egg-laying and provisioning of nest	first egg, then sawflies	first caterpillars, then egg

The way in which instinctive patterns of behavior are used to shed light on the evolution of behavior is shown by the work of Dr. Howard Evans of Harvard University on the prey-carrying of wasps. In a solitary wasp such as the digger wasp *Ammophila,* discussed earlier, the female digs a shallow nest in sand or soil, closes the nest, and then proceeds to capture insects. These prey are paralyzed by stinging and then are carried to the nest. After the nest is reopened, the prey is inserted and an egg is laid on the side of the abdomen. The prey serves as food for the developing larva.

Primitive wasps such as spider wasps, Pompilidae, capture only spiders, which are seized in the mandibles and dragged backwards into the nest. Some spider wasps have evolved the ability to straddle the spider and therefore can move forward by walking or by a series of short hops, but the prey-carrying behavior of spider wasps has certain disadvantages. Before the female can reopen her nest to deposit her prey she must release it and free her mandibles which are required in nest digging. While the female is thus occupied, foraging ants may carry off the prey, or parasitic wasps or flies may deposit their eggs on it. Either possibility can be disastrous for the young larva.

Digger wasps, Sphecidae, and vespid wasps have evolved the ability to hunt smaller insect prey which are transported in flight, held by the mandibles and legs. These insects can therefore be gathered from greater distances than large spiders which must be dragged over the surface, and nest provisioning can be accomplished more rapidly. In four groups of digger-wasp species, the prey is carried slung under the body and held by either the middle or the hind legs, sometimes both. In two groups of digger-wasp species, the prey is held by the end of the abdomen, either impaled on the sting or grasped by a kind of clamping, freeing the legs completely. Now the female need not release her prey while she reopens her nest, permitting fewer opportunities for predators or parasites to attack.

The phylogeny of the major groups of wasps has been investigated on the basis of the structural characteristics of the larvae and adults. Accompanying this evolution of structure has been the evolution of more efficient means of transporting prey to the nest, described above, of nest-building behavior, and of the selection of different kinds of prey.

Learning is a physiological process which results in changes in behavior as a result of experience. These changes are usually adaptive and of sufficient duration so that we exclude the effects of fatigue, sensory accommodation or injury, as well as maturational change. Because of the importance of learned behavior to man, the learning process has been intensively investigated, principally by psychologists, who recognize several types of learning: trial and error learning, habituation, conditioning, insight learning, reasoning, and so on.

In the 1930's Dr. Konrad Lorenz of Germany, an eminent student of animal behavior, divided an egg clutch of a Greylag goose into 2 groups. One group was incubated by the goose; the other was placed in an incubator. Upon hatching, the first group followed the mother; the second group, upon seeing Lorenz, followed him. When the two batches

FIGURE 78. The evolution of wasps.

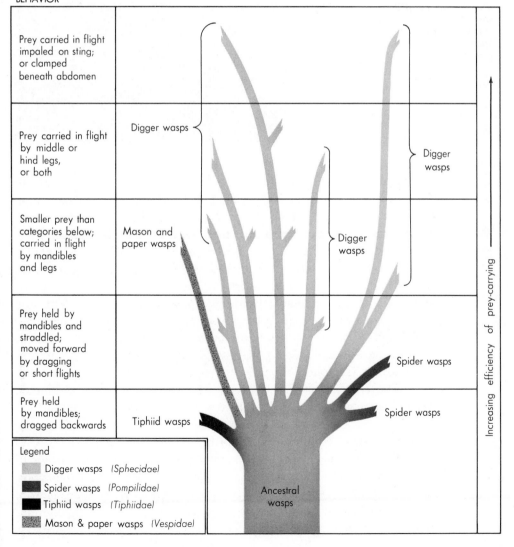

BEHAVIOR

Prey carried in flight impaled on sting; or clamped beneath abdomen		
Prey carried in flight by middle or hind legs, or both	Digger wasps	Digger wasps
Smaller prey than categories below; carried in flight by mandibles and legs	Mason and paper wasps	Digger wasps
Prey held by mandibles and straddled; moved forward by dragging or short flights		Spider wasps
Prey held by mandibles; dragged backwards	Tiphiid wasps	Spider wasps

Increasing efficiency of prey-carrying

Legend

Digger wasps *(Sphecidae)*

Spider wasps *(Pompilidae)*

Tiphiid wasps *(Tiphiidae)*

Mason & paper wasps *(Vespidae)*

Ancestral wasps

FIGURE 79. The imprinted ducklings are following the first object which they saw after hatching. Instead of their mother, they are following Dr. Konrad Lorenz, a well-known student of animal behavior.

of goslings were marked and placed together, they each followed the first moving object they had seen upon hatching. This phenomenon, called imprinting, is a type of learning which falls close to the margin of the continuum between instinctive behavior and learned behavior.

Further work has verified and extended Lorenz's results. Laboratory investigations with mallard ducks indicate that there is a critical age, a matter of hours after hatching, during which ducklings can be imprinted. While mallard ducklings respond to the size, form, and color of the object, ducklings of wood ducks, which nest in trees, are imprinted to the parent bird in response to sounds made by the parents. Imprinting appears greatest in those species in which the young can move about and follow the parents shortly after birth or hatching.

The classification of behavior into categories, such as kineses, taxes, instincts and learned behaviors, and the intensive analysis of a large number of each kind of behavior, will eventualy permit further generalizations about these kinds of behavior.

The question, "How does behavior contribute to the survival of the species?" provides another way of examining animal behavior. Here, we group behaviors that appear to be serving similar *functions,* such as orientation, communication, habitat selection, feeding, elimination, sanitation, defense, reproduction, care of young, and so forth. As examples of this approach to the study of behavior, we shall consider the questions "How do animals orient in space?" or, more simply, "How do animals find their way?" and "How do animals communicate?"

Orientation in Animals

In 1793, the Italian biologist, Spallanzani, blinded several night-flying bats and found that they not only flew as well as before, but could also catch food, for the stomachs of blinded and recaptured bats were packed with flying insects. However, when Spallanzani plugged their ears, these bats collided with buildings, with trees, and with other obstacles. Evidently bats orient in space and locate objects by using their ears, but how?

FIGURE 80. Between 13 and 16 hours after hatching, the critical period, ducklings are most susceptible to imprinting. After that time, imprintability declines rapidly.

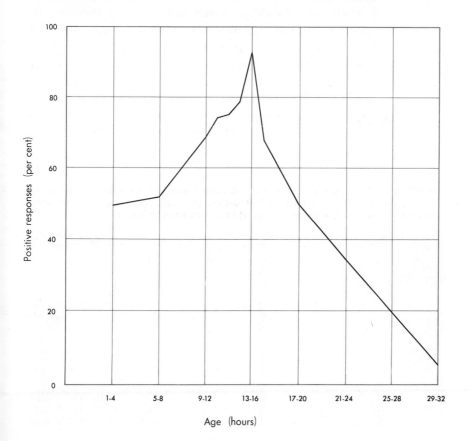

The problem remained unsolved until the 1940's when G. W. Pierce, at Harvard, using instruments sensitive to high-frequency sound, found that night-flying bats emit loud noises in the form of short, frequent bursts in the high frequency range of 10,000 to 150,000 cps (cycles per second). (By contrast children can detect sounds up to approximately 20,000 cps although many adults cannot detect frequencies higher than 10,000 cps.) When their mouths were covered, the bats tested could not orient. Evidently bats orient by some technique requiring emission and reception of high frequency sounds.

FIGURE 81. Many bats possess the remarkable ability to fly at night, avoiding obstacles and catching tiny flying insects.

Professor Donald Griffin of Harvard has demonstrated that the detection of echoes, or *echolocation*, is the simplest, and therefore the preferred explanation of the way bats employ these sounds in orientation. As a bat flies along at night, it emits pulses of high frequency sound which strike objects at a distance in front of the bat. The time required for the echo to return to the bat's ear is proportional to the distance of the object from the bat at the time the sound was emitted. By detecting these echoes, bats are continually informed not only of the presence of large obstacles but also of tiny objects such as flying insects.

FIGURE 82. When a bat is released into a darkened room across which wires are strung, the bat is able to fly back and forth without touching the wires. The man at the right holds a microphone to detect high frequency cries which the bat emits as it flies.

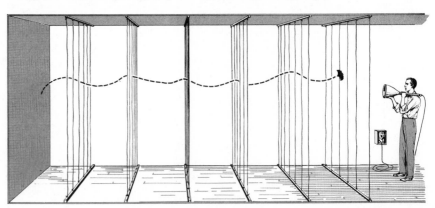

This ability to detect objects at a distance is very highly developed in bats. When released into a completely darkened room with wires strung in several directions, bats could fly about and avoid collision with walls, ceiling, or wires. When the wire diameter was reduced, Professor Griffin found that bats could avoid wire as thin as 0.12 mm in diameter! This extraordinary behavior is only possible because of specializations in the structure of certain organs: the bat's sound-producing organ, the larynx, and the ear. (Look at Figure 83 below.) Examination of the bat's brain shows great development and relative enlargement in those areas concerned with perception of sound and a high development of the neuromotor system related to hearing. There is a corresponding reduction of the sensory and motor systems not related to sound perception.

Since this discovery, biologists have found evidence of orientation by echolocation in porpoises, in cave-dwelling birds, and in the shrew, a rodent that lives underground. Surprisingly, certain night-flying moths, typical prey of bats, are capable of detecting the high-frequency cries of bats. They respond to these cries by performing power dives, passive dives, or a series of loops and turns which gradually carry them to the ground. (See Figure 84, page 300.) These maneuvers aid moths

FIGURE 83. The echolocation ability of bats is only possible because of anatomical and physiological specializations which accompany the behavior.

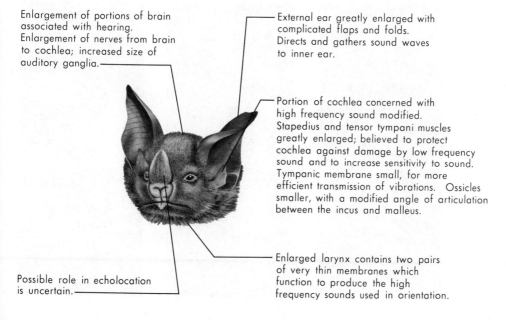

Enlargement of portions of brain associated with hearing. Enlargement of nerves from brain to cochlea; increased size of auditory ganglia.

External ear greatly enlarged with complicated flaps and folds. Directs and gathers sound waves to inner ear.

Portion of cochlea concerned with high frequency sound modified. Stapedius and tensor tympani muscles greatly enlarged; believed to protect cochlea against damage by low frequency sound and to increase sensitivity to sound. Tympanic membrane small, for more efficient transmission of vibrations. Ossicles smaller, with a modified angle of articulation between the incus and malleus.

Possible role in echolocation is uncertain.

Enlarged larynx contains two pairs of very thin membranes which function to produce the high frequency sounds used in orientation.

in escaping from insect-hunting bats. Mites which attack the hearing organs of these moths are never found on more than one hearing organ.

Worker fire ants are recruited and guided to a source of food by means of an *odor trail*. The trail-marking substance is laid by the workers that first discover the food. It consists of a volatile chemical secreted by a small abdominal gland, Dufour's gland. The trail substance is released in minute amounts when the sting is touched to the ground. Ants encountering this attractant simply follow the trail until the food source is reached. An extract of Dufour's gland can be used to create artificial trails, and worker fire ants will follow these artificial trails in circles, loops, or other patterns.

The western newt, mentioned earlier, probably recognizes its breeding site by means of the specific odor of the stream. When the newt's olfactory receptors are destroyed, the newt cannot home correctly.

Many animals—including migratory birds, bees, sea turtles, and some fish—are able to orient by means of the direction of the sun. Certain fish, such as the African river fish, *Gymnarchus,* can orient in

FIGURE 84. In response to high frequency sounds, such as the cries of bats, certain nocturnal moths drop to the ground. In the diagrams below, ultrasonic sound was emitted at the arrow, and the down course followed by three moths is shown. Moth flights include a passive dive (left), power dive (center), and double roll and dive (right). (Modified from photos of K. D. Roeder.)

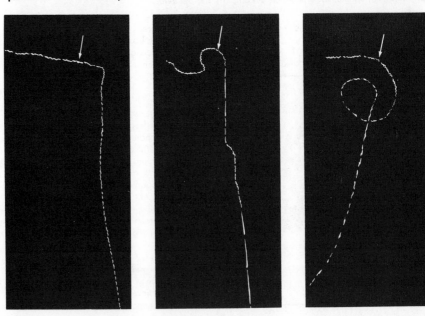

muddy waters by detecting distortions in an electrical field which the fish generates about itself. These distortions are caused by the presence of fish or other obstacles.

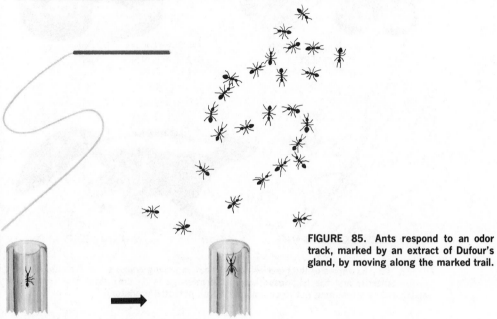

FIGURE 85. Ants respond to an odor track, marked by an extract of Dufour's gland, by moving along the marked trail.

Communication Among Animals

The complex social organization of groups of animals, such as bees, monkeys, and many kinds of birds, demands from these animals an ability to transmit information among members of their group. This information may concern food sources, presence of enemies, boundary of territories, readiness to mate, and so forth; however, even animals whose relationships with other members of their species are very limited, oysters and barnacles for example, need to communicate essential information such as that which signals the release of gametes. Therefore, ability of animals to communicate with other members of the species occurs very widely. This inter-animal communication, however, is not limited to members of the same species. A parent killdeer feigning a broken wing to lure a predator away from its nest and the walkingstick, an insect whose structure and posture, when disturbed, resembles a twig, are attempting to communicate with their predators. Duck hunters with their decoys, fishermen using lures or flies, and flowers, scented to attract pollinating insects and birds, are also involved in interspecies communication.

Monarch   Viceroy
MIMICRY

Hog-nosed snake
DEATH FEIGNING

Walkingstick
CRYPTIC RESEMBLANCE

Skunk
WARNING COLORATION

FIGURE 86. Examples of interspecies communication among animals. In the case of the Viceroy butterfly and the hog-nosed snake, the message is an untruthful one, while the walkingstick is attempting not to communicate with potential predators.

Animals communicate when one animal produces a signal, measurable in the language of physics or chemistry, which affects the behavior of another animal. Some animals can also communicate with themselves, as do dogs when they encounter their own scent on a post or tree. We may classify the kinds of animal communication on the basis of the receptor involved: chemical or olfactory, visual, acoustical, tactile or mechanical, and electrical. There is some overlap between acoustical and mechanical types of communication. Signals can also be classified as transient, exemplified by the bark of a dog, or persistent, such as the black "moustache" of a male yellow-shafted flicker.

As an example of the ways in which animals communicate, we shall examine the behavior of a honeybee scout returning to the hive with information about the distance, direction, and scent of a source of food. More than 20 years ago, Dr. Karl von Frisch of Germany, using glass-sided observation hives, noticed that returning scout bees performed a kind of dance on the vertical honeycomb within the hive. Worker bees eagerly followed the scout, touching its body with their antennae. From information received during the dance, the other bees are able to locate the food source, but precisely what are the signals by which this information is communicated?

By training scout bees to feed at food dishes containing scented sugars at increasing distances from their hive, von Frisch noticed that for food sources up to approximately 80 meters scout bees perform a "round dance." After perceiving this dance and the scent of the food clinging to the scout bee's hairs, nest bees are able to locate the food source after searching in all directions close to the hive. Evidently, the round dance contains the information, "There is a source of food close to the hive." The vigor and duration of the dance denotes the quality of the food.

For food sources beyond 80 meters, von Frisch noticed that the dance suddenly switched to a figure-eight pattern, or "tail-waggling dance." By placing food not only at varying distances from the hive but also in different directions, von Frisch was able to determine that the tail-waggling dance instructed nest bees not simply how far to fly for food, but in what direction. Since the tail-waggling dance is performed on the vertical honeycomb of the darkened hive, the scout bee obviously cannot "point" toward the food. Instead, the dancing bee indicates the angle at the nest which is identical to the angle between the food source and the sun. In describing this angle, bees use a simple transposition. The direction toward the sun is indicated by the top of the vertical honeycomb; the direction away from the sun by the

FIGURE 87. Scout bees perform food dances on the vertical honeycomb of the hive. The round dance (right) is performed when a food source is within 80 meters. This dance contains no information concerning direction. The waggle dance (left) conveys information as to both distance and direction of food sources from the hive.

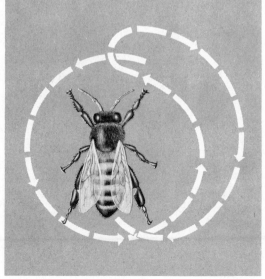

bottom. Thus, toward the sun is up, or away from gravity; away from the sun is down, or toward gravity. The transposition is therefore from a phototactic to a geotactic orientation. To indicate a food source 45° to the right of the sun's position, the scout bee orients the tail-waggling dance by moving in an angle 45° to the right of the vertical.

Scout bees can perform a correctly oriented tail-waggling dance even when the sun is obscured by clouds. If a patch of clear sky is visible, the direction of polarized light vibration indicates the sun's direction. If the sky is completely overcast, bees can, nevertheless, perceive the sun's position by the slight increase in the amount of ultraviolet light which penetrates the clouds.

The rhythm of the dance changes with increasing distance of the food source from the hive. By measuring the frequency of tail-waggling runs during each dance, von Frisch has determined that the number of runs per 15 seconds decreases as the food source is placed farther and farther away. More recently, biologists have found that during the tail-waggling dance, scout bees emit bursts of low-frequency sounds, each burst composed of approximately 32 pulses of sound per second. With increasing distance of food from the hive, both the length of the burst and the number of pulses per burst in-

FIGURE 88. The waggle dance informs hive bees about the direction of food sources by indicating on the vertical honeycomb the angle between the sun, hive, and food. Toward the sun is indicated by dancing upward, away from the sun by dancing downward.

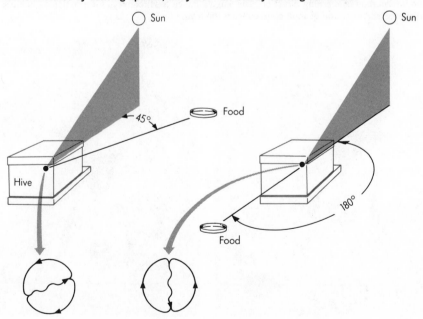

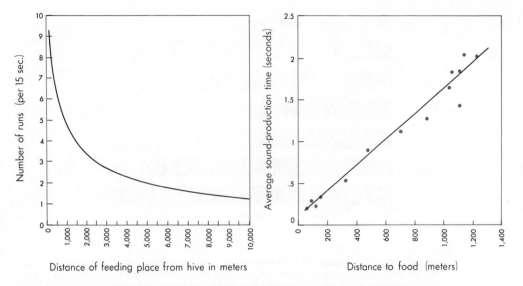

FIGURE 89. With increasing distance of food from the hive, the speed of the waggle dance decreases (left), and the duration of sound bursts made by scout bees increases (right).

crease. Thus, nest bees are furnished with both auditory and tactile signals which are correlated with food distance.

Variations in the pattern of food dancing exist among the races of honeybees. The Italian race, for example, indicates direction of food sources even for short distances from the hive by a distinct "sickle dance." For each of the races investigated, both the distance indicated by the round dance and the relationship between tail-waggling frequency and distance vary.

How could such a remarkable and precise behavior pattern have evolved? When von Frisch first announced his discoveries—the ability of bees to communicate distance and direction by means of food dances, to detect and orient by means of polarized light, and to transpose a light orientation to a gravity orientation—they were not known in other animals or in closely related forms. The pattern is essentially an elementary form of map-making and map-reading. In stingless bees, for example, the scout bee is capable of alerting and recruiting nest bees for foraging, as is the honeybee, but a pilot bee must guide the recruits to the food source over a scent trail which the scout bee has established.

Recent investigations, however, have shown that certain flies have a simple, light-oriented food dance, that ants and a species of dung beetle show a transposition from a phototactic orientation to a

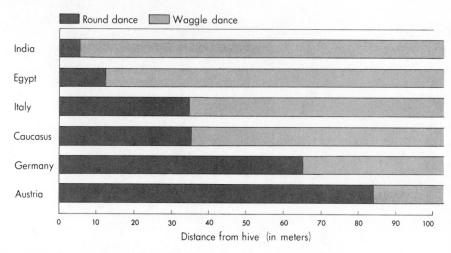

FIGURE 90. The distance of food from the hive, at which round dancing ceases and waggle dancing begins, differs among the separate races of honeybees. This difference suggests a genetic basis for the determination of the pattern of bee dances.

gravity orientation, and that several arthropods can detect and orient to polarized light. Therefore, we can surmise that the evolution of communication in bees has involved the selection and gradual refinement of behavior traits and sensory capabilities already present among the arthropods.

The use of chemical substances to transmit information is widespread among animals, including most mammals. Mammals employ scent in a number of ways: as sex attractants in rats; as repellents in skunks; to mark territories as in dogs; for recognition, as in elk which can thus identify their own young. Chemical substances used in animal communication are called *pheromones,* and a number of these pheromones have been isolated and chemically identified. The pheromone of the female gypsy moth, *gyplure,* is used as an attractant to enable males to locate females. It has been synthesized and is effective in gypsy-moth control.

While most pheromones act as stimuli eliciting behavioral responses, some alter physiological mechanisms which in turn alter behavior. In the migratory locust, mentioned earlier, adult male, gregarious-phase locusts secrete a volatile substance which accelerates maturation in young locusts. Termite and honeybee queens secrete a pheromone which inhibits production of additional queens.

The use of pheromones for communication presents special difficulties not present in acoustical or visual communication. Olfactory signals tend to persist; they have a long fade-out time and are therefore unsuitable for conveying information about position, change, or mood. Yet, their very persistence makes pheromones ideal for trail-marking, and they are so used by ants and certain bees. Furthermore, olfactory signals are capable of only limited modulation so that a multiplicity of signals, as in bird calls, is not possible for animals with only one or two pheromones.

The outstanding advantage to their use lies in the fact that some pheromones are capable of stimulating receptors at exceedingly low concentrations (approaching a few hundred molecules) and of acting at considerable distances from their point of origin. Thus, males of certain moth species have been lured to females from distances of two and one-half miles. German police dogs can distinguish between two trails left by dogs 30 minutes apart. The skin of a wounded minnow releases a pheromone called *schreckstoff* which causes other minnows to disperse, thus avoiding the predator. A 2-microgram skin fragment containing approximately 0.001 micrograms of *schreckstoff* is enough to disperse an entire school of minnows from a feeding place.

Other modes of communication among animals, such as posture, gait, and expression, convey important visual signals. A wolf's facial expression and the position of his tail inform other wolves about the mood of the individual (aggressive, submissive, and so forth). An elk threatens his attacker by holding his head high and his ears back, while stamping with his front feet. An elk's frozen stance denotes surprise, and a distinct, high stepping, marching gait denotes warning to other members of the herd, perhaps of the approach of a predator.

FIGURE 91. Posture, gait, and motion supply important signals in interspecies communication. In wolves, the position of the tail, the facial expression, and the movement of the ears convey information about the mood of the animal.

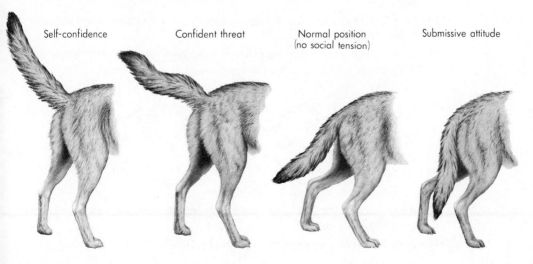

Self-confidence   Confident threat   Normal position   Submissive attitude
                                               (no social tension)

Certain species of mormyrids and gymnotids, as well as tropical South American and African freshwater fish, possess weak electric organs which are believed to function in communication as well as in orientation. When a second mormyrid was placed into a tank containing an already established resident, both fish released synchronous discharges, indicating that they had detected each other's presence.

QUESTIONS FOR DISCUSSION

1. Why is it difficult to define behavior precisely?
2. Define the term *stimulus*. Do all stimuli arise outside of an animal body? If not, give examples of internal stimuli. Can an animal perceive a stimulus but not respond to it? What is your basis for concluding this?
3. In addition to the gene and the behavioral types mentioned in the text, what other biological "entities" are recognized and used by biologists?
4. Why is the simplest explanation of bat orientation, echolocation, the preferred explanation? Would *hypothesis* be a suitable substitute for the word explanation as used here?
5. Animals find their way by means of landmarks, the position of the sun, and the detection of echoes. What other means of orientation might be used?
6. What reasons can you give for susceptibility of imprinting being greatest in the species in which it is found? What hazards exist for a species in which the young are capable of being imprinted?
7. What other functional categories of behavior can you suggest?
8. In what ways do you think behavior might affect the future evolution of a species of animal?
9. What kinds of structural and physiological specialization would you expect to accompany the exceptional olfactory activity of animals, such as rats, dogs, and pigs?
10. Give 3 examples of animal communication that would correspond to "lying" in human communication.
11. Draw a diagram of the relationship of food, hive, and sun which would exist if scout bees performed the following waggle-dance patterns.

 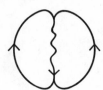

12. What would be the pattern of dance expected when the sun was *directly* overhead above the hive?
13. Does the human species communicate by means of odors?

SECTION **14** ANALYSIS OF BEHAVIOR

In the preceding material we saw that it was possible to generalize about the behavior patterns of animals and to group similar-appearing kinds of behavior, such as kineses and taxes, and patterns that were alike in function. From the total repertoire of behavior traits of a single species, we now wish to select a single pattern and to analyze this pattern in detail.

A careful description of the behavior is required first. Precisely what does the animal do? Is there a sequence or pattern of acts? Is this sequence invariably maintained, or is there variation in the order of the actions of the pattern?

How does the expression of the pattern vary with time of day? with the season? with the life cycle of the animal?

Is the pattern modified by the earlier experience of the animal? What role does learning play in the expression of the behavior?

What is the nature of the stimulus which elicits the behavior? To what, precisely, is the animal responding? Does the environmental setting of the animal affect the nature of the response? What receptors are involved in the response? Where are they located?

In what way does the internal state of the animal effect the response? What is the role of nutrition, hormones, and other chemicals in effecting the response?

Is the ability to express the pattern inherited? If so, in what way does the genetic mechanism control the behavior?

What is the function of this pattern in the life of the animal?

How did the pattern evolve? How does the expression of the pattern affect the evolution of the animal?

For our analysis, we shall select unlearned behavior—orientation movements in planaria.

INVESTIGATION 33
Planaria Sense and Respond to Gravity

The planarian is a common, freshwater, free-living flatworm. If you have not yet studied it and are unfamiliar with it, you should consult a general zoology text for a description of this animal including its position in the animal kingdom.

For our purpose you should know that planarians are multicellular and that they have a primitive nervous system consisting of two cerebral ganglia and a "ladder" type nerve cord.

The responses of planaria are often referred to as taxes.

MATERIALS (per team)

1 13 × 100 mm test tube and cork stopper
1 planarian (brown planarian, *Dugesia tigrini*)
1 medicine dropper
1 dissecting needle
1 marking pencil
1 test-tube rack or small bottle

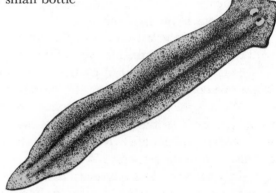

FIGURE 92. A planarian—a freshwater, free-living flatworm.

PROCEDURE

Obtain a small test tube (around 13 × 100 mm) and cork stopper. From the planaria culture bottle, suck up one worm into a medicine dropper and *gently* release the water containing the animal into the test tube. Unless this transfer is made quickly, the planarian may fasten itself to the inside wall of the medicine dropper. It may be dislodged by gently prodding it with a dissecting needle inserted through the open end of the dropper and squirting the water *gently* out as the worm is loosened.

Dechlorinate tap water by allowing it to stand several days in an open container or by treating it with a commercial dechlorinating compound. Add this water to the test tube until it is very close to being filled. Now, insert the cork stopper; at the same time, attempt to exclude all air or to trap as small an air bubble in the test tube as possible.

With a wax crayon (or other marking device), make a mark at the midpoint of the test tube. Also, take a small piece of aluminum foil and mold a light-tight cap that will cover half the length of the test tube. It is easier to mold this over the rounded end of the test tube than to attempt to do so over the corked end. Place the molded cap aside for the moment.

Place the test tube (corked end up) in a test-tube rack, or in a small bottle that will serve to hold it erect on the desk before you. Observe the activity of the planarian and clock the length of time that it spends in the top half (above the mark in the middle of the test tube) and in the bottom half of the test tube. Be sure that the test tube is as evenly illuminated with diffuse light from all sides and down its complete length as possible. Why? Continue your observations and your timing for at least 10 to 15 minutes to be certain that you have obtained an answer to the question being investigated. Record your data and conclusion.

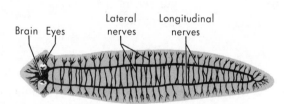

FIGURE 93. The nervous system of a planarian.

When you are certain that you have obtained an answer to the question, hold the test tube horizontally until the flatworm has moved to the middle of the tube (approximately). Then place the test tube upside down (with cork stopper down) and replace it in the rack or bottle. Again, observe and time the reactions of the planarian with the test tube in this position for 10 to 15 minutes. Record your data and conclusion.

QUESTIONS FOR DISCUSSION

1. Are the results obtained from observing a single planarian reliable?
2. Are the results obtained by the whole class more or less reliable?
3. Why, or why not?
4. Are the experimental results repeatable?
5. On what evidence do you base your answer to the question above?
6. Do the class results indicate that the planarian exhibits a positive (toward the stimulus source) or negative (away from the stimulus source) geotaxis?

7. What physiological basis can you suggest to explain the planarian's reaction to gravity?

8. Of what value are these reactions to the planarian?

INVESTIGATION 34
Planaria and Light

MATERIALS (per team)

    1 13 × 100 mm test tube and cork stopper
    1 piece foil wrap 12 × 12 cm.
    2 glass slides
    1 medicine dropper
    1 dissecting needle
    1 planarian

PROCEDURE

Hold the test tube containing the flatworm horizontally until the worm is approximately in the middle of the tube. Then, lay the tube on a white sheet of paper on the desk before you. If the cork elevates that end of the test tube slightly, place one or two (the second stacked on top of the first) glass slides under the opposite end of the test tube in order to make the tube as nearly horizontal as possible. Why?

Observe the behavior of the planarian with the test tube in this position and time the period it spends in each end of the test tube. Continue for approximately 10 minutes or until you are certain that you have ascertained whether or not the flatworm shows a preference for one or the other end of the test tube. Be careful that the test tube is evenly illuminated with diffuse light (room light) along its entire length. Why?

At a time when the planarian is approximately in the middle of the tube and is headed toward the corked end, slip the aluminum-foil cap over the rounded end of the test tube and replace it horizontally on the desk in front of you as before.

Observe the behavior of the worm and clock the length of time it spends in each end of the test tube. Continue these observations for at least 10 to 15 minutes to be certain that you have obtained an answer to the question being investigated. Record your data and conclusion.

QUESTIONS FOR DISCUSSION

1. Why was it necessary to prop up the rounded end of the test tube with one or two glass slides?
2. Why was it necessary to be careful about even illumination of the test tube along its entire length?
3. Are the experimental results repeatable?
4. Do the class results indicate that the planarian exhibits a positive or negative phototaxis?
5. Would you expect the response to be the same under a great variety of different light intensities? Design a simple experiment by which you might gather experimental evidence designed to answer this question.
6. What parts of the structure of the planarian are involved in its response to light?

INVESTIGATION FOR FURTHER STUDY

1. Consider that, in its natural habitat, the planarian is continually being stimulated both by gravity and by light of different intensities. How does the flatworm adjust to this stimulation?
   Design a simple experiment whereby you might test the relative effects of these two sources of stimulation upon the planarian.

INVESTIGATION 35
Planaria and Food

MATERIALS (per team)

- 1 planarian
- 1 watch glass
- 1 medicine dropper
- 1 small piece of fresh liver
- 1 dissecting needle
- 1 hand-magnifying glass or dissecting microscope

PROCEDURE

Obtain a Syracuse watch glass or similar container. Fill it approximately half full with conditioned tap water. Using a medicine dropper, obtain a planarian from the culture bottle and release it gently into the water in the watch glass.

Observe the behavior of the worm for 5 minutes. A hand-magnifying glass or low-power dissecting microscope will make the observations easier and more interesting. The watch glass should be on a square of white paper and should be as evenly illuminated with as diffuse light as possible.

At the end of the above period of observation, obtain two or three very small pieces of fresh, raw meat and place them, separated by a small distance, as close to the center of the water in the watch glass as possible. The pieces should be approximately 1 to 2 mm square. Raw liver is preferable, but raw beef will do. Lean meat (not fatty) should be used.

Observe the behavior of the planarian for 10 to 15 minutes and record your observations.

QUESTIONS FOR DISCUSSION

1. Why were you instructed to observe the behavior of the planarian for 5 minutes before placing the meat in the watch glass?
2. Describe the behavior of the worm during this period of time.
3. Did the planarian sense the meat and respond to it?
4. What was the response, if any?
5. Was the stimulus visual, tactile, chemical, or of some other type?
6. On what evidence do you base your answer to the above question?
7. Would it be possible to design an experiment to check on the type of stimulation? How would you go about it?
8. What parts of the planarian's anatomy are used in sensing and reacting to food?

INVESTIGATION 36
Planaria and Chemicals

MATERIALS (per team)

1 disk filter paper saturated in NaCl
1 disk filter paper saturated in acetic acid
7 cm. of 5-mm bore Tygon transparent tubing
1 pair scissors
1 medicine dropper
2 planarians

PROCEDURE

Obtain a 7-cm length of 5-mm bore Tygon transparent plastic tubing. Using scissors, cut this tubing in half lengthwise so that you will have two troughs each 7 cm long. If the tubing is curved, be sure to cut with the curve so that each half will lie flat on the desk. Place the two troughs on a sheet of white paper close together so that they will be under as nearly identical conditions as possible. Again, it is essential that the troughs should be evenly illuminated with diffuse light.

Using a medicine dropper, place a planarian and culture water in each of the two troughs, as shown in Figure 94. The "pond" of culture water in each trough should approximate 4 cm in length. Once each of the two troughs contains a planarian, you should spend a few moments observing the general behavior of the animals and in allowing them to become accustomed to this new situation.

Following this period of observation, test their sensitivity to salt and acetic acid by obtaining a strip of filter paper (approximately 2 mm wide and 6 cm long) that has been soaked in a saturated salt (NaCl) solution and allowed to dry. Obtain a similar strip of filter paper that has been soaked in glacial acetic acid and allowed to dry.

FIGURE 94. Planaria observation troughs.

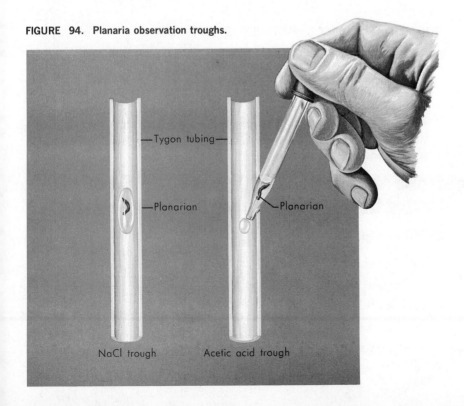

Be careful to keep these separate. Cut each strip into small sections—each 4 mm long—and keep the piles separated. Test one animal for a response to the salt and the other animal for a response to acetic acid by placing the soaked filter paper at the *extreme* edge of the water in each of the troughs. Then, holding it down with one of the prongs of the forceps, slide it toward the water until it makes contact. This should be done with the pieces of filter paper one at a time. Now repeat this procedure with the water in the second pool, using acetic-acid-soaked filter paper. Record your observations.

QUESTIONS FOR DISCUSSION

1. What is the purpose in using plastic troughs?
2. Why was it important to have the troughs evenly illuminated along their entire length?
3. What results did you obtain?
4. Is the experiment repeatable?
5. What were the results of the class as a whole?
6. How can you account for any differences in results that may be obtained by various teams?
7. Do planaria exhibit a positive or a negative chemotaxis?

INVESTIGATIONS FOR FURTHER STUDY

1. Set up a procedure whereby you can determine whether the planarians were reacting to an increase in the amount of chemicals added or to an increase in chemicals coming in contact with them because of the diffusion rates of the chemicals.
2. Set up an investigation to determine the degree of learning in planaria.
3. Modify this experiment, using an additional trough of NaCl and acetic acid on opposite ends of the same trough.

INVESTIGATION 37
Planaria and Electrical Shock

MATERIALS (per team)

    2 planarians

    2 troughs from Investigation 36

    1 $1\frac{1}{2}$-volt source

    1 medicine dropper

    1 dissecting needle

PROCEDURE

Thoroughly rinse one of the two troughs you have been using and obtain a "fresh" planarian and culture water. As before, place the flatworm in the trough with a "pond" of culture water 3 cm in length. Observe the worm to make certain that it is exhibiting normal behavior.

Using a $1\frac{1}{2}$-volt direct-current power supply and waiting until the planarian is in the middle of the "pond" of culture water in the trough, place the copper wires in opposite ends of the culture water in the trough.

Observe the behavior of the planarian. Repeat the experiment often enough to be certain of your results. You may reverse the positions of the wires, and you should also try out the procedure at times when the worm is moving away from the electrode to which it appears to react positively as well as at times when it is moving toward this electrode.

Record your observations.

QUESTIONS FOR DISCUSSION

1. What was the initial reaction of the planarian to the placing of the wires in the water?
2. What was its ultimate reaction?
3. How did your results compare with those of the class as a whole?
4. Were there any individual differences in behavior among the various worms utilized?
5. From the results obtained by your class, would it be safe to generalize about all species of flatworms? About all species of planaria?
6. From the results obtained by your class, would it be safe to generalize about the behavior of planaria when subjected to electrical stimulation of different voltages? If you have time and a source of stimulation, try subjecting the worms to 6 volts or greater.

INVESTIGATION 38
Observing Natural Animal Behavior

You now have had some experience in observing and analyzing animal behavior. In this Investigation you will select your own animal and carefully observe its behavior patterns. The objective is to learn more about animal behavior. The procedure is to be determined by you.

The animals selected by the class can be found in a city park, zoo, vacant lot, on a farm, or in a wood lot; they even can be the organisms in your classroom. The animal selected should be more complex than the planarian so that more complex animal-behavior patterns can be observed. Also the animal selected should be one that can be found day after day. Observations should be made over a week's time, with at least 3 hours spent in careful study.

In this Investigation you might find it helpful to work in pairs— one student making observations, the other recording these observations. And you might switch these assignments each day.

QUESTIONS FOR DISCUSSION

1. What kinds of behavior patterns were these—kinesis, taxes, reflex, instinct, or learned?
2. Of what value is this behavior pattern?
3. Were the behavior patterns of value to the individual, the species, or both?
4. Which senses were involved in the behavior pattern?
5. Were stimuli received by the animals without a reaction? If so, why wasn't there a reaction?
6. Would other animals have reacted in the same way these animals did?
7. Discuss other questions as indicated by your instructor.

Genes and Behavior

To what extent is the behavior of animals regulated by genes? We know that the expression of behavior depends on the structure and the physiology of animals, including both receptors and effectors. These characteristics are under genetic control. However, it is equally clear that both the development of structure and the functioning of an animal are continuously dependent on oxygen, nutrients, and the nature of the environment. Evidently, we can expect both genetic composition and environment to affect the expression of animal behavior.

It is erroneous to conclude that learned behavior is environmentally determined, while unlearned behavior is genetically determined. What is inherited is a framework of possibilities, or limits, within which changes in the environment, or experience, affect the expression of a behavioral trait to a greater or lesser degree. A chick or duckling has the genetically determined potentiality to be imprinted during a critical period shortly after hatching. The object upon which it will be imprinted is determined by the environment.

The relative contribution of heredity and of environment to the expression of behavior can be expected to differ from trait to trait. The variability in behavior observed within a population of animals permits the determination of the role of genes in affecting behavior or the heritability of behavior traits. Where there is uniformity in the expression of a trait within a population or between closely related species, we cannot conclude that the effects of the environment are negligible. Experimentation must precede conclusion.

Evidence for the heritability of behavior comes from observations of strain differences, from response to selection for a particular trait, and from observation of the effects of specific genetic changes.

Inbred strains of mice have been maintained in biological laboratories for use in studies which require animals of similar genetic background. Comparison of the performance on tests of exploratory activity among 15 mouse strains shows a 23-fold difference between the highest and lowest average score. This difference in behavior among mouse strains is especially impressive inasmuch as these strains were established without conscious concern for behavioral traits. In rats, the onset and amount of hoarding differs between strains; in rabbits, the nature of maternal care differs from strain to strain.

Selection experiments with *Drosophila* have been successful in producing two strains: one showing negative response to gravity (geotaxis); the other, positive response to gravity. Tests were initiated on a population that had a neutral response to gravity but contained some individuals which responded positively, and some negatively, to gravity. In the apparatus used, the flies responded by climbing upward or by climbing downward in a chamber in which all other environmental factors were as uniformly distributed as possible.

After the initial testing, negative-responding flies were selected and bred, and positive-responding flies were selected and bred. Offspring from these two populations were repeatedly tested, selected, and inbred. At the end of twenty generations of flies reared and tested under identical conditions, the two populations illustrated in Figure 95 had been developed.

Complex traits, such as the ability of rats to learn a maze, also respond to selection. By testing, selecting, and inbreeding rats that score few errors and repeating this procedure with rats that score many errors on a series of maze trials, it has been possible to develop a "maze-bright" strain and a "maze-dull" strain within a few generations from an initial foundation population. Conditions of rearing and testing were

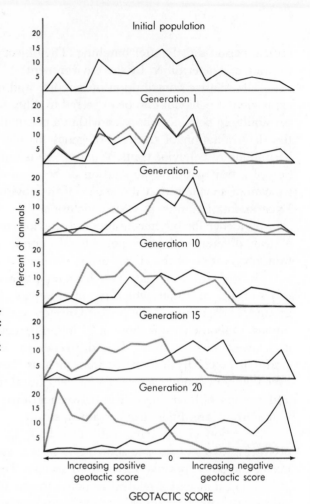

FIGURE 95. Selection for positive, geotactic response and negative geotactic response produced two distinct *Drosophila* populations with the desired trait by generation 20.

maintained unchanged throughout the experiment. The "maze-bright" rats did not perform better on all types of learning. (See Figure 96.)

The ability of behavior traits to respond to selection, such as geotaxis in *Drosophila* and maze learning in rats, is important evidence for the effect of genes upon behavior. As the genotype changes with selection, the variability within the population in the expression of the trait also changes.

Much of our knowledge of the effects of individual genes upon behavior results from studies with *Drosophila*. In this fly, it has been possible to determine that the gene *w* (white eye) decreases copulation frequency, while the gene *y* (yellow body color) reduces the strength and duration of wing vibration, which is important in male courtship. Another gene *e* (ebony body color), reduces the mating activity of males in the light but not in the dark. All these genes have morphological effects,

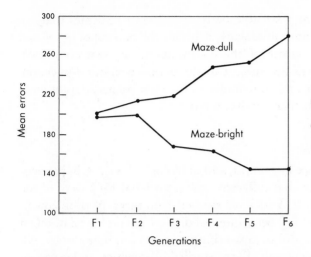

FIGURE 96. Selection for low-error maze performance and high-error maze performance from an initial population containing both low-error and high-error individuals produced the two populations illustrated in a few generations.

in addition to their action upon behavior, and the action of the gene is not independent of the conditions of the environment.

Genes act to affect behavior through several patterns. Some traits, as *phenylketonuria,* a type of mental retardation in man, result from the action of a single pair of genes. Other traits, such as the opening of brood cells on the comb and removal of diseased brood by honeybee workers, appear to be under the control of two pairs of genes, one pair affecting opening, the other the removal of the larvae or pupa (Figure 97). Most traits that have been investigated, however, appear to be influenced by genes at several loci, so-called polygenic inheritance.

The mechanism or the pathway through which genes affect behavior is currently under intensive investigation. In phenylketonuria, homozygous individuals lack an enzyme essential for the metabolism of the amino acid phenylalanine. The chemical products which accumulate from the enzyme deficit appear to be responsible for the mental retardation. A single-gene trait in mice, characterized by a peculiar motion, called "waltzing," results from a brain lesion. For most traits, the paths between genes and the behaviors they regulate remain to be discovered.

FIGURE 97. In response to the disease, American foulbrood, worker bees uncap cells (left) and remove infected larvae (right). This behavior response is controlled by two pairs of independently inherited genes.

We conclude that a sizeable body of evidence indicates that behavior, like structure and physiology, is under the control of genes and is influenced by the nature of the environment. As gene-controlled traits, behavior patterns are subject to the action of natural selection. Thus behavior evolves and contributes to the continuous adaptation of animal populations to their environment.

SUMMARY

Your concepts of science in general, and of biological science in particular, are probably somewhat different now from what they were at the beginning of this course. You have studied a number of problems in which biologists have long been interested. You have seen that, when an investigator is successful in answering one question, that answer will immediately raise a new one. Early studies in genetics and cytology gave us the concept of the gene as the determiner of heredity. But what is the gene and how does it work? It was found that chromosomes contained a great deal of DNA. What relation, if any, does DNA have to transmission of hereditary traits? Much later, Watson and Crick worked out the structure of DNA and showed how it might duplicate itself exactly. But how could DNA determine the hereditary traits? Evidence was obtained that DNA in the nucleus could determine the arrangement of the bases as RNA was synthesized and that certain arrangements of the bases on RNA would cause amino acids to be joined in certain orders during protein synthesis in the cytoplasm. But why do the same cells produce proteins (enzymes) which are different at one time from those at other times?

You have also seen that most of the results obtained in the laboratory are variable and that there is often the probability that sampling errors are involved. And you have seen that it is very difficult to "prove" any hypothesis; we simply cannot test all alternatives.

How are the knowledge and understandings which scientists have uncovered to be used for the best purposes? In fact what are the best purposes? These are often questions which the scientist cannot answer. Some may involve moral issues. Scientists claim no more insight into these problems than that possessed by other intelligent and educated persons.

In the next part of this book, an essay titled "The Tragedy of the Commons" considers questions of this sort. You may note that more knowledge and understanding are needed. But you might consider the problems as they are now, and as they might be, if we had the added knowledge.

CONCLUSIONS
AND BEGINNINGS

PART 3

SCIENCE AND SOCIETY

SECTION **15** RESPONSIBILITIES OF
THE SCIENTIST

Scientists have an obligation to explain their ideas and investigations
to those who support their work. The responsibility of the scientist to
inform the public and the responsibility of the public to learn about
science is the subject of the following contribution by a noted educator,
Dr. L. A. DuBridge, former President of California Institute of Technology.[1] Dr. DuBridge attempts to answer the questions: (1) What
goals do scientists seek? (2) What do scientists work for and how do
they achieve their ends? (3) What is a desirable relationship for scientists with the everyday world in which they, too, participate? (4) How
can obstructions to understanding between scientists and nonscientists
be removed?

[1] DuBridge, L. A. "The Inquiring Mind." *Engineering and Science,* vol. 18, October,
1954. California Institute of Technology, Pasadena, Calif. pp. 11–14.

Evaluate Dr. DuBridge's opinions but do not be dictated to by them. Formulate independent ideas as to the role of science and scientists in society.

THE INQUIRING MIND[2]

by L. A. DuBRIDGE

In 1798 a monk by the name of Thomas Robert Malthus published a paper with a long and complex title which attempted to analyze man's future on this planet. Examining past experience and bringing to bear on this experience the brilliant logic of an analytical mind, he came to some rather dire conclusions about the future. It was quite obvious to him that men had to eat; that the only major source of food was the arable land; that the area of such land was limited. Therefore, there was a limit to the potential food supply, and hence to the population that could exist on the earth.

On the other hand, he noted that the human population tended to grow at an ever-increasing rate. Any sort of voluntary birth control, it seemed to him, would be either unnatural or immoral. Therefore, the only possible future was one in which the population eventually outgrew the food supply, and thereafter death by starvation, disease and war would take over to balance a birth rate which knew no control.

Clearly, a world in which most of the people would assuredly die of one of these causes was not a very pleasant one to contemplate.

However, here we are 156 years after the Malthusian prediction, and the portion of the world that we live in does not face the Malthusian death sentence. Our population is expanding at a rate never dreamed of in Malthus' time. There are four times as many people on the earth now as then. At the same time, here in the United States at least, we have far more trouble with food surplus than with shortage. We buy potatoes and dye them blue, butter and let it spoil, wheat and give it away, in our desperate effort to avoid the economic consequences of growing more food than we can eat.

Surely Malthus was the most mistaken man in history. Or was he?

Actually, as Harrison Brown points out in his recent book (from which I shall now borrow heavily), *The Challenge of Man's Future*, Malthus' reasoning and logic were entirely correct. His only misfortune was that his observations and assumptions were later rendered obsolete by unforeseeable new developments. What were these new developments? They were of two kinds—technological and social. On the technological side men learned how to raise more pounds of food to the

[2] Reprinted by permission from *Engineering and Science*.

acre, learned to get more nutritive value to the pound, and learned how to transport food quickly from areas of surplus to areas of shortage. On the social side, great segments of the human race came to regard voluntary birth control not as a sin but as a virtue.

Now I think it is quite evident that without this latter factor—voluntary population control—the Malthusian disaster can be only postponed, and not finally prevented, by any advances in technology. We must admit that the supply of land is limited, that the productivity of land can *not* be expanded beyond all limit. But population, if not controlled, does expand without limit, and sooner or later—in 50, 250, 500 or 5000 years—a population which is doubling every 75 years or so is bound to outrun any given food supply.

This makes it clear that the primary need of the world is to insure that in all parts of it the population recognizes the need for growth that is controlled by voluntary action rather than through starvation. Clearly, this is not primarily a job for science and technology, but rather for education.

But science and technology do have some terribly important tasks to perform in this field. First, there is the task of improving the technology of producing, processing and preserving food so that the food supply will keep pace with population for the 25, 50 or 100 years required to complete the educational job. Second, there is the task of improving standards of living over a larger part of the world—for increased education goes only with increased living standards and increased disposable wealth. Finally, science and technology have the task of providing the necessary tools so that any segment of the population that has overcome the starvation limit can then proceed to help men and women lead happier and richer lives.

Now I claim that these constitute quite substantial and immensely challenging tasks. Another way of expressing them is to say simply that if men are to attain those social, moral and spiritual goals which we of the Christian nations believe desirable, then science and technology must provide the physical tools to make their attainment feasible.

This being about as important a goal as I can think of, it behooves those of us who are working in the fields of science and technology to ask ourselves how we are doing. Have we properly visualized our task and our goals? Have we properly analyzed and evaluated the steps which need to be taken, the prerequisites for progress? Are we putting first things first and do we know which things *are* first? Are we creating within science and technology itself, and within the community at large, the conditions most likely to nurture progress and success?

Now it would be presumptuous of me to attempt to answer these questions or to try to solve the problems they suggest. But I can presume to raise the questions and ask you to think about them, in the hope that if enough people think about them, we may some day get them answered.

The goals we seek

It seems to me obvious from the way in which I have stated the problem that it is important that we keep in mind the goals we seek. As I have suggested, these goals are not merely more food, more products, more gadgets. Our goal in the last analysis is a moral goal—more happiness for individual human beings, expressed in whatever terms their own philosophy of life dictates.

I emphasize and repeat this matter of ultimate goals precisely because it is so obvious to us that it is often forgotten. We become so absorbed in our gadgets, our machines, our new foods, new medicines, our new weapons, that only too often we think of them as ends in themselves—forgetting what they are *for*.

Now if we ourselves—if we scientists—forget the ends in our absorption with the means, that is bad enough; for then our work loses its meaning. But it is even more dangerous if we let the public believe that our machines and our mechanisms are ends in themselves. For then our work, which in the end depends upon public support, will surely be destroyed. And it will be destroyed by the public even though the public itself, rather than the scientists, would be the principal losers.

Let us bring this closer home. It is a paradoxical fact that, in these days of the mid-20th century, science and technology are being simultaneously praised to the skies and dammed with religious fervor; they are being handsomely supported and heartily kicked. Scientists are publicly acclaimed as a group and privately slugged as individuals.

Why is this?

Clearly, we have not told our story adequately. Our physical achievements are evident. But, because they are physical, we are accused of being materialists. Because the tools of science are powerful, their power is feared and those with the power are suspected of evil motives. Because weapons have been produced to help men fight in their own defense, it is assumed that they also make men *want* to fight. So we see that as we brag about our knowledge but are silent about our aims, then the public will come to ignore our knowledge and denounce our aims.

What scientists work for

So my first plea is that scientists shall throw off their reticence in speaking of their feelings and come out boldly and unashamedly to say, "We are working for the betterment and happiness of human beings—nothing less and nothing more."

But, in spite of the romanticism of the poet, we know full well that for most human beings *happiness* is not attained solely by sitting under a tree with a loaf of bread and a jug of wine. And even if it were, someone has to bake the bread and bottle the wine. The poet was right in suggesting that the essential elements of happiness consist of food, shelter, companionship and leisure. He only forgot to mention that these must be achieved by effort, and that the effort itself may bring happiness, too.

In any case, we are forced at once to consider how human effort can be most effectively employed to provide the physical elements for happiness and also the leisure to enjoy them. Nor are we content—as were those of medieval and ancient times—to have *many* people exert the effort and a few people enjoy the leisure. We have proved that *all* may work and *all* may play.

Now what is it that has made it possible for us today to think of a modest amount of happiness coupled with a reasonable amount of work as a possible goal for *all* people, rather than just a few? The answer is, clearly, that a series of *intellectual achievements* have enabled men to enlarge, to expand, and to dream of achieving a moral goal.

What are the intellectual achievements?

I think it is fair to say that the essential cause of the difference in the physical and the moral outlook of the western world in the 20th century, as compared to the 10th is simply that, along some time between those dates, men invented a new process of thinking.

Men had, of course, always thought, always observed, always speculated, always wondered, always asked questions, always explored. But along about 1700 men began to do these things in a new way. Men began to realize that by making observations carefully and analyzing them quantitatively, it could be shown that nature behaved in a regular manner and that these regularities could be discovered, reduced to mathematical form and used to predict future events.

This was an astonishing discovery. And as this new concept, outlined by Francis Bacon, was pursued—first by Galileo, then by Newton, then many others— a new world of understanding was opened to men's minds. Nature was partly comprehensible, not wholly mysterious and capricious. The falling stone and the moving planets became

suddenly not only understandable but miraculously and simply related. Men couldn't *affect* the motion of the planets, but they *could*
control the motion of the stone and of other objects.

And so, machines were invented, the concept of energy emerged,
steam was put to work—and suddenly, after thousands of years of
doing work only with the muscles of men and animals, men found
that a piece of burning wood or coal could take the place of many
slaves or horses or oxen.

From that time on, happiness and leisure for all men became
a possible goal, not a crazy dream.

A limitless quest

But that was only the beginning. The scientific method led from physics
to astronomy to chemistry to biology. A beachhead on the shores of
ignorance became a vast area of knowledge and understanding. Yet, as
the frontiers of knowledge advanced, the area of ignorance also seemed
to enlarge. Nature was not simple after all. A literal eternity of new
frontiers was opened up. The quest for understanding, we now see,
will, for finite man, be limitless.

I need not recount the way in which this new understanding has
spread—often slowly, often with startling rapidity—from one field to
another.

But I would like to direct your attention to the conditions that
are required for knowledge and understanding to grow and to spread.
Intellectual advancement does not come about automatically and
without attention. There have been throughout human history only
a few places and a few periods in which there have been great advances
in knowledge. Only under certain special conditions does the inquiring mind develop and function effectively. Can we identify these conditions? Certainly we must try.

The first condition, of course, is that at least a few people must
recognize the value of the inquiring mind. Here we all take for granted
that new advances in understanding come only from the acts of creative
thinking on the part of individual human beings. We know that, and
we respect and admire the men who have shown the ability to think
creatively. But we mustn't get the idea that our admiration for original
thought is shared by all people.

Even in this country, the man who thinks differently is more
often despised than admired. If he confines his new thoughts to the
realms of abstruse theoretical physics or astronomy, he may not be
molested. For then he will be speaking only to those who understand

him. But if he wanders into biology or medicine, into psychology or sociology or politics, then he should beware.

Now in recognizing the virtues of thinking differently, we do not mean that we must encourage the idiot, the criminal or the traitor. Honest, truly intellectual inquiry is perfectly easily recognizable by those who have some training in the field. But just here we run into difficulty. Those who are incompetent to judge may nevertheless render judgment and pass sentence on those with whom they disagree, or whom they fear.

One of the great unsolved problems of a democracy is how to insure that, in intellectual matters, judgments are left to those who are competent, and the people will respect that competence. But when uneducated fanatics presume to choose and to censor textbooks, when government officials impose tests of political conformity on the scholars that may leave or enter a country, and when the editors of a popular magazine set themselves up to judge who had the proper opinions of nuclear physics, then the inquiring mind finds itself in an atmosphere not exactly conducive to maximum productivity.

Fortunately, for the past 100 years in Western Europe and in the United States the impediments to creative scholarship have been less important than the great encouragements. In the past 10 years the physical conditions necessary for research in the sciences have enormously improved. More opportunities have been created to study, to travel, to carry on research, than ever before existed.

The needs of the inquiring mind

But physical conditions are not enough. Big, beautiful laboratories do not themselves produce research—only the men in them can think. And if conditions are such as not to attract men who think or such as to impede their thinking, then the laboratory is sterile. Such laboratories, as you well know, do exist. There is no use storming and raging at the perverseness of scientists who refuse to work when conditions are not just to their liking. We don't call a rose bush perverse if it fails to bloom when deprived of proper water and soil. A community or a nation which wishes to enjoy the benefits that flow from active inquiring minds needs to recognize that the inquiring mind is a delicate flower, and if we want it to flourish we are only wasting our time if we do not create those conditions most conducive to flowering. The cost of doing so will be well repaid.

The inquiring mind then needs, first of all, some degree of understanding and sympathy within the community. And if there are those who cannot understand, then at least they must be insulated by those who do, so that they do the least harm. As someone has said, we can stand having a few idiots in each community—as long as we don't put them on the school board.

As I have already suggested, it is not enough for the scholar or the scientist to wring his hands and wish that there were fewer idiots or that they had less influence. He must also, to the extent of his ability, explain to those who can understand what he is doing and why. We now see that an intelligent and informed segment of public understanding is essential to the progress of scholarly endeavor.

Scientist and government

This leads me to another subject which has become timely to the scientist and to the citizen in recent years; that is, the relation of the scientist and the government. This is obviously a very large subject which I cannot attempt to explore here. But as the scholar needs an informed community to support him, so he owes an obligation to that community.

The prime obligation of the scholar, of course, is to pursue scholarship. That is, he must seek answers to important questions, observe carefully, analyze accurately, test rigidly, explain imaginatively, and test and test again. Then he must publish his results, fully, fearlessly, objectively, and defend them enthusiastically unless or until the facts prove him wrong. Through such intellectual struggle does the truth emerge.

But in these days the results of science impinge so heavily on public affairs that the public—in particular the government—needs the scientist's help in so many ways. Obviously, the government needs the direct services of thousands of scientists and engineers to carry on work in public health, standards of measurement, agriculture, conservation of resources and in military weapons, to name a few.

But when there is developed a new weapon, a new treatment for a disease, a new way of using public resources, does the scientist's responsibility end there? I think not. There are so many ways in which important matters of public policy are affected by these new scientific achievements that scientists must stand by as advisers at least to interpret, explain, criticize and suggest on policy matters.

Scientific advice

We would not think, of course, of allowing a new law affecting public health to be passed without asking a physician's advice on whether it is wisely conceived. Yet I am sure state and federal legislatures *have* thought of it—in the various antivivisection bills, for example. Fortunately, (for this purpose at least) the medical profession has great influence and can make its opinions heard. And most of the public respects its doctors.

But when national security matters are being discussed which involve the nation's strength in atomic weapons, it is clear that those in charge of forming policy will need to have much help on questions of what atomic weapons really are, what they do individually, and what would be the effects of setting off the whole stock pile. I am not saying that such scientific advice is not sought (though I think it is not always adequately used). But I do say that scientists need to be ready to help. Yes, they may need to be ready to intrude with their advice even if it is not asked for.

This problem has, of course, caused much recent trouble and misunderstanding. Many prominent citizens, including many politicians and editors, apparently feel that scientists should stick to the laboratory and let public policy matters be handled by others. Now no one argues that *decisions* on public matters must be made by the properly constituted responsible officials. But *advice* and *information* on scientific aspects of the problem are often essential and must come from scientists.

It is often true that the scientific aspects of a problem are so important that they overshadow all else—and the scientist's advice becomes adopted as a decision. But in other cases, other factors may appear important and the scientist's advice may be wrong, or may not be taken. Even the scientist, being human and being a citizen, will take non-scientific matters into account in rendering his advice. He may be just as competent to do this as anyone else. Being a scientist does not disqualify a person from being an intelligent citizen. But the possibilities of disagreement and misunderstanding are very great.

A risky course

A very great and admittedly loyal scientist is right now being persecuted partly because, though he gave advice of surpassing value on many, many occasions, he gave on one occasion advice which some (but by no means all, then or now) believe was wrong. The sad part of this case is not so much the harm to the individual, as the harm to the country that will result if scientists cannot give honest advice to their government officials, or will be no longer asked for advice, or listened to. Dire disaster could indeed follow from such a course pursued in the thermonuclear age.

I fervently believe that the world has been remade the past century—remade physically, socially, and spiritually—by the work of the inquiring scholars. These scholars have sought new knowledge and new understanding; they have sought to use this understanding to produce those things that men needed—or thought they needed—to improve their health, their comfort, their happiness, their security.

Scholars will continue these activities and the world will continue to change. Their efforts must be aided; for though what they do may yield dangers, the dangers are far greater if they do less. And since what they do affects the world, affects you and me and our community and our country, we should have these inquiring and active minds around all the time to direct their attention to the most difficult of all problems—how to help men make better use, in their relations with each other, of the great new areas of knowledge which can yield so much to make men happier and better.

END OF ARTICLE

SECTION **16** THE ROLE OF CONTROVERSY
IN SCIENCE

Controversy . . . . a discussion of a question in which opposing opinions clash; . . . *Webster's New World Dictionary, College Edition.*

Science has been emphasized as a process of inquiry, but because the conclusions derived from inquiry may often be diverse, science involves controversy as well as inquiry. This is so true that *constructive controversy* is a major mechanism in the growth of science.

History shows that scientific achievement along with its resulting technological advances has grown under the freedom of controversy and has become restricted (or extinct) by the restraint of free debate.

However, the society which permits free debate must be an educated society if it is to choose wisely among the alternatives offered it. Survival itself demands a society objectively familiar with the ways in which science and scientists affect it.

When we look at the past and attempt to evaluate the knowledge and attitudes of previous centuries, we find many examples of superstition and ignorance masquerading as science, and scientific inquiry not only being neglected, but even prohibited. Compare that darkness with the brilliance of our times—but what of our times? Are we really able to rationalize current controversies with scientific candor? Are we honestly wiser than our ancestors in the exercise of rational thought and emotional control? Are we able to make the decisions that will affect our future and the future of all others as yet unborn, or will our descendants look back and remark on the "ignorance of the past"?

We have the opportunity to become a better educated society through the use of constructive controversy and through understanding its contribution to the progress of science.

How does controversy contribute to the rational progress of science? The conclusions of science depend on the results of its experiments. Experiments constantly attempt to establish the correctness or fallacy of ideas. Thus, experimentation is the way science approaches controversy. While well-designed experiments may settle one controversy, often new knowledge derived from these experiments engenders new controversy, and this becomes, in turn, a part of the growth of science.

The following essays represent two examples of current scientific controversy. No guide questions or suggested points of discussion will be given. Read them, individually deliberate their contents, and then collectively and freely discuss these opinions. It will be strange if your discussions do not bring out ideas and solutions which would not even have been suggested by the authors.

The Use of Insecticides

In 1962, the late Rachel Carson's book, *Silent Spring,* appeared on the nation's bookshelves. *Silent Spring* became a bestseller overnight, for it introduced an important scientific controversy: the use of chemicals, specifically insecticides and herbicides, to protect certain plant life from insects or other plants. Miss Carson skillfully and convincingly leads the reader to the conclusion that the continuation of life on earth may well depend on halting the use of insecticides. Her book was acclaimed almost unanimously by book reviewers.

For years before the talented Miss Carson's debut into the field of insecticides, however, other biologists had been debating the dangers and merits of these chemicals used in such huge quantities throughout the country. Many felt that the use of insecticides was not an "all-or-none" proposition. Trained scientists believed that perhaps the use of insecticides could be justified under certain conditions, but just what conditions justify this use presented an area for a many-sided controversy. Some favored the application of insecticides on a rather broad scale and others preferred a very restricted use of these chemicals. Undoubtedly all shades of opinion between the "all-or-none" extremes could be found in the scientific community.

Biologists' continuing interest in insecticides is well illustrated in an article, and subsequent letter to the editor in reply, which appeared in two issues of *The AIBS Bulletin* during 1960.[1] In the April issue, Dr. George Decker, Head, Section of Economic Entomology of the Illinois Natural History Survey, takes a strong stand for a realistic attitude toward the necessity of insecticides in our modern world. Dr. Samuel A. Graham, School of Natural Resources, University of Michigan apparently favors a somewhat more restricted use of insecticides than does Dr. Decker. We have reprinted both scientists' articles for your examination.

In both extracts from *The AIBS Bulletin,* we witness scientific literature functioning as a medium for free and open controversy. Neither scientist takes as extreme a view of the evils of insecticides as does Miss Carson. Neither sees the issue as "black or white," for both are trying to settle on a shade of grey that could represent adequately the many pros and cons of this very controversial subject. Perhaps an ultimate test of discretion is to be able to distinguish, and to compromise capably, moderate views. Certainly very little insight is required to distinguish an extremist position from its opposite, such as the indiscriminate use of insecticides as opposed to a complete abandonment of their usage.

Read the articles by these two scientists and discuss the merits of each point. You should be familiar with Carson's *Silent Spring* for yet a third view on the use of insecticides.

INSECTICIDES IN THE
20th CENTURY ENVIRONMENT[2]

by GEORGE C. DECKER

When the first white men came to North America, they found a race of rather primitive men living in reasonable harmony with a relatively stable environment. Under those conditions, this continent supported a population of about one million persons and provided in excess of

---

[1] For reference, the *Bulletin* was renamed *Bioscience* in January, 1964.

[2] Condensed from a paper presented at a Symposium sponsored by the Ecological Society of America. "Ecological Problems of Pest Insects," given at the AIBS-sponsored meetings in Bloomington, Indiana. August, 1958. Reprinted from *The AIBS Bulletin,* April, 1960. pp. 27–31.

2000 acres per capita. Then, as now, literally dozens of insects attacked every crop that grew and neither man nor beast escaped their ravages. In the years that followed, with agriculture on a subsistence basis and a seemingly endless supply of land available, there was plenty for all, and farmers raised only feeble objections to share cropping with the insects. Later, as urban populations increased, each farmer was called upon to meet the food and fiber requirements of an ever-increasing number of individuals and to do so on an ever-decreasing number of acres per capita. This trend continued until at present we have only a little over ten acres per person, seven of which are classified as farm land but only two of which are devoted to crop production.

The early American farmers had little choice but to rely upon nature to control their insect enemies. Then, as losses mounted and the standards of perfection demanded by an increasingly more discriminating consuming public rose, farmers began to clamor for governmental aid and scientific guidance to solve their insect control problems. The early state and federal entomologists were essentially naturalists, and they preached a gospel of biological and cultural insect control methods. For years such measures dominated all entomological endeavor, but finally when natural controls proved wholly inadequate, entomologists reluctantly turned to chemicals, and thus we entered an age of chemical insect control.

The Advent of Insecticides

The large-scale practical use of insecticides is in reality one of the important technological developments of the 20th century. While it is true that numerous nondescript concoctions of lye, lime, soap, turpentine, brine, vinegar, fish oil, and even some tobacco, pyrethrum powder, mineral oil, and arsenic were reportedly used as insecticides prior to the year 1800, effective use of agricultural insecticides had its origin with the use of Paris green to control the Colorado potato beetle in 1867. For the next seventy-five years, arsenical compounds played an ever-increasing role in insect control. Thus, considering the many insecticidal uses for white arsenic, sodium arsenite, lead arsenate, calcium arsenate, and other arsenical salts, it is not surprising to find that from 1939 to 1948 the domestic consumption of white arsenic averaged over 35,000 tons per year. It must be noted also that the arsenicals were not the only chemicals used as insecticides in the pre-DDT era. To obtain a fair estimate of the extent of insecticide usage in the early 1940's, we would have to add about 15 million pounds of cryolite and related fluorine compounds, some 15 million pounds of pyrethrum flowers, 8-10 million pounds of rotenone-bearing roots and powders, and at least one million pounds of nicotine. Then to all of this we must add literally millions of gallons of petroleum oils, unestimated quantities of tars, cresols, fish oil, and many lesser products.

With the advent of DDT for agricultural use in 1945 and the large array of chlorinated hydrocarbon and organophosphate insecticides that followed in quick succession, many of the older materials suffered a rapid decline in popularity and they were largely replaced by the more effective synthetic organic insecticides. It appears that the actual tonnage of primary insecticidal chemicals produced for domestic consumption each year may not have changed materially. However, with the use of newer and more effective materials at much lower rates of application, the acreage treated has increased several fold in the last decade. Moreover, prior to the advent of DDT the use of insecticides was for the most part confined to fruit, vegetable, cotton, and a few miscellaneous crops of high-per-acre value, they are now used quite extensively on several field crops, pastures, meadows, and forests.

When DDT and at least a dozen other new chemicals became available for general use, a number of competent and distinguished scientists expressed concern that widespread use of these materials might create a public health problem. Immediately a number of publicity seekers and misguided individuals seized upon the idea that the public was being poisoned, and the country was deluged with an amazing flood of scare stories. Then, as the general public began to show some concern, the witch hunt got underway in earnest. As absurd charges and counter-charges were hurled back and forth in congressional committees and in the press, the scientists settled down to a detailed analysis and factual study of the problem. The public health aspects were reviewed by several scientific bodies, notably the World Health Organization (1, 12), the U. S. Public Health Service (7), and the Food Protection Committee of the National Research Council (4, 6). The general conclusions drawn in each instance were: (a) The large-scale usage of pesticides in the manner recommended by manufacturers or competent authorities and consistent with the rules and regulations promulgated under existing laws would not be inconsistent with sound public health programs, and (b) although the careless or unauthorized use of pesticidal chemicals might pose potential hazards requiring further consideration and study, there was no cause for alarm.

These encouraging conclusions notwithstanding, the fact that insecticides may be misused remains a matter of concern to a considerable segment of the American public. This is true particularly of conservationists who quite correctly insist that many forms of wildlife are subjected to certain potential hazards not shared by man and his domestic animals.

Clarification of Terms

As a prelude to any attempted evaluation of the hazards which may be inherent in the use of insecticides, a few frequently unused [*sic*] terms must be clearly defined. To avoid endless confusion, a careful distinction must be drawn between the terms hazard and toxicity. As the Food Protection Committee of the National Research Council has repeatedly pointed out: "Toxicity is the capacity of a substance to produce injury; hazard is the probability that injury will result from the use of the substance in the quantity and in the manner proposed." Therefore, to be at all reliable, an estimate of the hazard involved in the use of any substance must be based upon a knowledge of its inherent toxicity and upon the details of its proposed use.

It is also imperative to recognize and understand the equally clear-cut difference between the terms acute and chronic toxicity. In general, *acute toxicity* refers to toxic effects (either lethal or profound) occurring immediately after, or at least definitely attributable to, a single exposure by any route of administration, but usually referring to ingestion. *Chronic toxicity* refers to toxic effects resulting from repeated

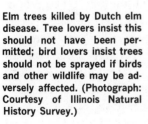

Elm trees killed by Dutch elm disease. Tree lovers insist this should not have been permitted; bird lovers insist trees should not be sprayed if birds and other wildlife may be adversely affected. (Photograph: Courtesy of Illinois Natural History Survey.)

or multiple exposures extending over a period of time and is generally considered to be accumulative in nature.

As noted earlier, the hazards to man and his domestic animals resulting from approved uses for insecticides in public health and agricultural insect control programs have been extensively studied and the general sum and substance of all research in this field has failed to indicate any significant public health hazards. It is generally conceded that safety factors ranging from 10 to 100 fold have been included in most recommendations and at times such factors have been superimposed one upon another until the possibility that an actual hazard may exist is fantastically remote. Whereas man and his domestic animals are afforded much protection by the labeling requirements of the Insecticide, Fungicide, and Rodenticide Act of 1947, which specifies definite time-lapse intervals between insecticide applications and the harvest or consumption of the crop, it is generally recognized that most forms of wildlife can hardly be expected to avail themselves of such precautions. It is also recognized that fish, reptiles, amphibians, and numerous arthropods are more susceptible to insecticide poisoning than are mammals and birds, and in fact there have been a number of cases where fish were killed following the direct or inadvertent contamination of streams and lakes.

As a matter of fact, entomologists, toxicologists, and wildlife biologists alike recognize that some species of wildlife are subjected to hazards not encountered by man or by his domestic animals and thus they present special problems requiring special research attention. For example, it is well known that in general the food consumption of animals is more or less inversely proportionate to their size; thus birds, rabbits, and other forms of wildlife receiving the same diet as much larger animals actually ingest larger amounts of pesticidal chemicals in terms of mg/kg of body weight. Then too, in many treated areas the intake of insecticides by various forms of wildlife may be by ingestion, inhalation, or absorption, and under some conditions certain species may be subjected simultaneously to exposure by all three routes of administration. Finally, it is obvious that relatively small species moving in, around, and under a vegetative cover are subjected to greater and more prolonged exposure than are larger domestic animals standing well above the contaminated vegetation.

In view of the foregoing factors, it is not surprising that there have been numerous instances where wildlife of varied types has been adversely affected by insecticides. At the same time, the preponderance of evidence to date indicates that such incidents can usually be traced to carelessness, to accidents, to instances of outright experimentation, or to non-agricultural programs involving anticipatable hazards. Considering the thousands of tons of insecticides that have been used—and misused—in the control of agricultural pests during the last 50 years, it can be concluded that to date the impact of insecticide usage on wildlife has neither been great nor disastrous. This holds true for orchards, market gardens and cotton fields, which represent the most extensive and intensive agricultural usage of insecticides. Actually, considered from the broad point of view, the impact of agricultural and public health insecticide use on wildlife has been insignificant when compared with many of the other everyday activities of man.

In contrast to these remarkably safe agricultural practices, the existence of some very specialized insect control programs that call for per-acre insecticide dosage rates as much as 5 to 10 times those required in most agricultural uses must be recognized. These do involve calculated risks insofar as wildlife is concerned. The spraying of elms for the control of phloem necrosis or the Dutch elm disease, and the fire ant control program are examples. In such cases the interested parties must, or at least should, with the aid of competent experts, weight the facts pro and con and then adopt a course of action that will best serve overall interests.

The Balance of Nature

Not infrequently insecticides are accused of upsetting the balance of nature, when in many instances it would be more accurate to say they were used to suppress an organism already out of balance. Actually, man himself has been the primary factor in upsetting the so-called natural balance. When he cleared the forest, plowed the prairies, drained a marsh, or dammed a stream, he altered an entire environmental complex and set up an entirely new set of opposing forces which if left uninhibited would establish an entirely new biotic equilibrium. Presumably it is obvious to all that nature's balance is not a static condition, but is rather a fluid condition changing from day to day, and at the same time constantly moving forward in response to the forces of changing geologic time and advancing culture. Insecticides represent only one minor element of these dynamic forces.

In nature, every living organism is engaged in the most ruthless kind of competition with every other organism upon which its interests impinge. Man is a part of that environment. That he has been eminently successful is evidenced by the fact that the human population of this country has risen from less than one million to over 170 million in some fifteen to twenty generations. It now becomes apparent that to clothe and feed this vastly increased population, man must maintain his position of dominance, and our agricultural production must continue to increase even at the expense of the further displacement of native plants and animals. It may be news to some that these pressures are now intensified by the fact we have at last absorbed and brought into production most of the lands suitable for agricultural production. As a matter of fact, 1954 marked a turning point in our history because then for the first time the withdrawal of agricultural land for use as home or industrial sites, airports, and highways exceeded the land reclaimed, and the number of acres in farms showed a decline. While America is presently blessed or, as some say, plagued, by overproduction, with populations increasing and the area of farm land tending downward, it will be a matter of only a few years until agricultural scientists and farmers will have to make ever-increasing use of new advances in agricultural technology, including even greater use of insecticides and other pesticides, to meet the nation's food and fiber requirements.

Population and Land Use

In a recent discussion of population dynamics and land use problems, H. B. Mills (9) said: "When we think of the future of recreational lands in all of this competition for land use, we cannot get away from zoning. We must think of areas where people will live, where they will work, where their roads will go, where they will produce their food, where they can find outdoor relaxation, and where wildlife can live and prosper. We may not like the idea of such regimentation; it will be an expression of restriction of personal freedom due to increased population density."

In 1954, the United States Department of Agriculture (11) estimated that to offset the pest losses in agricultural production, an extra 88 million acres must be cultivated, and that losses subsequent to harvest equal the production of an additional 32 million acres. Is it not possible that the benefits to be derived by diverting these acres to soil and wildlife conservation would far exceed any foreseeable damage that might accrue from the approved use of insecticides? The recent amendment of the duck stamp law is a step in the right direction, but this is apparently not enough. There is still some talk of a land bank, and the disposition of submarginal lands is still an open question. But 10 years hence may be too late. Is it possible that while dissipating our energies pondering potential hazards that may never materialize, we are missing the opportunity of a lifetime to secure and preserve an adequate representation of our native fauna and flora to be forever maintained in suitable conservation and recreational areas?

Difficulties of Evaluation

Those familiar with the excellent reviews on wildlife-insecticide relationships by Brown (2), Rudd and Genelly (10), and Cope and Springer (3) must be aware of the fact there is a considerable volume of excellent field and laboratory research, much of which tends to pinpoint areas of considerable concern where further research is needed. Nevertheless, at times it is difficult to evaluate properly many of the criticisms directed against insecticides and certain insect control practices. For example, some authors frequently note that a certain insecticide such as aldrin or heptachlor, is 5, 10, or even 20 times as toxic as DDT and then, after commenting in a matter-of-fact way that it may persist for several years, they imply that the hazard involved is tremendous, when actually the hazard may be insignificant because the dosage is low and the residue is rapidly lost. All too often the results of some laboratory or field experiments are cited as if they were typical of conditions to be encountered following the legitimate use of the insecticide in the field. Then, too, one frequently encounters vague references to mass applications of insecticides without any further detail. Some attempt should be made to distinguish between the treatment of a million acres of cotton with X insecticide at 2 ounces per acre, and an equal area of marsh or timberland at 2 pounds per acre. A million acres is a lot of ground, but, after all, it represents considerably less than $\frac{1}{10}$ of one per cent of the land area of the United States.

Pre-testing Insecticides

It is frequently asserted that new insecticides are introduced and released for large-scale insect control operations without adequate pretesting, but apparently these critics are unaware that as early as 1948, Dr. A. J. Lehman (8) clearly and with amazing accuracy set forth the toxicological characteristics of DDT and related compounds. Then, too, the very fact that the review by Rudd and Genelly (10) contains a bibliography in excess of a thousand titles would indicate that the characteristics of insecticidal chemicals are not entirely unknown. As a matter of fact, there are experts who would testify that few if any chemicals known to man have undergone the toxicological scrutiny to which DDT has been subjected. It is true that all questions and problems have not been entirely resolved "beyond a shadow of a doubt," but that is not surprising, for after all, science is seldom if ever static and there are few instances where a scientist can claim he has established all of the answers finally, conclusively, and irrevocably.

Actually, there are many problems related to insecticidal chemicals that cannot be anticipated or resolved in advance in any laboratory or small plot experiments. As a matter of fact, in establishing the principles to be observed in the evaluation of new insecticides, the Food Protection Committee of the National Research Council (5) acknowledged this basic truth when it said: "Complete knowledge of many factors pertaining to pesticide usage, performance, and ultimate safety can be developed only through actual use in large scale performance tests. Hence, any system proposed for regulating the distribution of new materials should provide for their orderly release with recognized steps between strictly controlled small plot experiments and full scale commercial operations." This being the case, one must anticipate certain adverse reactions and responses in the early large-scale usage of a pesticide. The correct procedure is to make sure such instances are detected and corrected as soon as possible. While the occurrence of such incidents is disturbing and regrettable, it seems reasonably certain the adverse effects will rarely if ever be disastrous or permanent.

Not infrequently the wildlife specialist and the conservationist feel that they are placed at a disadvantage in presenting their case because in a large degree the values with which they deal are intangible and thus the reconciliation of divergent points of view which often elicit strong emotional responses becomes doubly difficult. Even here, there are at least some entomologists who can lend a sympathetic ear, for insect control also has its intangible considerations which can likewise induce violent emotional eruptions.

Many of the most dreaded diseases of man are insect-borne, and there are those who regard the continued control of insect disease vectors as essential to the maintenance of successful public health programs. The hundreds of phone calls that besiege public health officials when nuisance mosquitoes become annoying lend ample testimony to the public interest in and the demand for local mosquito control programs.

Then, too, a large segment of the American public is profoundly interested in the protection of shade trees, ornamental plantings, and forests. Literally thousands of individuals, recalling with profound regret the demise of the chestnut tree because no effective disease control measures were available, vociferously demand that every possible effort be made to assure that the same fate does not befall the American elm, the most common and perhaps the most beloved shade tree in America. In the battle for the elms we approach the tragedy of civil war, for here we have entomologist pitted against entomologist, conservationist against conservationist, and neighbor against neighbor. Last but not least, the fact that the Congress of the United States and a score of state legislatures, normally reluctant to appropriate funds, except in response to great pressure, have approved plans and appropriated funds for several large-scale insect control programs, testifies to the public interest in phases of insect control which do not affect its diet and only indirectly its pocketbook.

Considering All Sides

Any fair and impartial appraisal of the impact of insecticides on wildlife must give equal consideration to both the good and bad side effects that may occur, and if we are to be honest we must look for the good as diligently as we look for the bad. Unfortunately, harmful side-effects are usually readily apparent, whereas indirect beneficial results are apt to pass totally unobserved. Nevertheless, the impartial observer may be pardoned if he considers the loss of a few song birds attributable to the spraying of elm trees less harmful than the permanent loss of such trees, with the accompanying disappearance of nest sites.

Unfortunately, the ultimate effect of insecticide usage on animal life cannot always be measured in terms of initial mortality of individual species. In reality it must reflect and encompass the long-term effects on both plant and animal life. Since the latter are largely dependent upon plants for both food and shelter, is it not possible that the destruction of timber, range vegetation, or cultivated crops by insects may produce a chain reaction that will ultimately affect all of the forms of life in the area?

Conclusions

There is no question but that the future food, fiber, and public health needs of this country will assure the continued, if not, indeed, the greatly expanded use of insecticides for generations to come.

Despite the widespread use of insecticides totaling billions of pounds in the last decade, profound fears that the large-scale usage of modern pesticidal chemicals would seriously upset the balance of nature and result in disastrous losses of wildlife have not been realized. To quote from the most comprehensive and complete study of the problem (10), "Considered in its broadest scope, at the present time pesticides seem to be only minor influents in nature compared to other factors in land and water development and use."

Favorable results notwithstanding, the many diverse and complex problems of wildlife conservation in a chemical world must be kept under continued surveillance. Particular attention should be devoted to the welfare of rare species of restricted habitat, to the impact of new chemicals as they are introduced, and to pest control programs involving the widespread application of insecticides over large contiguous areas.

Since most of the unfortunate incidents, problems, and differences of opinion that have arisen or are likely to arise, involve insecticide usage (unintentional or otherwise) that was not covered by label approval, it appears that the solution of the problem at hand rests in the detection, isolation, proper evaluation, and eventual elimination of malpractices rather than indulgence in wholesale condemnation of insecticides and insecticide usage *per se*.

BIBLIOGRAPHY

(1) Barnes, J. M. 1953. Toxic hazards of certain pesticides to man. World Health Organization Monograph Series N. 16. 129 pp.

(2) Brown, A. W. A. 1951. Insect control by chemicals. John Wiley & Sons, Inc., London. 817 pp.

(3) Cope, O. B., and P. F. Springer. 1958. Mass control of insects: The effects on fish and wildlife. Bulletin of the Entomological Society of America. 4(2):52–56.

(4) Food Protection Committee. 1951. Use of chemical additives in foods. National Research Council. Food and Nutrition Board. 23 pp.

(5) ———. 1952. Safe use of chemical additives in foods. National Research Council. Food and Nutrition Board. 26 pp.

(6) ———. 1956. Safe use of pesticides in food production. National Research Council. Publication 470. 16 pp.

(7) Hayes, Wayland J., Jr. 1954. Agricultural chemicals and public health. U. S. Dept. of Health, Education, and Welfare Public Health Reports. 69(10):893–898.

(8) Lehman, A. J. 1948. The toxicology of the newer agricultural chemicals. Bul. Assoc. Food & Drug Officials. 12(3):82–89.

(9) Mills, H. B. 1959. The importance of being nourished. Trans. Ill. Acad. Sci. 52 (1, 2).

(10) Rudd, Robert L., and Richard Genelly. 1956. Pesticides: Their use and toxicity in relation to wildlife. State of California Department of Fish and Game, Game Bulletin No. 7. 209 pp.

(11) United States Department of Agriculture. 1954. Losses in agriculture. Agricultural Research Service. 190 pp.

(12) World Health Organization. 1956. Toxic hazards of pesticides to man. Tech. Report Series No. 114. 51 pp. END OF ARTICLE

LETTER[1]

by Dr. SAMUEL A. GRAHAM

July 26, 1960

George Decker's article on insecticides and their use in the 20th Century, (*AIBS Bulletin* April, 1960) is an excellent presentation of the agricultural entomologist's viewpoint on this controversial subject. One can find little to criticize either in the logic of his argument or the general conclusions reached, in so far as they relate to the production of human foods. There are, however, some apparent implications inferring blanket approval of the widespread application of insecticides which deserve comment, in view of the current controversy between bird lovers and control agencies. Knowing Dr. Decker as I do, I feel sure that these implications were not intended, and were the natural result of time and space limitations that prevented adequate coverage of a very complex subject. This letter to the editor is written, not to criticize but rather to amplify and clarify.

We must agree that insecticides will continue to be used, perhaps in increasing quantities, in the artificially simplified environment of our fields, gardens and orchards. This seems inevitable. We have no desire to eat wormy apples, and most of us object to too many aphids in our spinach. Furthermore, one would encounter difficulty in persuading the potato grower to desist from using insecticides when he has seen per acre production rise from 75 to 500 or even 1000 bushels as a result of his use of agricultural chemicals, of which insecticides and fungicides are outstandingly important.

But when we depart from the fields and orchards, conditions are by no means comparable. The widespread application of poisons cannot be justified by the fact that the use of these chemicals on farms is necessary. Most of the controversy about the use of toxicants is centered around projects involving large areas, often under multiple

[1] Reprinted from *The AIBS Bulletin*, October, 1960. pp. 5–7.

ownership. Usually the projects which are questioned are conducted by public agencies, municipal, state, or federal, and therefore are not subject to regulations that apply to projects conducted by individuals or corporations.

Decker points out correctly that DDT has received a tremendous amount of research attention and he seems to imply that all insecticides have been subjected to similar careful scrutiny before they are approved for use. Actually, DDT is the only chlorinated hydrocarbon insecticide that has been adequately evaluated in terms of side effects on many organisms that are either of direct value to man, or appreciated for their beauty or aesthetic value. The widespread use of toxicants cannot be justified until small scale tests have proved their worth and their shortcomings. That is the reason that forest entomologists, while experimenting with other materials on a limited scale, have used only DDT on extensive projects.

We know that DDT used at dosages greater than one or possibly two pounds per acre may cause injury to certain terrestrial organisms. $\frac{1}{2}$ pound per acre or even less may cause serious damage to fish. Why unexpected mortality of fish from low dosages sometimes happens is difficult to explain, especially when the mortality is delayed until several months or more after spraying. Possibly the death of organisms on which they feed results in malnutrition of fish, but more likely the toxicant is concentrated in the bodies of organisms used by fish for food. We know this to be true in the case of earthworms feeding on materials contaminated with insecticides of the chlorinated hydrocarbon family, and that robins feeding upon such worms are poisoned (Barker, 1958). The cherry orchardist may be happy about the reduction in the robin population, but others may not be.

The question of when, how, and where to use insecticides on a widespread scale is not easy to answer. Decker suggested that benefits to be derived should be weighed against the damage that may be done. Obviously this evaluation should not be entrusted to individuals or organizations which will do the actual control work and thereby stand to benefit. Sincere though they may be, such people are likely to be biased in viewpoint, and unwise decisions may result. The recklessness with which insecticides have sometimes been applied by public agencies, especially municipalities, for the control of the Dutch elm disease is a good illustration of carelessness. Dosages have ranged from one pound to over 100 pounds of DDT per acre. In one locality with which I am personally acquainted a private operator sprays the elms at so much per tree per treatment, and instead of one treatment when the trees are dormant and birds are least likely to be injured, three treatments are applied, thus tripling the income of the operator, but causing needless injury to birds. The residents, whose trees are sprayed without their direct authorization, are charged personally for the work.

Decker inadvertently infers by the photograph of elms in his article that economic entomologists approve without reservation the Dutch elm disease spraying program, as it is conducted. This is not altogether true because we know that the disease cannot be controlled by spraying alone, as Decker and his associates have elsewhere pointed out. (I.N.H.S.) The difficulties encountered in controlling the Dutch elm disease are evident on the campus of The University of Illinois, where, in spite of spraying, the elms have died in great numbers, not only from the elm diseases but from the scale insects which almost always increase following the application of DDT. Control of the Dutch elm disease can only be accomplished by intelligent use of insecticides combined with sanitation. Furthermore, sanitation does not mean merely the prompt removal of dead wood, as we are inaccurately told by control agencies, but rather the prompt removal of dying or recently killed branches and dying trees, regardless of the cause of mortality, before the elm beetles have had a chance to breed in them. The beetles cannot breed in trees or branches that have been dead long enough for the inner bark to have become darkened, and without the bark beetles the Dutch elm disease would be without its primary means of dissemination. These facts have almost never been fully publicized by those responsible for the Dutch elm disease control on the local level.

The actual decision to spray or not to spray is all too often in the hands of inadequately-trained men, who can only see the pros. The Dutch elm disease situation is only one illustration of this. We need some provisions that will assure wise and dispassionate decisions on whether or not the broadcast use of an insecticide is justified in a specific instance.

Decker may seem to imply that insecticides must always be the "backbone" of insect control, although he recognizes that cultural practices and natural enemies of insects are helpful. Some of us feel that other means of insect control are even more important than insecticides, especially in forests, and judging from the research program under his supervision Decker himself is among these. Some research even in orchards, has shown that in certain localities proper selection and timing of insecticidal applications can increase the effectiveness of natural enemies in limiting the multiplication of some pests. The result is a material reduction in the amount of toxicants needed. (Pickett, 1956). Much more ecological work along this line should be encouraged.

Admitting, as we must, that the use of toxicants is essential to man's well-being, we need not assume that broadcasting these poisons is the best way to apply them. Widespread application from the air is spectacular and attractive because it makes news, but it inevitably entails a certain amount of risk. Fortunately we have some researchers looking for safer ways of using toxicants; among them are F. S. Arant and Sidney Hays in Alabama who are developing a technique for poisoning the imported fire ant that seems safe. The "catch" is that the technique must be applied by individuals who wish to control the ants on their property, and not by the widespread and spectacular treatments under federal aid, so attractive to some public agencies.

Unfortunately space does not permit the enumeration of other work leading toward the safer application of insecticides, but fortunately considerable effort is being directed along this line. These efforts should be encouraged.

I have raised a question concerning the wisdom of the Dutch elm disease program as currently conducted. Now let us look at the fire ant situation. We know that the toxicants used in the broadcast applications to "eradicate" this insect are highly destructive to some birds and mammals. That this is recognized by proponents of the program is indicated by the warnings issued. These include, covering of fishponds, keeping pets and small children off treated ground for a certain length of time after treatment, covering leafy vegetables in the garden, and preventing pets from drinking from pools of water standing after a rain.

Only a major emergency could justify a widespread program that requires such precautions. Can the fire ant be so classified? Many entomologists think not. In fact it appears from personal conversation with one entomologist who has studied the insect intensively, that the fire ant is chiefly a nuisance. On the credit side it is highly predatory upon other insects such as the cotton boll weevil and boll worm. Furthermore, it appears to be less destructive to ground nesting birds than is our native fire ant, which it seems to be replacing, and with which we have lived for generations. These facts cause us to wonder whether or not the pros and cons were carefully weighed before the fire ant program was undertaken.

Also, the program to eradicate the gypsy moth is open to question. Would it be practicable to eradicate this insect? Most forest entomologists who are acquainted with the situation believe not. I do not question the sincerity of those who proposed the project, but I do question their judgment and their competence to weigh the pros and cons. In support of this viewpoint, we in Michigan have been trying unsuccessfully to eradicate the gypsy moth from a small area around

Lansing since 1954. If, after five or six years of spraying, a small infestation such as this has not been eradicated, what chance is there that the New York-New England program can succeed?

Some entomologists seriously question whether or not the eradication of *any* thoroughly established insect can be accomplished with means now at our disposal. The case of the Mediterranean fruit fly in Florida is often cited as an example of successful eradication, but actually the results there signify little if we accept Hopkin's (1938) conclusion that the climate in Florida is unsuited to the year in year out survival of the insect.

Still other widespread control projects supported by public funds might be mentioned, some justifiable and some not. But from the examples mentioned, the evidence is clear that enough mistakes have been made to raise questions concerning the infallibility of some publicly-supported insect control agencies.

My conclusion is this and I hope that Dr. Decker will agree: that the use of insecticides is a necessity for production of foods in the quantity and quality that we require. However, numerous widespread projects involving the broadcasting of insecticides that have been endorsed enthusiastically by the public-supported control agencies, are open to question. Apparently the decision to spray or not to spray cannot be safely left to these control agencies. The temptations of empire building are too great. The pros and cons should be weighed by persons with broad training and experience, who can evaluate all available information dispassionately, thus reaching a decision that will be in the best long-term interest of mankind and as nearly unbiased as possible. The viewpoint of forest entomologists on the broadcasting of insecticides deserves special comment because it is a sensible one. It is this: All agree that the application of insecticides over large areas must be regarded as an emergency treatment, comparable to extinguishing a fire or removing a man's appendix. Control projects involving the broadcasting of insecticides should not be entered upon lightly.

REFERENCES

Arant, F. S., Kirby L. Hays, and Dan W. Speak. 1958. Facts about the imported fire ant, Highlights of Agricultural Research. The Alabama Polytechnic Institute, Auburn, Ala. 5: #4.

Baker, Roy J. 1958. Notes on some ecological effects of DDT sprayed elms. Jour. Wildlife Mgt. 22:269–274.

DeWitt, J. B. 1956. Chronic toxicity to quail and pheasants of some chlorinated insecticides. Jour. Agric. and Food Chem. 4: 863–866.

Hays, Kirby L. 1958. The present status of the imported fire ant in Argentina. Jour. Econ. Ent. 51: 111 & 112.

Hopkins, A. D. 1938. Bioclimatics. U.S.D., Misc. Pub. 280, p. 64.

George, John L. and Robert T. Mitchell. 1947. The effects of feeding DDT-treated insects to nestling birds. Jour. Econ. Ent. 40: 782–789.

Illinois Natural History Survey Publications—by several authors, Jour. Econ. Ent. 47: 624–627—Plant Reporter, 44: 163–166-etc.

Pickett, A. D., et al. 1956. Progress in harmonizing biological and chemical control of orchard pests in eastern Canada. Proceedings of the Tenth International Congress of Entomology, 3. 169–174.

Wallace, G. J. 1959. Insecticides and birds. Audubon Mag. 61: 10–12.

END OF ARTICLE

Too Many People?

Next to the problem of controlling nuclear energy, perhaps the most urgent controversy of our age is the exploding world population. There are many opinions on what to do about this problem, and the only positive statement that can be made is that the problem remains. Inquiring minds, free to express opinions, must provide solutions.

Two essays by contemporary scientists follow as an illustration. While the general theme of these two articles is similar, there are differences in approach to the problem under discussion as well as in the degree of technicality employed by the two authors to communicate their ideas to the reader.

Note also that the article by Darwin was written in 1956, some years ago—as we measure time in this age of rapid scientific and technological advance—while the work by Hardin is much more recent.

FORECASTING THE FUTURE[1]

by SIR CHARLES DARWIN

We none of us can help hoping that when anyone undertakes to prophesy the future, the facts will prove him wrong. I share this taste

[1] Reprinted by permission from *Engineering and Science,* April, 1956. pp. 22–36.

myself, and yet it may appear that I too am starting to prophesy. In fact I am going to try and do something much more modest. Forecasting is the word used for the predictions that the meteorologists make about the probable future weather, and this is the analogy I am going to follow. Through the reports he receives the meteorologist knows better than the rest of us what is happening in other parts of the world, and though he is very conscious that there are a great many things he does not know, with the information and experience that he has, he is in a good position to forecast the *probabilities* of future weather.

Sir Charles Darwin.

The present director of the British Meteorological Office, Sir Graham Sutton, recently wrote an article which describes the situation admirably. In making his forecast the meteorologist is doing the same sort of thing that a player does when he bids his hand at the game of bridge. If he were required to predict what tricks he would take with absolute certainty, he would not get very far; for example, if he had the ace and king of a suit he would only be *absolutely* certain of two tricks if that suit were trumps.

In fact, he does not declare that he will get two tricks, but he makes the estimate that he will probably get say eight or nine tricks. He reckons that this is the probability; he knows that one or two of his strongest cards may possibly fail to win the tricks he expects, but then he knows that this will most likely be compensated by tricks from some of his other cards he was not so confidently counting on. He estimates probabilities, and if he is an experienced player he is usually not far from right in a general way, even though some of his details may be wrong.

That is the sort of prediction that the meteorologist makes about the weather, and it is the sort of prediction that I am going to try and make about the future prospects of the world.

I want to work out this analogy with meteorology rather further. There are two separate branches of that subject, called respectively weather-forecasting and climatology. In forecasting, the meteorologist uses all the detailed knowledge of conditions in the world at the present moment and applies to them the laws of mechanics and also a good deal of personal experience and personal judgment, and from all this he says what things will be like twenty-four hours hence, and he usually gets it fairly right. He also tries to do forty-eight hours, but has a good deal less confidence about that, because as time goes on the things he does not know get proportionately more and more important.

The subject of climatology is quite different. In this there is no forecasting of what things will be like tomorrow, but instead there are general statements such as that this place will be a desert, that place a tropical jungle, while yet another one has a climate which will support good agriculture most of the time. It is much less detailed but a much more general subject, and it is one that must always be in the back of the mind of the forecaster when he makes his predictions.

I am going to try and make a forecast for the fairly close future, say fifty or a hundred years, but before coming to that I must say something about what I call the climatology of my subject, because that really is a deeper part of it. I will begin this by taking a simplified example. Suppose that somewhere in the ocean there is an island that is completely isolated from contacts with other parts of the world. I am told, in a general way, such things as what its climate is, how hot it is, how much rainfall it has, and what the soil is like. I am also told a little about the inhabitants and their state of culture—say that they know about the use of metals, but have only rather inferior food crops.

With only this information I could say a great deal about the life of the island; for instance, I could make a very fair estimate of the numbers of its population. To do this I should take as my principle that the normal way that any living species of animal survives is by producing too many offspring, of which only a fraction survive. With many lower animals the excess is often enormous, with a million produced of which only one may survive, but the same rule holds for the higher animals, too; the excess production is much less, but it is still there.

The same rule applies to man. The families on the island will mostly each produce several children and the parents will do all they can to keep their children alive and to bring them up. Now, simply to replace the numbers of the two parents, two children would be enough, but most peasant families surely produce more than two children, so that there is a tendency for the population to increase.

What is it that determines the total population then? The whole island will have come under cultivation, and it will be yielding all the food it can. Through the uncertainties of the weather, in some years there will be good harvests and in some years bad, and the peasants will accumulate a certain amount of reserve food against the bad harvests. But sometimes there will be two or three bad years running, and then they will get short of food, and perhaps two or three times in a century there may be four bad years running, and then there will be real famine. It will be these occasional famines that will determine the average number of people on the island.

This is not the sort of thing we see now anywhere in the world, but, for example, it was what used to control the population of India until about a hundred years ago. All this may seem rather obvious, but it is worth noting that we can say with some confidence that one of the most important features in the life of the island will be famines at the rate perhaps of three a century, and it is these famines that will mainly determine the number of people on it.

Now, suppose that the island has settled down into this state, but that its perfect isolation is broken by a ship which is wrecked on its coast and in which there happens to be a cargo of potatoes or some such crop. The new crop will give a much better yield than any of the previous food crops of the island, and it will be gradually adopted by the inhabitants. Every acre of ground will now yield twice as much food as it did previously.

Man is a rather slow breeder, so that the most conspicuous thing first to be noticed is that there is plenty of food for everyone. The bad old days of famines have disappeared and the population starts to increase. The historians of the island will record that it is a Golden Age, with an easy life very different from that of their parents. They will probably have a very human failing; they will forget about the cargo of potatoes, and they will claim how clever the present inhabitants are in overcoming the difficulties of life that used to afflict their ancestors.

This Golden Age will go on for a century or two, while the population increases to double its previous numbers, but at the end of that time the old troubles will begin all over again, because now again the yield of the crops will only be about enough to provide food for the new numbers of the population. There will be the old trouble over occasional successions of bad harvests which will produce famines again, and this will limit the population in the same old way. Something very like this was what happened in Ireland in the 1840's.

I have developed this imaginary example at some length, because it has a most important application to the present condition of the whole world. The world is just now in a highly abnormal condition, as is shown by the consideration of the increasing numbers of humanity. We are living in a Golden Age, which for man may well be the most wonderful Golden Age of all time. The historians have made fairly reliable estimates of the numbers of world population at different periods of history, and these numbers reveal it rather clearly. At the beginning of the Christian era the population of the world was about 350 million. It fluctuated up and down a bit, and by A.D. 1650 it was still only 470 million. But by 1750 it had risen to 700 million, and now it is 2500 million. That is to say that for 1700 years it was fairly constant, and then in 200 years it has suddenly quadrupled itself.

The increase of world population is still going on at a rate of doubling itself in a century, but it is a most menacing thing to think about. Year in year out the increase is at a rate of about one percent, and this means that every day there are 80,000 more people on the earth. That is the daily difference between the number of babies born and the number of people dying. Even those who are not conscious of this fact are unconsciously used to it, and accept it as natural, but it quite obviously cannot go on forever like this, and the most crucial question for us all is how long it *can* go on.

An abnormal state of affairs

This will be the main thing I shall want to discuss, but to see how abnormal the present condition is, I will imagine for a moment that it was the normal condition and I will look at the consequences that would follow. If the population were going to be able to double itself in each century, it would only be two thousand years before it was a million times what it is now, and two thousand years is only a short time in the period of human history. As a matter of simple arithmetic, if the population were a million times what it is now, there would be just about standing room on the land surfaces of the earth, but not room for the people to lie down! This would obviously be a fantastically impossible state of affairs, but it illustrates what an abnormal state the world is in just now with its population increasing at this rate.

It is obvious what has produced this present abnormal state of the world. There have been two chief causes. One of them was the discovery of the New World, much of it barely inhabited, which has provided enormous areas for possible expansion, in particular for the white races. The other is the development of science, through which it has been possible for man to find ways of producing a great deal more food, and in particular of transporting it from the places where it is produced to the places where it is needed. The Scientific Revolution, which began about three hundred years ago, must rank as one of the two really great episodes in human history; the only thing comparable with it in importance is the Agricultural Revolution. This happened in about 10,000 B.C., when man learned how to become a food grower instead of merely a food collector.

The climatology of humanity

I want to give more consideration to what I have called by analogy the climatology of humanity. As I have shown, the present time is very abnormal, and so present conditions cannot be of much help in this. Are there any deeper principles that can be used? I think there are sufficient of them for us to be able to say a good deal about it. The first point is that the climate—and here I mean the actual climate—of the earth has been fairly constant for something like a thousand million years at least. It is eminently reasonable, then, to expect that we can count on it for say at least one more million years. Here is one constant datum we can use in our estimates.

A second thing is the finite size of the earth, and the fact that its whole surface is now fairly well known. This knowledge, of course, is quite a new thing; even a century ago there were great areas in Africa and South America that were hardly known, and they might have held something quite unexpected. There may, of course, still be many things to be discovered; there might possibly be other gold fields like the South African one, or perhaps great ore-fields of other, more practically valuable metals, but we can now be fairly confident that there is not room on the earth for anything, at present unknown, on a scale that would materially alter the possibilities of our ways of life.

The third principal we can use is much the most important. It is human nature. The characteristics of mankind are conveniently, though only roughly, divided into two parts, which have—as I think, rather clumsily—been called nature and nurture. Nurture signifies the environment in which people grow up and live, and it is, of course, what determines most of their day-to-day behavior. It is thus immensely important in making the short-term forecast, but the conditions of life have varied enormously from century to century, and they will surely continue to do so, and therefore nurture gives little reliable help in estimating what the long-term character of human life will be.

The matter is quite different when we consider nature. Here, as we know from the study of many types of animals heredity plays a predominating part, and so for as long as any of us can really care about—say a hundred thousand years at least—we must accept that man will be just like what he is now, with all his virtues and all his defects. There is simply no prospect at all of any millennium in which pure virtue triumphs, because that is not in the nature of the species Homo Sapiens. In so far as heredity determines man's behavior, we can take this as a constant in making our predictions about his destiny.

The most important human characteristics, for my present purpose, are the deepest instincts which human beings have. These are the instincts which are directed towards the perpetuation of the species. One of them is the fear of death, shared by such a vast proportion of humanity that even under the most dreadful catastrophes very few people do actually commit suicide. This instinct serves to help in keeping the individual alive.

Equally important are the instincts serving to reproduce the species. In man and in the higher animals this characteristic falls into two rather separate parts, the sexual instinct and the parental instinct. Among the animals these two instincts suffice to perpetuate the species, and until very recently the same has been true of man. Things have, however, been changed by the developments of methods of birth-control, which have revealed a curious gap in our equipment of instincts.

Most people feel the sexual instinct with a force almost as great as the fear of death, and most people, when they have got children, have a very intense instinct to care for them and bring them up, but a good many people lack the desire to have children in advance; or at any rate, if they have the instinct, it is very much weaker than the other two. The parental instinct seems to be evoked mainly by the presence of the children, and thus it has come about that the sexual instinct can be satisfied without leading to the consequence it ought to have of ensuring the creation of a next generation. This third instinct, coming between the sexual and the parental, may be called the procreative instinct; it is much weaker than the other two, and indeed seems to be absent in a good many people.

Long-range forecast

The really important condition essential for human life was first fully described by Thomas Malthus in 1799, in his celebrated book, *An Essay on Population*. In this he drew attention to the necessity of a balance between the numbers of a population and the food it will require. He pointed out, with numerous examples, that there is a tendency for population to increase in geometrical ratio, whereas the area from which they will derive their food cannot possibly increase in this ratio.

Malthus could not be expected to have foreseen the consequences of the Scientific Revolution, which was going for a time entirely to upset the balance between the two sides of his account. During the 19th century it was possible to take the view that the disasters foretold by him had not occurred and that, therefore, his principles had been proved wrong.

This comfortable view overlooked the fact that all through that century population was, in fact, increasing geometrically, just as he had said, but for a time this was being balanced by the opening up for agriculture of barely inhabited regions in the New World, from which the newly invented railways and steamboats could convey the food to the places where it was needed.

It was the developments of the Scientific Revolution that for a time upset Malthus's balance, but now once again the balance is coming into effect, because we are now very fully conscious of the finiteness of the earth. There are few more regions that can be opened out for agriculture, and once again we have to face the problem of how our rapidly increasing populations are to be fed.

Population and food production

I have noticed that most people, when for the first time they face the population problem, at once think about the possibilities of producing more food. They first think perhaps of the fields we all notice here and there that are not being properly cultivated. Then they may think of improved breeds of plants that will produce two or three crops a year instead of only one. Then there is the possibility of cultivating the ocean. And there is the Chlorella, an alga which might be grown on a sort of moving belt in a factory; it can produce proteins perhaps ten times more efficiently than the garden vegetables do, but unfortunately at a hundred times the cost. Finally with the rapid progress in our knowledge of chemistry, it is not to be excluded that one day the food-stuffs necessary for life will be synthesized in factories from their original elements, carbon, nitrogen, phosphorus and so on.

All these things are possible, and I do not doubt that some of them will be done, but to accomplish them is no help, because of the central point made by Malthus, that there has to be a balance between food production and population numbers. Until population numbers are controlled, it will always continue to be true that, *no matter what food is produced there will be too many mouths asking for it.* New discoveries in the way of food production may make it possible for many more people to keep alive, but what is the advantage of having twenty billion hungry people instead of only three billion?

In the light of these considerations it seems to me that the food problem can be left to look after itself and that all attention must be given to the other side of the balance. Can anything be done about it? Frankly, though perhaps for a short term something might be done, in the long run I doubt it. My reason is this. Nature's control of animal populations is a simple, brutal one. In order to survive, every animal produces too many for the next generation, and the excess is killed off in one way or another. It is a method of control of tremendous efficiency, and during most of his history it has also applied to man. To replace a mechanism of this tremendous efficiency it is no use thinking of anything small; the alternative we must offer, if we want to beat nature, must also be tremendous.

The difficulty is even greater than it appears at first sight, because there would be an instability about any alternative scheme deliberately adopted. Thus, suppose some really good solution was found, and was adopted by half the world. For a generation or two this half would prosper. Its numbers would stay constant and the people would not be hungry, but all the time the numbers in the other half of the world would be increasing, so that in the end they would swamp the first half. That is the terrible menace of the matter; there is a strong survival value in being one of those who refuse to limit population.

The most easily imagined solution would be the establishment of some *world-wide* creed prohibiting large families, but when we reflect how many rival religious creeds there already are, all largely subsisting on account of their mutual differences, there seems little hope for any universal creed which would permanently limit population in this way.

It is very much to be hoped that a great deal of thought will be given to this matter on the chance that someone may hit on a solution, but I must repeat that nature's method of limiting population is so brutally tremendous that it can never be replaced by any such triviality as the extension of methods of birth control. It calls for something much more tremendous if there is to be any prospect of success.

Short-range forecast

I have said all I want to say about what by analogy I called climatology, and I will turn to weather forecasting; that is to say, I will attempt to forecast what will happen in the near future of say 50 or 100 years. I would remind you of the description of forecasting that I gave at the start, that it is like declaring a hand at bridge, where one makes a general estimate on incomplete data and one only expects to be right in general and not in detail. The weather forecaster can only do his work by receiving a great deal of information coming from all over the earth, and I need similar information for my forecasting. I have derived this from a fairly wide variety of sources. One of the most useful sources was a book entitled *The Challenge of Man's Future,* by Prof. Harrison Brown of Caltech. As a geochemist his study of the prospects of shortages in the future supply of various minerals led him on to study other shortages facing the world. A second book, *The Future of Energy*, by P. C. Putnam, deals very usefully with a narrower subject, the rate of exhaustion of our present fuel supplies and the various possible alternatives to them. Another very valuable source of information came from attendance at the UNO Conference on Population which was held in Rome in 1954. I may also refer to a book, *World Population and Resources,* recently composed in England by the organization known as P.E.P.

Cautious estimates

As I have already shown, we have been living during the past hundred and fifty years or so in a period of history of quite unique prosperity. Expert demographers estimate that our present two and a half billion population will have become four billion by A.D. 2000 and six billion by A.D. 2050. These estimated increases will be fairly equally distributed among the different races and among the social classes in each. For example, one of the most rapidly increasing groups at present consists of the moderately well-to-do Americans, who are increasing at a rate faster than the peoples of India or Japan. I may say that these estimates should be regarded as cautious ones.

The first thing we may think of which might reduce the numbers is war, but most war is not nearly murderous enough to have any effect. Thus we should count as a really bad war one in which five million people were killed, but this would only set back the population increase for less than three months, and that hardly seems to matter. I doubt if even an atomic war would have any serious influence on the estimate, unless it led to such appalling destruction of both the contestants that the economy of the whole world was so entirely ruined that barbarism and starvation would ensue. There is perhaps some hope that man will be wise enough not to embark on such a war, but anyhow I shall refuse to consider it in my forecast.

Some people may feel that methods of birth control might upset the whole forecast. This is a most important matter, which must be considered. The proponents say a contraceptive may be discovered which would put in our hands the possibility of completely controlling population numbers. It is very possible that such a discovery may be made, and I hope it will, but I do not think it seriously affects the forecast. This is because of the time scale in human affairs. Even if we already possessed the full knowledge of what I may call the "contraceptive pill," a good deal of time would be taken in building factories to make it on a scale large enough to provide pills for the whole world population and the world-wide distribution would take some arranging; but there are other more serious troubles which would also have to be overcome.

It is hardly likely that the physiologists could be absolutely confident that such a drastic medicine would have no collateral effect at all, and to verify this, many years of experiment on a smaller scale would be necessary. For example, it would take two or three decades to verify that when the habitual users of the pill did decide to have children, those children would grow up into normal adults. It would be necessary to verify that there were no unforeseen collateral effects, such as a premature aging of the habitual user, or perhaps a special liability to some disease. I may quote as a parallel the liability of people exposed to X-rays to develop cancer a good many years later.

Furthermore there would need to be an enormous educational campaign, and the number of educators would have to be so vast that it would take all of a generation to train them, and therefore two generations for them to produce their results.

On all these counts I think it is safe to say that no large-scale effects could possibly be seen under two generations or so, and therefore the contraceptive pill—which in fact we have not got yet—would have little influence in affecting the forecast for fifty years, though it might for a hundred. But things are unlikely to be even as favorable as this; there are religious doctrines that might prohibit the use of the pill and there is a tremendous stock of unreasoning emotion in such intimate matters that would make a lot of unforeseeable difficulties.

A population of four billion

In the light of these considerations I see no escape from the estimate that by A.D. 2000 the world population will be four billion.

It is time to turn to the other side of the Malthusian account. Malthus only thought of actual food production as the balancing item, but since his day there are a lot of other things to be included which he could not have foreseen—such things as the supply of energy and the metals which are essential for the city life which alone can carry large populations.

First, the agriculturists at the 1954 Rome Conference on Population claimed that a doubling of food production can probably be achieved, but to do so everything has got to be exactly right. There must be no creation of dust bowls by the exhaustion of poor soils, and the stores of artificial fertilizers must not be distributed freely, but must be controlled so that they are only used in the places where they will give the most advantage. I am not competent to discuss this matter, but I do wonder how far this strict control will be possible.

In connection with agriculture I may refer to a thing of the recent past which is at least suggestive. Between 1947 and 1953 the world's agriculture made the most tremendous strides; in these seven years it increased by 8 percent, a truly wonderful performance, which we owe largely to the brilliant work of the scientific agriculturists. *But*—during those seven years the world's population increased not by 8 percent, but by 11 percent, so that the world was hungrier at the end than at the beginning. So, as I have said, I forecast there will be four billion people in fifty years from now, but I forecast that they will be hungrier than the two and a half billion we have now.

Now, to turn to other matters, Malthus needed only to think about agriculture, but we have to consider the provision of a lot of other things, because since his day the enormously increased numbers can only exist by living in large cities, and these demand all sorts of equipment like good roads, railways, water supply, electricity and so on. If some of these things could not be supplied it would be quite impossible to maintain the large numbers we have. So we must add to the right-hand side of Malthus's balance sheet things like energy and metals, and consider whether the supply of these will be adequate to keep us going for the next fifty years.

The prospects for energy

As to energy, as far as we can see the prospects are not too bad. There are only three sources which can provide power in quantities sufficient to be important. They are the "fossil fuels" coal and oil, nuclear energy, and the direct use of sunlight. Notice that water-power is not in the list; this is because the total quantity yielded, if all the rivers of the whole earth were fully exploited, would be only 12 percent or so of even the present energy developed.

At present, of course, practically all the power comes from coal and oil, and it is being used up at an ever-increasing rate. It is not possible to estimate the reserves with any great accuracy because it would be necessary to take some standard of the ease with which the coal can be won; for example, would it ever be worth while to mine a seam only a foot thick? But an estimate very definitely on the optimistic side predicts that the coal will be all gone in 500 years. Since it took some 500 million years to make the coal, it may be said—speaking only very loosely, of course—that we are living on our capital at the rate of a million to one. Is it surprising that we can create wonderful prosperity for a short time? Oil is won much more easily than coal, and it is expected it may at most last for a century.

The prospects for nuclear energy are good, but the construction of nuclear power stations will inevitably take a good many years. It has been estimated that at the end of 30 or 40 years something like a quarter of the power developed in Britain will come from uranium instead of coal. Even at the present rates of consumption of power this would still mean a very large demand for coal, and as the demand in fact is growing year by year, there seems little prospect of the coal situation improving. Indeed, I would not be surprised if there was going to be a rather awkward period for us in about 50 years, when the expense of winning the remaining coal has increased a good deal, while there are still not enough nuclear power stations.

These difficulties apply specially in Great Britain. In America the situation is much easier in respect to coal. It is being consumed at an almost fantastic rate here, but there would seem to be enough easily mineable coal to last you a century. I have called the rate fantastic, and this can be justified by the following consideration. In the history of the world man has burnt up a very considerable amount of coal in all, but half of this total has been burnt in the United States since 1920.

Favorable prospects

As far as we can judge in these rather early days there is not likely to be any shortage of uranium for many centuries, and there is also always the possibility that the fusion of deuterium into helium may be made to occur slowly instead of, as now, only in the form of a super-bomb. The prospects for the supply of energy are therefore rather favorable, but it must be noticed that it may make very considerable changes in our ways of life. Nuclear power units are likely to have to be very large, and this may mean that there will have to be far fewer small units such as motor-cars. This suggests that in the nuclear age the population will be concentrated in the great cities even more than it is now.

The energy arriving at the earth day by day in the form of sunlight is quite enormous, and if it could be turned into mechanical power it would supply many times over the needs of mankind. A square yard facing the sun receives energy at a rate of about a horse power, but this implies that a great area would be required in order to make any reasonable power station. It may well be that improving techniques will solve this problem, but there is certainly a long way to go. Indeed it is rather humiliating to know that at the present time the most efficient way of collecting solar energy is to plant a row of trees, let them grow, cut them down and burn them.

If the provision of energy is not necessarily going to be a great difficulty, the same cannot be said of many other raw materials, in particular many of the metals, though even the supply of such a common thing as fresh water is going to be a formidable problem. Of course, strictly speaking, the metals, unlike coal, are indestructible; once won they can be used again and again, but in fact there is always some wastage due to wear or to actual loss, and this wastage must be allowed for. There has been the same enormous increase in the extraction of metals as of coal; in fact, of all the metal mined from the earth, half has been dug up in the last 30 years.

The possession of metals in great quantity seems to be essential for industrial development. It would appear likely that there simply is not enough of many of them, such as lead or tin or copper, to permit the underdeveloped countries to become industrialized on a scale at all equivalent to that of the highly developed ones. It is true that substitutes can often be found, but usually they will be inferior; for example, an electric transformer could be made with aluminum wires to replace the copper, but it would be less efficient. The underdeveloped countries which are trying to improve their industrial power are already handicapped in two respects. They lack capital, and they lack engineering experience, and to these difficulties must be added a third, the expected world shortage of constructional materials. So I forecast that at the end of this century industrialization will not have spread very greatly over the less developed parts of the world.

My general conclusion then is that in fifty years the population of the world will be four billion. They will be a rather hungry four billion, busily engaged in straining the resources of the earth to yield enough food, but they will not have succeeded very much in their present ambitions about becoming more industrialized.

I regard the forecast for a century with a great deal more doubt. The demographers forecast six billion for the year 2050, but my own guess is that the world will not have succeeded in yielding enough food for this, and that by then the world will have begun to go back into what I earlier called its normal state, the state in which natural selection operates by producing rather too many people, so that the excess simply cannot survive.

A gloomy picture

I fear this is a gloomy picture, and I ought to say that there are many people who forecast quite the opposite. They are the technological enthusiasts. They claim that whenever a shortage has declared itself the technologists have produced a substitute and that things will go on forever like that. To me they do not seem to appreciate the overwhelming importance and difficulty concerning the population numbers, and that is why I must disagree with them. If they are right and I am wrong the world can look forward longer than I expect to a continuance of the present era of prosperity.

I hope that they will prove right, and that I shall be proved wrong, but I must repeat my opinion that the central problem is that of world-population. I do not see any happy solution of this, but I earnestly hope that if many people face the difficulties, someone may possibly be inspired to find an acceptable solution.

END OF ARTICLE

THE TRAGEDY OF THE COMMONS*

The population problem has no technical solution;
it requires a fundamental extension in morality.

by GARRETT HARDIN†

At the end of a thoughtful article on the future of nuclear war, Wiesner and York (1) concluded that: "Both sides in the arms race are . . . confronted by the dilemma of steadily increasing military power and steadily decreasing national security. *It is our considered professional judgment that this dilemma has no technical solution.* If the great powers continue to look for solutions in the area of science and technology only, the result will be to worsen the situation."

I would like to focus your attention not on the subject of the articles (national security in a nuclear world) but on the kind of conclusion they reached—namely that there is no technical solution to the problem. An implicit and almost universal assumption of discussions published in professional and semipopular scientific journals is that the problem under discussion has a technical solution. A technical solution may be defined as one that requires a change only in the techniques of the natural sciences, demanding little or nothing in the way of change in human values or ideas of morality.

In our day (though not in earlier times) technical solutions are always welcome. Because of previous failures in prophecy, it takes courage to assert that a desired technical solution is not possible. Wiesner and York exhibited this courage; publishing in a science journal, they insisted that the solution to the problem was not to be found in the natural sciences. They cautiously qualified their statement with the phrase, "It is our considered professional judgment. . . ." Whether they were right or not is not the concern of the present article. Rather, the concern here is with the important concept of a class of human problems which can be called "no technical solution problems," and, more specifically, with the identification and discussion of one of these.

It is easy to show that the class is not a null class. Recall the game of tick-tack-toe. Consider the problem, "How can I win the game of

* Reprinted from Science, Dec. 13, 1968, Vol. 162, pp. 1243–1248 published by the American Association for the Advancement of Science, Washington, D.C. by permission of the publisher.
† The author is professor of biology, University of California, Santa Barbara. This article is based on a presidential address presented before the meeting of the Pacific Division of the American Association for the Advancement of Science at Utah State University, Logan, 25 June 1968.

tick-tack-toe?" It is well known that I cannot, if I assume (in keeping with the conventions of game theory) that my opponent understands the game perfectly. Put another way, there is no "technical solution" to the problem. I can win only by giving a radical meaning to the word "win." I can hit my opponent over the head; or I can drug him; or I can falsify the records. Every way in which I "win" involves, in some sense, an abandonment of the game, as we intuitively understand it. (I can also, of course, openly abandon the game—refuse to play it. This is what most adults do.)

The class of "No technical solution problems" has members. My thesis is that the "population problem," as conventionally conceived, is a member of this class. How it is conventionally conceived needs some comment. It is fair to say that most people who anguish over the population problem are trying to find a way to avoid the evils of overpopulation without relinquishing any of the privileges they now enjoy. They think that farming the seas or developing new strains of wheat will solve the problem—technologically. I try to show here that the solution they seek cannot be found. The population problem cannot be solved in a technical way, any more than can the problem of winning the game of tick-tack-toe.

What Shall We Maximize?

Population, as Malthus said, naturally tends to grow "geometrically," or, as we would now say, exponentially. In a finite world this means that the per capita share of the world's goods must steadily decrease. Is ours a finite world?

A fair defense can be put forward for the view that the world is infinite; or that we do not know that it is not. But, in terms of the practical problems that we must face in the next few generations with the foreseeable technology, it is clear that we will greatly increase human misery if we do not, during the immediate future, assume that the world available to the terrestrial human population is finite. "Space" is no escape (2).

A finite world can support only a finite population; therefore, population growth must eventually equal zero. (The case of perpetual wide fluctuations above and below zero is a trivial variant that need not be discussed.) When this condition is met, what will be the situation of mankind? Specifically, can Bentham's goal of "the greatest good for the greatest number" be realized?

No—for two reasons, each sufficient by itself. The first is a theoretical one. It is not mathematically possible to maximize for two (or more) variables at the same time. This was clearly stated by von Neumann and Morgenstern (*3*), but the principle is implicit in the theory of partial differential equations, dating back at least to D'Alembert (1717–1783).

The second reason springs directly from biological facts. To live, any organism must have a source of energy (for example, food). This energy is utilized for two purposes: mere maintenance and work. For man, maintenance of life requires about 1600 kilocalories a day ("maintenance calories"). Anything that he does over and above merely staying alive will be defined as work, and is supported by "work calories" which he takes in. Work calories are used not only for what we call work in common speech; they are also required for all forms of enjoyment, from swimming and automobile racing to playing music and writing poetry. If our goal is to maximize population it is obvious what we must do: We must make the work calories per person approach as close to zero as possible. No gourmet meals, no vacations, no sports, no music, no literature, no art. . . . I think that everyone will grant, without argument or proof, that maximizing population does not maximize goods. Bentham's goal is impossible.

In reaching this conclusion I have made the usual assumption that it is the acquisition of energy that is the problem. The appearance of atomic energy has led some to question this assumption. However, given an infinite source of energy, population growth still produces an inescapable problem. The problem of the acquisition of energy is replaced by the problem of its dissipation, as J. H. Fremlin has so wittily shown (*4*). The arithmetic signs in the analysis are, as it were, reversed; but Bentham's goal is still unobtainable.

The optimum population is, then, less than the maximum. The difficulty of defining the optimum is enormous; so far as I know, no one has seriously tackled this problem. Reaching an acceptable and stable solution will surely require more than one generation of hard analytical work—and much persuasion.

We want the maximum good per person; but what is good? To one person it is wilderness, to another it is ski lodges for thousands. To one it is estuaries to nourish ducks for hunters to shoot; to another it is factory land. Comparing one good with another is, we usually say, impossible because goods are incommensurable. Incommensurables cannot be compared.

Theoretically this may be true; but in real life incommensurables *are* commensurable. Only a criterion of judgment and a system of weighting are needed. In nature the criterion is survival. Is it better for a species to be small and hideable, or large and powerful? Natural selection commensurates the incommensurables. The compromise achieved depends on a natural weighting of the values of the variables.

Man must imitate this process. There is no doubt that in fact he already does, but unconsciously. It is when the hidden decisions are made explicit that the arguments begin. The problem for the years ahead is to work out an acceptable theory of weighting. Synergistic effects, nonlinear variation, and difficulties in discounting the future make the intellectual problem difficult, but not (in principle) insoluble.

Has any cultural group solved this practical problem at the present time, even on an intuitive level? One simple fact proves that none has: there is no prosperous population in the world today that has, and has had for some time, a growth rate of zero. Any people that has intuitively identified its optimum point will soon reach it, after which its growth rate becomes and remains zero.

Of course, a positive growth rate might be taken as evidence that a population is below its optimum. However, by any reasonable standards, the most rapidly growing populations on earth today are (in general) the most miserable. This association (which need not be invariable) casts doubt on the optimistic assumption that the positive growth rate of a population is evidence that it has yet to reach its optimum.

We can make little progress in working toward optimum population size until we explicitly exorcize the spirit of Adam Smith in the field of practical demography. In economic affairs, *The Wealth of Nations* (1776) popularized the "invisible hand," the idea that an individual who "intends only his own gain," is, as it were, "led by an invisible hand to promote . . . the public interest" (5). Adam Smith did not assert that this was invariably true, and perhaps neither did any of his followers. But he contributed to a dominant tendency of thought that has ever since interfered with positive action based on rational analysis, namely, the tendency to assume that decisions reached individually will, in fact, be the best decisions for an entire society. If this assumption is correct it justifies the continuance of our present policy of laissez-faire in reproduction. If it is correct we can assume that men will control their individual fecundity so as to produce the optimum population. If the assumption is not correct, we need to reexamine our individual freedoms to see which ones are defensible.

Tragedy of Freedom in a Commons

The rebuttal to the invisible hand in population control is to be found in a scenario first sketched in a little-known pamphlet (6) in 1833 by a mathematical amateur named William Forster Lloyd (1794–1852). We may well call it "the tragedy of the commons," using the word "tragedy" as the philosopher Whitehead used it (7): "The essence of dramatic tragedy is not unhappiness. It resides in the solemnity of the remorseless working of things." He then goes on to say, "This inevitableness of destiny can only be illustrated in terms of human life by incidents which in fact involve unhappiness. For it is only by them that the futility of escape can be made evident in the drama."

The tragedy of the commons develops in this way. Picture a pasture open to all. It is to be expected that each herdsman will try to keep as many cattle as possible on the commons. Such an arrangement may work reasonably satisfactorily for centuries because tribal wars, poaching, and disease keep the numbers of both man and beast well below the carrying capacity of the land. Finally, however, comes the day of reckoning, that is, the day when the long-desired goal of social stability becomes a reality. At this point, the inherent logic of the commons remorselessly generates tragedy.

As a rational being, each herdsman seeks to maximize his gain. Explicitly or implicitly, more or less consciously, he asks, "What is the utility *to me* of adding one more animal to my herd?" This utility has one negative and one positive component.

1) The positive component is a function of the increment of one animal. Since the herdsman receives all the proceeds from the sale of the additional animal, the positive utility is nearly $+1$.

2) The negative component is a function of the additional overgrazing created by one more animal. Since, however, the effects of overgrazing are shared by all the herdsmen, the negative utility for any particular decision-making herdsman is only a fraction of -1.

Adding together the component partial utilities, the rational herdsman concludes that the only sensible course for him to pursue is to add another animal to his herd. And another; and another. . . . But this is the conclusion reached by each and every rational herdsman sharing a commons. Therein is the tragedy. Each man is locked into a system that compels him to increase his herd without limit—in a world that is limited. Ruin is the destination toward which all men rush, each pursuing his own best interest in a society that believes in the freedom of the commons. Freedom in a commons brings ruin to all.

Some would say that this is a platitude. Would that it were! In a sense, it was learned thousands of years ago, but natural selection favors the forces of psychological denial (8). The individual benefits as an individual from his ability to deny the truth even though society as a whole, of which he is a part, suffers. Education can counteract the natural tendency to do the wrong thing, but the inexorable succession of generations requires that the basis for this knowledge be constantly refreshed.

A simple incident that occurred a few years ago in Leominster, Massachusetts, shows how perishable the knowledge is. During the Christmas shopping season the parking meters downtown were covered with plastic bags that bore tags reading: "Do not open until after Christmas. Free parking courtesy of the mayor and city council." In other words, facing the prospect of an increased demand for already scarce space, the city fathers reinstituted the system of the commons. (Cynically, we suspect that they gained more votes than they lost by this retrogressive act.)

In an approximate way, the logic of the commons has been understood for a long time, perhaps since the discovery of agriculture or the invention of private property in real estate. But it is understood mostly only in special cases which are not sufficiently generalized. Even at this late date, cattlemen leasing national land on the western ranges demonstrate no more than an ambivalent understanding, in constantly pressuring federal authorities to increase the head count to the point where overgrazing produces erosion and weed-dominance. Likewise, the oceans of the world continue to suffer from the survival of the philosophy of the commons. Maritime nations still respond automatically to the shibboleth of the "freedom of the seas." Professing to believe in the "inexhaustible resources of the oceans," they bring species after species of fish and whales closer to extinction (9).

The National Parks present another instance of the working out of the tragedy of the commons. At present, they are open to all, without limit. The parks themselves are limited in extent—there is only one Yosemite Valley—whereas population seems to grow without limit. The values that visitors seek in the parks are steadily eroded. Plainly, we must soon cease to treat the parks as commons or they will be of no value to anyone.

What shall we do? We have several options. We might sell them off as private property. We might keep them as public property, but allocate the right to enter them. The allocation might be on the basis of wealth, by the use of an auction system. It might be on the basis of merit, as defined by some agreed-upon standards. It might be by lottery. Or it might be on a first-come, first-served basis, administered to long queues. These, I think, are all the reasonable possibilities. They are all objectionable. But we must choose—or acquiesce in the destruction of the commons that we call our National Parks.

Pollution

In a reverse way, the tragedy of the commons reappears in problems of pollution. Here it is not a question of taking something out of the commons, but of putting something in—sewage, or chemical, radioactive, and heat wastes into water; noxious and dangerous fumes into the air; and distracting and unpleasant advertising signs into the line of sight. The calculations of utility are much the same as before. The rational man finds that his share of the cost of the wastes he discharges into the commons is less than the cost of purifying his wastes before releasing them. Since this is true for everyone, we are locked into a system of "fouling our own nest," so long as we behave only as independent, rational, free-enterprisers.

The tragedy of the commons as a food basket is averted by private property, or something formally like it. But the air and waters surrounding us cannot readily be fenced, and so the tragedy of the commons as a cesspool must be prevented by different means, by coercive laws or taxing devices that make it cheaper for the polluter to treat his pollutants than to discharge them untreated. We have not progressed as far with the solution of this problem as we have with the first. Indeed, our particular concept of private property, which deters us from exhausting the positive resources of the earth, favors pollution. The owner of a factory on the bank of a stream—whose property extends to the middle of the stream—often has difficulty seeing why it is not his natural right to muddy the waters flowing past his door. The law, always behind the times, requires elaborate stitching and fitting to adapt it to this newly perceived aspect of the commons.

The pollution problem is a consequence of population. It did not much matter how a lonely American frontiersman disposed of his waste. "Flowing water purifies itself every 10 miles," my grandfather used to say, and the myth was near enough to the truth when he was a boy, for there were not too many people. But as population became denser, the natural chemical and biological recycling processes became overloaded, calling for a redefinition of property rights.

How To Legislate Temperance?

Analysis of the pollution problem as a function of population density uncovers a not generally recognized principle of morality, namely: *the morality of an act is a function of the state of the system at the time it is performed*

(*10*). Using the commons as a cesspool does not harm the general public under frontier conditions, because there is no public; the same behavior in a metropolis is unbearable. A hundred and fifty years ago a plainsman could kill an American bison, cut out only the tongue for his dinner, and discard the rest of the animal. He was not in any important sense being wasteful. Today, with only a few thousand bison left, we would be appalled at such behavior.

In passing, it is worth noting that the morality of an act cannot be determined from a photograph. One does not know whether a man killing an elephant or setting fire to the grassland is harming others until one knows the total system in which his act appears. "One picture is worth a thousand words," said an ancient Chinese; but it may take 10,000 words to validate it. It is as tempting to ecologists as it is to reformers in general to try to persuade others by way of the photographic shortcut. But the essence of an argument cannot be photographed: it must be presented rationally—in words.

That morality is system-sensitive escaped the attention of most codifiers of ethics in the past. "Thou shalt not . . ." is the form of traditional ethical directives which make no allowance for particular circumstances. The laws of our society follow the pattern of ancient ethics, and therefore are poorly suited to governing a complex, crowded, changeable world. Our epicyclic solution is to augment statutory law with administrative law. Since it is practically impossible to spell out all the conditions under which it is safe to burn trash in the back yard or to run an automobile without smog-control, by law we delegate the details to bureaus. The result is administrative law, which is rightly feared for an ancient reason—*Quis custodiet ipsos custodes?*—"Who shall watch the watchers themselves?" John Adams said that we must have "a government of laws and not men." Bureau administrators, trying to evaluate the morality of acts in the total system, are singularly liable to corruption, producing a government by men, not laws.

Prohibition is easy to legislate (though not necessarily to enforce); but how do we legislate temperance? Experience indicates that it can be accomplished best through the mediation of administrative law. We limit possibilities unnecessarily if we suppose that the sentiment of *Quis custodiet* denies us the use of administrative law. We should rather retain the phrase as a perpetual reminder of fearful dangers we cannot avoid. The great challenge facing us now is to invent the corrective feedbacks that are needed to keep custodians honest. We must find ways to legitimate the needed authority of both the custodians and the corrective feedbacks.

Freedom To Breed Is Intolerable

The tragedy of the commons is involved in population problems in another way. In a world governed solely by the principle of "dog eat dog"—if indeed there ever was such a world—how many children a family had would not be a matter of public concern. Parents who bred too exuberantly would leave fewer descendants, not more, because they would be unable to care adequately for their children. David Lack and others have found that such a negative feedback demonstrably controls the fecundity of birds (*11*). But men are not birds, and have not acted like them for millenniums, at least.

If each human family were dependent only on its own resources; *if* the children of improvident parents starved to death; *if,* thus, over-breeding brought its own "punishment" to the germ line—*then* there would be no public interest in controlling the breeding of families. But our society is deeply committed to the welfare state (*12*), and hence is confronted with another aspect of the tragedy of the commons.

In a welfare state, how shall we deal with the family, the religion, the race, or the class (or indeed any distinguishable and cohesive group) that adopts overbreeding as a policy to secure its own aggrandizement (*13*)? To couple the concept of freedom to breed with the belief that everyone born has an equal right to the commons is to lock the world into a tragic course of action.

Unfortunately this is just the course of action that is being pursued by the United Nations. In late 1967, some 30 nations agreed to the following (*14*):

> The Universal Declaration of Human Rights describes the family as the natural and fundamental unit of society. It follows that any choice and decision with regard to the size of the family must irrevocably rest with the family itself, and cannot be made by anyone else.

It is painful to have to deny categorically the validity of this right; denying it, one feels as uncomfortable as a resident of Salem, Massachusetts, who denied the reality of witches in the 17th century. At the present time, in liberal quarters, something like a taboo acts to inhibit criticism of the United Nations. There is a feeling that the United Nations is "our last and best hope," that we shouldn't find fault with it; we shouldn't play into the hands of the archconservatives. However, let us not forget what Robert Louis Stevenson said: "The truth that is suppressed by friends is the readiest weapon of the enemy." If we love the truth we must openly deny the validity of the Universal Declaration of Human Rights, even though it is promoted by the United Nations. We should also join with Kingsley Davis (*15*) in attempting to get Planned Parenthood-World Population to see the error of its ways in embracing the same tragic ideal.

Conscience Is Self-Eliminating

It is a mistake to think that we can control the breeding of mankind in the long run by an appeal to conscience. Charles Galton Darwin made this point when he spoke on the centennial of the publication of his grandfather's great book. The argument is straightforward and Darwinian.

People vary. Confronted with appeals to limit breeding, some people will undoubtedly respond to the plea more than others. Those who have more children will produce a larger fraction of the next generation than those with more susceptible consciences. The difference will be accentuated, generation by generation.

In C. G. Darwin's words: "It may well be that it would take hundreds of generations for the progenitive instinct to develop in this way, but if it should do so, nature would have taken her revenge, and the variety *Homo contracipiens* would become extinct and would be replaced by the variety *Homo progenitivus*" (*16*).

The argument assumes that conscience or the desire for children (no matter which) is hereditary—but hereditary only in the most general formal sense. The result will be the same whether the attitude is transmitted through germ cells, or exosomatically, to use A. J. Lotka's term. (If one denies the latter possibility as well as the former, then what's the point of education?) The argument has here been stated in the context of the population problem, but it applies equally well to any instance in which society appeals to an individual exploiting a commons to restrain himself for the general good—by means of his conscience. To make such an appeal is to set up a selective system that works toward the elimination of conscience from the race.

Pathogenic Effects of Conscience

The long-term disadvantage of an appeal to conscience should be enough to condemn it; but has serious short-term disadvantages as well. If we ask a man who is exploiting a commons to desist "in the name of conscience," what are we saying to him? What does he hear? —not only at the moment but also in the wee small hours of the night when, half asleep, he remembers not merely the words we used but also the nonverbal communication cues we gave him unawares? Sooner or later, consciously or subconsciously, he senses that he has received two communications, and that they are contradictory: (i) (intended communication) "If you don't do as we ask, we will openly condemn you for not acting like a responsible citizen"; (ii) (the unintended communication) "If you *do* behave as we ask, we will secretly condemn you for a simpleton who can be shamed into standing aside while the rest of us exploit the commons."

Every man then is caught in what Bateson has called a "double bind." Bateson and his co-workers have made a plausible case for viewing the double bind as an important causative factor in the genesis of schizophrenia (*17*). The double bind may not always be so damaging, but it always endangers the mental health of anyone to whom it is applied. "A bad conscience," said Nietzsche, "is a kind of illness."

To conjure up a conscience in others is tempting to anyone who wishes to extend his control beyond the legal limits. Leaders at the highest level succumb to this temptation. Has any President during the past generation failed to call on labor unions to moderate voluntarily their demands for higher wages, or to steel companies to honor voluntary guidelines on prices? I can recall none. The rhetoric used on such occasions is designed to produce feelings of guilt in noncooperators.

For centuries it was assumed without proof that guilt was a valuable, perhaps even an indispensable, ingredient of the civilized life. Now, in this post-Freudian world, we doubt it.

Paul Goodman speaks from the modern point of view when he says: "No good has ever come from feeling guilty, neither intelligence, policy, nor compassion. The guilty do not pay attention to the object but only to themselves, and not even to their own interests, which might make sense, but to their anxieties" (*18*).

One does not have to be a professional psychiatrist to see the consequences of anxiety. We in the Western world are just emerging from a dreadful two-centuries-long Dark Ages of Eros that was sustained partly by prohibition laws, but perhaps more effectively by the anxiety-generating mechanisms of education. Alex Comfort has told the story well in *The Anxiety Makers* (*19*); it is not a pretty one.

Since proof is difficult, we may even concede that the results of anxiety may sometimes, from certain points of view, be desirable. The larger question we should ask is whether, as a matter of policy, we should ever encourage the use of a technique the tendency (if not the intention) of which is psychologically pathogenic. We hear much talk

these days of responsible parenthood; the coupled words are incorporated into the titles of some organizations devoted to birth control.
Some people have proposed massive propaganda campaigns to instill
responsibility into the nation's (or the world's) breeders. But what is
the meaning of the word responsibility in this context? Is it not merely
a synonym for the word conscience? When we use the word responsibility in the absence of substantial sanctions are we not trying to brow-
beat a free man in a commons into acting against his own interest?
Responsibility is a verbal counterfeit for a substantial *quid pro quo*. It is
an attempt to get something for nothing.

If the word responsibility is to be used at all, I suggest that it be
in the sense Charles Frankel uses it (*20*). "Responsibility," says this
philosopher, "is the product of definite social arrangements." Notice
that Frankel calls for social arrangements—not propaganda.

Mutual Coercion Mutually Agreed upon

The social arrangements that produce responsibility are arrangements
that create coercion, of some sort. Consider bank-robbing. The man
who takes money from a bank acts as if the bank were a commons.
How do we prevent such action? Certainly not by trying to control his
behavior solely by a verbal appeal to his sense of responsibility. Rather
than rely on propaganda we follow Frankel's lead and insist that a bank
is not a commons; we seek the definite social arrangements that will
keep it from becoming a commons. That we thereby infringe on the
freedom of would-be robbers we neither deny nor regret.

The morality of bank-robbing is particularly easy to understand
because we accept complete prohibition of this activity. We are willing
to say "Thou shalt not rob banks," without providing for exceptions.
But temperance also can be created by coercion. Taxing is a good
coercive device. To keep downtown shoppers temperate in their use of
parking space we introduce parking meters for short periods, and
traffic fines for longer ones. We need not actually forbid a citizen to
park as long as he wants to; we need merely make it increasingly expensive for him to do so. Not prohibition, but carefully biased options
are what we offer him. A Madison Avenue man might call this persuasion; I prefer the greater candor of the word coercion.

Coercion is a dirty word to most liberals now, but it need not forever be so. As with the four-letter words, its dirtiness can be cleansed
away by exposure to the light, by saying it over and over without
apology or embarrassment. To many, the word coercion implies arbitrary decisions of distant and irresponsible bureaucrats; but this is not a

necessary part of its meaning. The only kind of coercion I recommend is mutual coercion, mutually agreed upon by the majority of the people affected.

To say that we mutually agree to coercion is not to say that we are required to enjoy it, or even to pretend we enjoy it. Who enjoys taxes? We all grumble about them. But we accept compulsory taxes because we recognize that voluntary taxes would favor the conscienceless. We institute and (grumblingly) support taxes and other coercive devices to escape the horror of the commons.

An alternative to the commons need not be perfectly just to be preferable. With real estate and other material goods, the alternative we have chosen is the institution of private property coupled with legal inheritance. Is this system perfectly just? As a genetically trained biologist I deny that it is. It seems to me that, if there are to be differences in individual inheritance, legal possession should be perfectly correlated with biological inheritance—that those who are biologically more fit to be the custodians of property and power should legally inherit more. But genetic recombination continually makes a mockery of the doctrine of "like father, like son" implicit in our laws of legal inheritance. An idiot can inherit millions, and a trust fund can keep his estate intact. We must admit that our legal system of private property plus inheritance is unjust—but we put up with it because we are not convinced, at the moment, that anyone has invented a better system. The alternative of the commons is too horrifying to contemplate. Injustice is preferable to total ruin.

It is one of the peculiarities of the warfare between reform and the status quo that it is thoughtlessly governed by a double standard. Whenever a reform measure is proposed it is often defeated when its opponents triumphantly discover a flaw in it. As Kingsley Davis has pointed out (*21*), worshippers of the status quo sometimes imply that no reform is possible without unanimous agreement, an implication contrary to historical fact. As nearly as I can make out, automatic rejection of proposed reforms is based on one of two unconscious assumptions: (i) that the status quo is perfect; or (ii) that the choice we face is between reform and no action; if the proposed reform is imperfect, we presumably should take no action at all, while we wait for a perfect proposal.

But we can never do nothing. That which we have done for thousands of years is also action. It also produces evils. Once we are aware that the status quo is action, we can then compare its discoverable advantages and disadvantages with the predicted advantages and disadvantages of the proposed reform, discounting as best we can for our lack of experience. On the basis of such a comparison, we can make a rational decision which will not involve the unworkable assumption that only perfect systems are tolerable.

Recognition of Necessity

Perhaps the simplest summary of this analysis of man's population problems is this: the commons, if justifiable at all, is justifiable only under conditions of low-population density. As the human population has increased, the commons has had to be abandoned in one aspect after another.

First we abandoned the commons in food gathering, enclosing farm land and restricting pastures and hunting and fishing areas. These restrictions are still not complete throughout the world.

Somewhat later we saw that the commons as a place for waste disposal would also have to be abandoned. Restrictions on the disposal of domestic sewage are widely accepted in the Western world; we are still struggling to close the commons to pollution by automobiles, factories, insecticide sprayers, fertilizing operations, and atomic energy installations.

In a still more embryonic state is our recognition of the evils of the commons in matters of pleasure. There is almost no restriction on the propagation of sound waves in the public medium. The shopping public is assaulted with mindless music, without its consent. Our government is paying out billions of dollars to create supersonic transport which will disturb 50,000 people for every one person who is whisked from coast to coast 3 hours faster. Advertisers muddy the airwaves of radio and television and pollute the view of travelers. We are a long way from outlawing the commons in matters of pleasure. Is this because our Puritan inheritance makes us view pleasure as something of a sin, and pain (that is, the pollution of advertising) as the sign of virtue?

Every new enclosure of the commons involves the infringement of somebody's personal liberty. Infringements made in the distant past are accepted because no contemporary complains of a loss. It is the newly proposed infringements that we vigorously oppose; cries of "rights" and "freedom" fill the air. But what does "freedom" mean? When men mutually agreed to pass laws against robbing, mankind became more free, not less so. Individuals locked into the logic of the commons are free only to bring on universal ruin; once they see the necessity of mutual coercion, they become free to pursue other goals.

I believe it was Hegel who said, "Freedom is the recognition of necessity."

The most important aspect of necessity that we must now recognize, is the necessity of abandoning the commons in breeding. No technical solution can rescue us from the misery of overpopulation. Freedom to breed will bring ruin to all. At the moment, to avoid hard decisions many of us are tempted to propagandize for conscience and responsible parenthood. The temptation must be resisted, because an appeal to independently acting consciences selects for the disappearance of all conscience in the long run, and an increase in anxiety in the short.

The only way we can preserve and nurture other and more precious freedoms is by relinquishing the freedom to breed, and that very soon. "Freedom is the recognition of necessity"—and it is the role of education to reveal to all the necessity of abandoning the freedom to breed. Only so, can we put an end to this aspect of the tragedy of the commons.

REFERENCES

1. J. B. Wiesner and H. F. York, *Sci. Amer.* **211** (No. 4), 27 (1964).
2. G. Hardin, *J. Hered.* **50,** 68 (1959); S. von Hoernor, *Science* **137,** 18 (1962).
3. J. von Neumann and O. Morgenstern, *Theory of Games and Economic Behavior* (Princeton Univ. Press, Princeton, N.J., 1947), p. 11.
4. J. H. Fremlin, *New Sci.*, No. 415 (1964), p. 285.
5. A. Smith, *The Wealth of Nations* (Modern Library, New York, 1937), p. 423.
6. W. F. Lloyd, *Two Lectures on the Checks to Population* (Oxford Univ. Press, Oxford, England, 1833), reprinted (in part) in *Population, Evolution, and Birth Control*, G. Hardin, Ed. (Freeman, San Francisco, 1964), p. 37.
7. A. N. Whitehead, *Science and the Modern World* (Mentor, New York, 1948), p. 17.
8. G. Hardin, Ed. *Population, Evolution, and Birth Control* (Freeman, San Francisco, 1964), p. 56.
9. S. McVay, *Sci. Amer.* **216** (No. 8), 13 (1966).
10. J. Fletcher, *Situation Ethics* (Westminster, Philadelphia, 1966).
11. D. Lack, *The Natural Regulations of Animal Numbers* (Clarendon Press, Oxford, 1954).

12. H. Girvetz, *From Wealth to Welfare* (Stanford Univ. Press, Stanford, Calif., 1950).

13. G. Hardin, *Perspec. Biol. Med.* **6,** 366 (1963).

14. U. Thant, *Int. Planned Parenthood News,* No. 168 (February 1968), p. 3.

15. K. Davis, *Science* **158,** 730 (1967).

16. S. Tax, Ed., *Evolution after Darwin* (Univ. of Chicago Press, Chicago, 1960), vol. 2, p. 469.

17. G. Bateson, D. D. Jackson, J. Haley, J. Weakland, *Behav. Sci.* **1,** 251 (1956).

18. P. Goodman, *New York Rev. Books* **10(8), 22** (23 May 1968).

19. A. Comfort, *The Anxiety Makers* (Nelson, London, 1967).

20. C. Frankel, *The Case for Modern Man* (Harper, New York, 1955), p. 203.

21. J. D. Roslansky, *Genetics and the Future of Man* (Appleton-Century-Crofts, New York, 1966), p. 177.

END OF ARTICLE

The Course in Retrospect

If this course has served its purpose, you now have an increased awareness that science is a major force in the destiny of living things. Through science, man will continue to accumulate knowledge, and with that knowledge he will acquire the ability to design the future according to his aspirations.

The BSCS Second Course in biology has not attempted to define science precisely; such a definition may be neither possible nor useful. Rather, the course was designed to broaden your understanding of science as a constantly growing and changing process which defies precise definition. If you elect a career in science, you will be enlarging continually on your understanding of what science is. If you are not going to be a scientist, we hope you can grasp the potentials of science now. If you can, your year's work has been worthwhile.

As a process, science includes the recognition of problems, the formation of hypotheses, and the designing of experiments. Scientific progress depends on carefully controlled experiments evaluated as objectively as possible. This evaluation thrives on publication. Published experiments can be studied by others, evaluated, and re-evaluated. Their methods and conclusions can be applied to new problems. Thus, an intellectual history of science has accumulated, and it continues to grow.

The evaluation of scientific progress continues as new insights are continuously being acquired and recorded. Even now, the accumulated knowledge of the past has resulted in new instrumentation and techniques which allow scientists to probe the inner structure of molecules and to speculate on the basic physical structures of life. More important in science than knowledge and instrumentation, however, is man's unique ability to reason. It is responsible for his past successes and failures and will certainly determine his destiny. Even though man's destiny must depend on his purpose, that purpose in turn must depend on his capacity for clear and rational thought. We hope this course has enabled each of you to increase that capacity.

APPENDICES

APPENDIX A

GENERAL LABORATORY REQUIREMENTS

A complete list of equipment and materials needed for a class of 24 students is given below. Orders for materials should be placed well in advance of their projected use. Orders for living material should be scheduled so that organisms arrive close to the time they are needed.

QUANTITY	GENERAL EQUIPMENT
2	Atomizers, hand ("Windex" sprayers, and so forth)
1	Autoclave or pressure cooker (22-qt)
1	Balance, analytical or torsion
6	Balances, triple-beam (0.01 g sensitivity)
6	Batteries, 1.5-volt
1	Brooder, chick
1	Burner, Fischer
6	Burners, Bunsen
1 roll	Cellophane, dark blue (MSC 300 or equivalent)
1 roll	Cellophane, dark green (MSC 300 or equivalent)
1 roll	Cellophane, dark red (MSC 300 or equivalent)
1	Centrifuge
6	Coleoptile cutters
1 lb	Cotton, nonabsorbent
6	Counters, mechanical (optional)
1	De-ionizing column (or source of distilled water)
6	Dissecting kits, each including
	1 Forceps (medium), medium points
	1 Forceps (fine), sharp points
	2 Needles, dissecting
	1 Scalpel

	1 Scissors (medium size), sharp point
	1 Scissors (small size), fine point
1	Incubator
24	Loops, inoculating
1	Meter, footcandle (optional)
12	Microscopes, compound monocular
12	Microscopes, stereo binocular
6	Needles, hypodermic ($\frac{3}{4}$ inch, 18 gauge)
6	Needles, hypodermic ($\frac{3}{4}$ inch, 22 gauge)
2 boxes	Paper, filter (10 cm)
25	Pins, insect, No. 1
1 large roll	Polyethylene film (.004–.006)
400 assorted	Polyethylene plugs, flask size
500 assorted	Polyethylene plugs, test-tube size
1	Pump, aquarium
1	Refrigerator
12	Rubber stoppers, 2-hole, No. 4
2	Rubber stoppers, 3-hole, No. 7
15 ft	Rubber tubing, 6 mm
24	Rulers, plastic (mm)
1	Stopwatch (optional)
12	Spatulas
1	Stove or hot plate
6	Syringes, hypodermic (0.1 ml accuracy)
1	Temperature-gradient box
10	Test-tube racks
12	Thermometers, centigrade (0–100° C range)
—	Tygon tubing (6 mm)
1	Ultraviolet-light source, germicidal (optional)
6	Viewing chambers

6	Volumeters
—	Watering pan, chick (1/group of chicks)

QUANTITY	GLASSWARE
24	Beakers, Griffin low-form (100-ml Pyrex or Kimax)
12	Beakers, Griffin low-form (500-ml Pyrex or Kimax)
6	Beakers, Griffin low-form (1000-ml Pyrex or Kimax)
12	Bottles, flat-sided
1 oz	Cover glass (No. 1)
1 oz	Cover glass (No. 2) rectangular
40	Dishes, culture (4-inch fingerbowls or round 8-oz polyethylene refrigerator dishes with covers)
24	Dishes, culture (10-inch fingerbowls or round 64 oz polyethylene refrigerator dishes with covers)
24	Flasks, Erlenmeyer (125-ml Pyrex or Kimax)
24	Flasks, Erlenmeyer (250-ml Pyrex or Kimax)
12	Flasks, Erlenmeyer (500-ml Pyrex or Kimax)
24	Flasks, Erlenmeyer (1000-ml Pyrex or Kimax)
1	Flask, side-arm (500-ml Pyrex or Kimax)
1	Funnel, Buchner
1 lb	Glass beads
10 ft	Glass rod, solid
1	Graduated cylinder (1000-ml)
6	Graduated cylinders (100-ml)
6	Graduated cylinders (50-ml)
6	Graduated cylinders (25-ml)
6	Graduated cylinders (10-ml)
12	Hemocytometers and cover glasses
1 box	Microscope slides, regular
24	Pipettes (1-ml), graduated in 0.1-ml units

6	Pipettes (10-ml), graduated in 0.1-ml units
6	Pipettes, Pasteur (with rubber bulbs)
72	Plates, petri (glass)
12	Serum vials
100	Test tubes (13 mm x 100 mm)
100	Test tubes (22 mm x 175 mm)
12	Tubes, centrifuge (15-ml)
12	Tubes, centrifuge (60-ml)
30 ft	Tubing, glass (6 mm)
12	Watch glasses

QUANTITY	CHEMICALS
1 pt	Acetic acid
1 qt	Acetone
1 lb	Agar
1 lb	Agar, nutrient
1 qt	Alcohol, 95% ethyl
0.5 pt	Alcohol, isoamyl
1 gal	Alcohol, methyl
5 g	Ammonium oxalate
100 g	Ammonium sulfate
1 lb	Ascarite, coarse
10 mg	Biotin (optional)
1 g	Boric acid (H_3BO_3)
20 g	Brain-heart medium
25 g	Calcium chloride
75 g	Calcium nitrate
10 g	Chloretone (1,1,1-trichloro-2-methyl-2-propanol) (MS-222 preferred)

10,000 I.U.	Chorionic gonadotrophin
1 pt	Chloroform (optional)
1 g	Copper sulfate
5 g	Crystal violet
5 g	EDTA
1 pt	Ether (anesthetic)
100 mg	Gibberellic acid
1 lb	Glucose
10 g	Glycerol (optional)
1 oz	Gram's stain
1 g	Indoleacetic acid
1 g	Indolebutyric acid
10 g	Iodine crystals
10 g	Iron chelate (iron ethylenediamine-tetraacetate)
$\frac{1}{4}$ lb	Lanolin
0.5 g	Lysozyme
150 g	Magnesium sulfate
5 g	Manganous chloride
1 g	Malachite green (optional)
0.5 g	Mercuric chloride
5 g	Methylene blue
1 qt	Mineral oil
1 g	MS-222 (Tricaine methane sulphonate)
1 g	Para-aminobenzoic acid (optional)
0.5 g	Penicillin
5 g	Peptone (optional)
0.5 g	Phenol
1.0 g	Phenolphthalein
10 g	Potassium chloride
100 g	Potassium hydroxide

100 g	Potassium iodide
75 g	Potassium nitrate
125 g	Potassium phosphate, monobasic
10 g	Potassium phosphate, dibasic
20 mg	Pyridoxine HCl (vitamin B$_6$) (optional)
1 g	Safranin
1 qt	Sesame oil
100 g	Sodium acetate
10 g	Sodium azide
10 g	Sodium carbonate
1 lb	Sodium chloride (noniodized table salt)
10 g	Sodium citrate
140 g	Sodium hydroxide pellets
0.5 lb	Sodium lauryl sulfate
1.0 g	Sodium molybdate
100 g	Sodium phosphate, monobasic
100 g	Sodium phosphate, dibasic
1 lb	Starch, soluble
0.5 g	Streptomycin
5 lbs	Sucrose, commercial grade
5 g	Thiourea
2 rolls	Tes-Tape
500 mg	Testosterone propionate
1 g	L-thyroxin
1 g	L-triiodothyronine
5 g	2-,3-,5-triphenyltetrazolium chloride
1 pkg.	Vitamin tablets, multiple
1 lb	Wax
15 g	Yeast extract
1 g	Zinc sulfate

QUANTITY	BIOLOGICAL MATERIALS
	Animals
26	Chicks (1- or 2-day-old cockerels of the same strain)
26	Chicks (1- or 2-day-old pullets of the same strain)
10*	Frogs, mature females
30–50*	Frogs, mature males (check number required for particular season)
75	Planaria, brown (*Dugesia tigrini*) (Another species may be used)
25	Tadpoles, small *Rana pipiens* or *clamitans* (if not saved from earlier Investigation)
35	Tadpoles, large (hindbud stage)

* **Note:** the number of frogs may be reduced if pre-injected females are used.

	Microorganisms
1 culture	*Serratia marcescens,* strains D1, 933, and WCF (optional)
6 pkgs	Dry yeast (active)

	Plant Items
2200	Alaska pea seeds
300	Corn grains
3 g	Great Lakes lettuce seeds
3 g	Grand Rapids lettuce seeds
640	Kentucky Wonder bean seeds
620	Little Marvel dwarf pea seeds
900	Oats (*Avena sativa*), variety Victory, seedlings
450	Pinto bean seeds
250	Sorghum RS610 seeds

QUANTITY	MISCELLANEOUS MATERIALS
1 roll	Aluminum foil

15 lbs	Baby-chick feed
—	Box and heat source for young chicks (if no brooder)
25	Boxes, corrugated (for growing flats, 3 inches to 5 inches high)
25	Boxes, sweater, or shirt (for germination trays)
1 roll	Cellophane or plastic tape
5 rolls	Cellophane or plastic tape (5 colors)
1 qt	Detergent, commercial
12	Flowerpots, 3 inches in diameter
—	Fluorescent tubes, daylight
1 bottle	Food coloring
1 gal	Hexol or Lysol
6	Light bulbs (50-watt)
2	Light bulbs (100-watt)
—	Light bulbs (150-watt)
5 rolls	Masking tape
1 qt	Milk, skim or fat-free
1 qt	Molasses
24 sheets	Paper, log, 5 cycle
—	Paper towels
1 set	Pens, marking
2 pkgs	Razor blades
1 lb	Washed sand (if no glass beads)
1 box	Saran wrap
1	Storage container, plastic (5 gal)
1 box	Toothpicks
1 cu yd	Vermiculite
5 gal	Water, distilled

APPENDIX B

LABORATORY REQUIREMENTS FOR EACH INVESTIGATION

In the following lists the quantities specified are requirements for each team unless they are preceded by an asterisk, which indicates that the quantity is for the class.

When quantities are not specified, review the Investigation and determine the amount required.

INVESTIGATION 1:   The Problem

1 pkg	Dry yeast
1	100-ml graduated cylinder
10	Test tubes (13 mm x 100 mm)
10	Test tubes (22 mm x 175 mm)
—	Test-tube racks
*1 pt	Commercial molasses
2	Erlenmeyer Flasks, 125- or 250-ml
*—	Distilled water

INVESTIGATION 2:   A Study of Variables

10	Test tubes capable of holding 10 ml
1	10-ml pipette graduated in 1-ml units
10 ml	10% sucrose solution
10 ml	Buffer, pH^4 (glacial acetic acid, sodium acetate)
1 ml	1% yeast-extract solution

5 ml	1% ammonium sulfate solution
5 ml	1% monobasic potassium phosphate solution
3 ml	Vitamin solution
3 ml	Soil extract
*—	Suspension of live yeast

INVESTIGATION 3:   Further Variables and Controls

10	Test tubes capable of holding 10 ml
*—	50% sucrose solution
1	Balance
*—	Turbidity standards
*—	Yeast suspension
*—	Supplements as in Investigation 2

INVESTIGATION 4:   A New Question

Materials are the same as those used in Investigations 1, 2, and 3.

INVESTIGATION 5:   Resting Cells

14	Test tubes (13-mm x 100-mm)
14	Test tubes (22-mm x 175-mm)
100 ml	4% yeast suspension (in formate buffer)
100 ml	60% sucrose solution
25 ml	Molasses
1	Graduated cylinder
14 ml	Acetate buffer (as in Investigation 2)
100 ml	Formate buffer (pH 3.5)
—	Supplements from Investigation 2

INVESTIGATION 6:   Making an Enzyme Preparation

2	Small dishes (or spot plate)
*—	Tes-Tape
*—	Acetone (or acetone preparation)
2 ml	00.5 M phosphate buffer, 4.5 pH (if no centrifuge)
0.5 ml	1% sucrose
—	Filter paper
*—	Toothpicks
1	Scissors
*—	Yeast
*1	Centrifuge (if available)

INVESTIGATION 7:   Measuring Rates of Respiration

1	Volumeter (complete)
1	Thermometer
100	Alaska pea seeds
1	Germination tray (sweater box and polyethylene liner)
1	100-ml graduated cylinder
—	Glass beads
3	150-ml beakers
—	Solution of vegetable dye (food color)
—	Cotton
—	Ascarite
—	Eyedroppers

INVESTIGATION 8:   Growth of a Yeast Population

1	Microscope
1	Hemocytometer (counting chamber)
2	150-ml flasks (99 ml sterile water)

2	150-ml flasks (49 ml sterile-culture medium)
1	Volumetric pipette, 1-ml accuracy
—	Mechanical counters (1/team, if available)
*1 pkg	Dry yeast (active) or yeast suspension
*—	12°C and 22°C incubators
*1	Balance

INVESTIGATION 9:   A Mixed Population

1.0 g	Soil
5	Sterile 9 ml water blanks
6	Sterile petri dishes
1	Flask with 50 ml sterile soil agar
1	Flask with 50 ml sterile nutrient agar
—	Sterile 1-ml pipettes (graduated in 0.1-ml units)
1	Spreading rod
—	70% alcohol
1	Bunsen burner

INVESTIGATION 10:   Isolation of a Pure Culture

2	Sterile slants of nutrient agar
1	Inoculating needle
1	Bunsen burner
—	Soil plates from Investigation 9
*—	Incubator

INVESTIGATION 11:   A Population Sequence

—	Microscope slides
1	Microscope
—	Methylene blue stain

*1 pt	Fat-free milk
0.5 g	Soil
1	Bunsen burner
1	Flat-sided bottle
1	Inoculating loop

INVESTIGATION 12:   Interaction in Populations

—	Plates (at 1/HT) from Investigation 10
2	Tubes sterile nutrient agar slants
—	Culture of *B. mycoides* (isolated in Investigation 9)
1	Bunsen burner
1	Inoculating loop
*1	Incubator

INVESTIGATION 13:   Isolation of Mutants from Bacteria

2	Sterile nutrient agar slants
*10 ml	Sterile water
3	Test tubes of sterile nutrient agar (20 ml/tube)
3	Sterile petri dishes
3	Sterile 1-ml pipettes
2 ml	Streptomycin solution (1 mg/ml)
—	Culture of *B. mycoides*
1	Glass-rod spreader
*—	70% alcohol
1	Inoculating loop

INVESTIGATION 14:   Resistance and Sensitivity to Chemical Agents

2	Sterile petri dishes
3	Sterile tubes nutrient broth

2	Tubes sterile nutrient agar (15-ml tube)
1	Sterile pipette (0.1-ml graduations)
—	Filter-paper disks
*—	Solutions of streptomycin, penicillin, mercuric chloride and phenol
1	Glass-spreading rod
*—	70% alcohol
*1	Incubator

INVESTIGATION 15:   Isolation of DNA

50 ml	Sterile nutrient broth
50 ml	Saline citrate buffer (0.6 gm NaCl plus 3.0 gm sodium citrate (2H$_2$O), adjusted to pH 7)
0.5 ml	0.1 M EDTA
0.5 ml	Lysozyme (10 mg/ml)
0.2 ml	45% alcohol saturated with sodium lauryl sulfate pH8
6 ml	Chloroform
0.3 ml	Isoamyl alcohol
6 ml	95% ethanol
2 ml	Sterile 10% NaCl in small test tubes
*1	Centrifuge
*—	Centrifuge tubes

INVESTIGATION 16:   Transformation

1	Tube containing 2 ml sterile 10% NaCl
10 ml	Enriched broth
2.5–3.0 ml	Streptomycin solution (75 mg/100 ml)
2	Sterile petri dishes
2	Tubes containing 15 ml sterile nutrient agar

1	5-ml sterile pipette
4	1-ml sterile pipettes
1	Broth culture of sensitive *B. mycoides* less than 7 days old
—	DNA saline from Investigation 15
1	Sterile glass-rod spreader
—	70% alcohol

INVESTIGATION 17:   Sensitive, Resistant, and Transformed Strains

Materials as in Investigation 14.

INVESTIGATION 18:   Preparation of a Scientific Paper

No laboratory materials.

INVESTIGATION 19:   Enzyme Activity in Germinating Seeds

—	Corn grains, soaked
1	Razor blade
3	Culture dishes
50 ml	Starch agar
*1 roll	Tes-Tape
25 ml	Tetrazolium solution
1	100-ml beaker

INVESTIGATION 20:   Isolation of an Enzyme

2	Plates of starch agar
2	Petri plates or culture dishes
*1 roll	Tes-Tape

25 ml	Tetrazolium solution
1	Razor blade

INVESTIGATION 21:   Testing for Seed Viability

—	Pea seeds
25 ml	Tetrazolium chloride
2	100-ml beakers

INVESTIGATION 22:   Effects of Light on Germination of Seeds

The materials vary with each design, but germination trays, seeds, and a light source are necessary.

INVESTIGATION 23:   The Effects of Different Wavelengths of Light on Germination of Seeds

—	Grand Rapids and Great Lakes lettuce seeds
—	Germination trays
—	Blue, green, and red filters or sheets of cellophane
—	Daylight fluorescent light source
—	Incandescent light source

INVESTIGATION 24:   Mineral Requirements of Sorghum Plants

16	Test tubes (22 mm x 175 mm)
40	Seeds of RS610 sorghum
—	Cardboard box and styrofoam lid
—	Mineral solutions (see Appendix C)
—	Cotton for plugging test tubes
—	Distilled water
—	Detergent

—	Paper toweling
—	Razor blade

PREPARATORY TECHNIQUE 1:   Obtaining Frog Pituitaries

1 to 10	Male frogs
—	Chloroform
—	Cotton
1	Fine scissors
1	Fine forceps
—	10% Holtfreter's solution

PREPARATORY TECHNIQUE 2:   Injecting Pituitaries

—	Hypodermic syringe and No. 18 needle
1 to 5	Female frogs
—	Plastic containers and lids
—	70% alcohol solution
—	Cotton

PREPARATORY TECHNIQUE 3:   Fertilization in Vitro

1	Female frog previously injected with pituitary glands
1	Male frog, mature
10	Plastic dishes and covers
—	Fine, sharp scissors
—	Fine forceps
—	Cotton
—	Medicine dropper
—	10% Holtfreter's solution

PREPARATORY TECHNIQUE 4: Establishing and Maintaining Frog-Embryo Cultures

—	Temperature-gradient box
10	Plastic dishes
—	Pond water or 10% Holtfreter's solution
—	*Elodea*
—	Refrigerator
—	Fertilized eggs from Preparatory Technique 3

INVESTIGATION 25: Development of the Frog Embryo

10	Plastic dishes with embryos from Preparatory Technique 4
—	Lettuce or *Elodea*

INVESTIGATION FOR FURTHER STUDY: Effects of Ultraviolet Irradiation on Frog Development

*1	Ultraviolet lamp (germicidal)
70 ml	10% Holtfreter's solution
1	Male frog
*1	Female frog (pituitary-injected)
9	Fingerbowls
—	Dissecting instruments
1	Petri dish
1	2.0-ml pipette

INVESTIGATION 26: Observation of Regenerating Tissues

—	Separate containers for tadpoles
*—	Insect pins

—	Tadpoles
3	Petri plates
*—	Anesthetic MS-222 or Chloretone in 10% Holtfreter's solution
—	Lettuce or *Elodea* for tadpole food
1	Viewing chamber
1	Microscope
1	New razor blade
*—	10% Holtfreter's solution

INVESTIGATION 27: Hormonal Control of the Development of Frog Embryos

7	Flat culture dishes or large plastic bowls (1 liter capacity)
7	Stones or bricks
1	1-liter graduated cylinder
—	Stock solutions of thyroxin, triiodothyronine, and iodine
35	Tadpoles
3	10-ml graduated pipettes
—	Pond water

INVESTIGATION 28: Behavior and Development in Chicks

5	1-day-old male chicks
5	1-day-old female chicks
*—	Brooder or several boxes of sawdust and heat source
*4	Hypodermic syringes graduated to 0.1-ml with No. 21 or 22 needle
*—	Solutions of hormones (testosterone propionate and chorionic gonadotrophin)

*___	Chick food and water trays
*___	Marking pencils
*___	Plastic tape (5 colors)
*___	Millimeter ruler
1	Scale or balance
*___	70% alcohol
*	Sesame oil
1	Centigram balance

INVESTIGATION 29:   The Effect of Light on the Growth of Seedlings

—	Germination trays
—	Alaska pea seeds
—	Sand or vermiculite
—	Paper toweling

INVESTIGATION 30:   A Biological Assay

—	*Avena sativa,* variety Victory, seedlings
—	Germination box with red cellophane and light shield
—	Indoleacetic acid
—	Sucrose
1	Coleoptile cutter
6	Petri plates
1	Millimeter ruler
1	10-ml pipette

INVESTIGATION 31:   Effect of Gibberellic Acid

—	Alaska and Little Marvel pea seeds
4	4-inch flower pots or other containers

—	Solution of gibberellic acid (100 mg per liter)
—	Germination tray
—	Sand and soil

INVESTIGATION 32: Effects of Growth-Regulating Substances on Plants

60 to 80	Pinto bean seeds
1	Germination tray
21	Bottles (baby-food jars)
—	Plant-growth regulators (indoleactic acid, indolebutyric acid, and gibberellic acid)
—	4% ethyl alcohol
—	Lanolin
—	Toothpicks
—	Dropper pipettes

INVESTIGATION 33: Planaria Sense and Respond to Gravity

1	13- x 100-mm test tube and stopper
1	Brown planarian (*Dugesia Tigrini*)
1	Medicine dropper
1	Dissecting needle
1	Marking pencil
1	Test-tube rack or small bottle

INVESTIGATION 34: Planaria and Light

1	13- x 100-mm test tube
*—	Foil wrap
2	Glass slides
1	Medicine dropper

1	Dissecting needle
1	Brown planarian

INVESTIGATION 35:   Planaria and Food

1	Brown planarian
1	Watch glass
1	Medicine dropper
*—	Fresh liver
1	Dissecting needle
1	Hand lens or dissecting microscope

INVESTIGATION 36:   Planaria and Chemicals

*—	Filter paper saturated in NaCl
*—	Filter paper saturated in acetic acid
*—	Tygon tubing
1	Scissors
1	Medicine dropper
2	Brown planaria

INVESTIGATION 37:   Planaria and Electrical Shock

2	Brown planaria
2	Tygon troughs from Investigation 36
1	$1\frac{1}{2}$ volt source
1	Medicine dropper
1	Dissecting needle

INVESTIGATION 38:   Observing Natural Animal Behavior

Laboratory equipment needed, if any, will be variable.

APPENDIX C

PREPARATION OF CHEMICAL SOLUTIONS

A. GENERAL INFORMATION

Solutions are used in many scientific investigations; therefore, it is necessary to become familiar with what certain solutions contain and with some of the techniques used in preparing them. In the laboratory one uses many different solutions in varying concentrations. The term *concentration* means the amount of one substance that is contained within a definite amount of another substance. There are numerous ways of expressing concentration, some of which follow:

1. Percent Concentration

The number of parts by weight or volume of one substance contained by 100 parts by weight or volume of solution.

2. Parts per Million

The number of parts of one substance contained in one million parts of solution; usually expressed by weight; 1 mg/liter = 1 ppm.

3. Molarity

The number of moles of one substance contained in one liter of solution. *Mol* means the molecular weight of a substance expressed in grams. *Liter* is a unit of volume in the metric system.

The terms *solute* and *solvent* are used to designate the components of a solution. The solute is the substance that is dissolved in the solvent. A *solution* is obtained when a solute is dissolved in a solvent.

The following conversion factors will aid you in understanding the discussion of a *dilution*, which is used to obtain weights of amounts of chemicals too small to be weighed on the balances in the laboratory:

1 gram (g) = 1000 milligrams (mg) = 1,000,000 micrograms (μg)
   1 mg = 1000 μg
   1 liter = 1000 milliliters (ml)

B. PREPARATION OF SOLUTIONS OF LOW CONCENTRATIONS BY SERIAL DILUTIONS

Scientists often use extremely small quantities of chemicals—amounts less than the finest balances can weigh accurately. For accurate measurement, then, it is necessary to start with a volume of a solution containing an amount of the chemical which can be weighed accurately. Assume you have prepared 1 liter of a uniform solution containing 1 g of solute. Since 1000 ml of this solution contains 1000 mg of the chemical, 1 ml contains 1 mg. One mg is less than can be weighed accurately on a triple-beam balance. Can you see how *volume measurements* may be used to obtain the weights of small quantities of a substance?

Even smaller weights of the chemical can be obtained easily by preparing more dilute solutions. Assume that 1 ml of the above solution (thus, 1 mg, or 1000 μg, of the chemical) is diluted in 999 ml of distilled water. We can now obtain 1 μg of the chemical by using 1 ml of this second solution. This is known as the serial-dilution technique.

Serial-dilution techniques may be applied to such diverse materials as chemicals, microbial cultures, serums, and so on, in the laboratory. In this case we have used 1000-fold dilutions, but any convenient dilution factor (2-fold, 4-fold, 10-fold, and so on) may be employed as the occasion demands. Solutions which contain a known weight per volume, and which are kept on hand to provide a definite quantity of material, are often called *stock solutions*.

C. STERILIZING SOLUTIONS

In biological work it is often necessary to use solutions which do not contain pathogens (disease-causing organisms). This is particularly true when one is working with microbes or with solutions which are to be injected into living tissue. Two common methods of heating solutions for such studies are autoclaving and pasteurizing.

Autoclaving. Autoclaving sterilizes (kills all life) in the material so treated. To autoclave, prepare the solution and place it in the container. Place a cap on the container, loosely so that steam can escape, and place it in an autoclave or pressure cooker. Bring the temperature of the autoclave to 120°C and maintain it for 20 minutes. If a pressure cooker is used, a pressure of 15 pounds will approximate this temperature. After 20 minutes, remove the heat source and allow the apparatus to cool. Allowing the pressure to go down suddenly may cause the caps or stoppers to blow off, so be certain to allow the apparatus to cool slowly by itself. Remove the solution from the autoclave or pressure cooker and secure the caps or stoppers immediately.

Pasteurizing. Pasteurization kills unwanted organisms but usually does not kill all life in the product. Prepare the solution and place it in the container. Cap it loosely so that air resulting from its expanding contents can escape. Place the container of solution in a pan of water on a heat source or in a constant-temperature water bath and bring the temperature of the water to 62°C. Maintain this temperature for 30 minutes, then remove the container from the water and secure the cap or stopper. This is the temperature and time often used for pasteurizing milk. Other products may require different temperatures and times to be pasteurized effectively.

D. PREPARATION OF SOLUTIONS

1. Tetrazolium-Test Solution

2,3,5-triphenyltetrazolium
chloride   0.5 g
Distilled water   10.0 ml

Solution should be used within 3 to 6 hours after its preparation.

2. Iodine-Potassium Iodide Solution

Iodine (crystals)   5.0 g
Potassium iodide   10.0 g
Distilled water   100.0 ml

Dilute this stock solution one part to ten parts water before using.

3. Growth-Regulator Solution

Specific regulator   100.0 mg
Ethyl alcohol (95% absolute)   40.0 ml
Distilled water   960.0 ml

This is for the preparation of a solution of 100 ppm of growth regulator (indoleacetic acid, indolebutyric acid, and gibberellic acid). For each separate solution, dissolve the regulator in the alcohol and dilute with water, stirring constantly.

4. Growth-Regulator Paste

Specific regulator	0.1 g
Lanolin	9.9 g

For a preparation of a 1.0% growth-regulator paste, melt lanolin in a small beaker placed in a water bath and add regulator, stirring for several minutes. (For a 0.1% paste, use 1.0 g of the 1.0% preparation and 9.0 g of lanolin.)

5. Thyroid Hormone Solutions

a. *Thyroxin Stock Solution.* Dissolve 0.1 g thyroxin in 10 ml of 0.1 M sodium hydroxide and combine this with 90 ml of distilled water. Combine 1 ml of this solution with 1 liter of distilled water to prepare the final stock solution which contains 1 μg of thyroxin per ml.

50 μg/liter solution: Combine 50 ml of thyroxin stock solution with 950 ml of pond water.

25 μg/liter solution: Combine 25 ml of thyroxin stock solution with 975 ml of pond water.

10 μg/liter solution: Combine 10 ml of thyroxin stock solution with 990 ml of pond water.

b. *Triiodothyronine Stock Solution.* Substitute triiodothyronine for thyroxin and follow the directions given for preparing the thyroxin stock solution.

c. *Iodine Stock Solution.* Dissolve 0.1 g of iodine crystals in 10 ml of alcohol, and combine this with 90 ml of distilled water into which has been dissolved 0.1 g potassium iodide. Combine 1 ml of this solution with 1 liter of water to prepare the final stock solution which contains 1 μg of iodine per ml of solution.

2 μg/liter solution: Combine 2 ml of iodine stock solution with 1 liter of pond water.

1 μg/liter solution: Combine 1 ml of iodine stock solution with 1 liter of pond water.

d. *Thiourea Culture Solution.* Prepare a 3.3 g/liter culture solution by dissolving 3.3 g of thiourea in 1 liter of pond water.

6. Holtfreter's Solution

a. *Standard Stock Solution.* Combine 3.5 g of sodium chloride, 0.05 g of potassium chloride, and 0.2 g of sodium bicarbonate with 1 liter of distilled or deionized water.

b. *10% Holtfreter's Solution.* Prepare this solution by combining 1 part standard stock solution with 9 parts distilled water.

7. Sex Hormones

a. *Testosterone Injection Solution.* Dissolve 250 mg of testosterone propionate in 10 ml of sesame oil. Place this solution in a serum vial and pasteurize it.

b. *Chorionic Gonadotrophic Injection Solution.* Chorionic gonadotrophin will be supplied in a sealed vial containing 500 Cortland-Nelson units (10,000 international units).

8. Physiological Saline

0.9% Saline Solution. Dissolve 9 g of sodium chloride in 1 liter of distilled water.

9. Preparation of Staining Materials

a. *Ammonium Oxalate-Crystal Violet Solution (for Gram's stain).* Make one solution by dissolving 2 g of crystal violet in 20 ml of 95% ethanol. Make a second solution by dissolving 0.8 g of ammonium oxalate in 80 ml of distilled water. Then mix the two solutions together and store.

b. *Gram's Iodine Solution.* Grind 1 g of iodine crystals and 2 g of potassium iodide in a mortar until they are finely dispersed. Then dissolve the mixture in 300 ml of distilled water.

c. *Safranin Counterstain.* Dissolve 0.25 g of safranin in 10 ml of 95% ethanol. Then add 90 ml of distilled water to bring the volume up to 100 ml of solution.

d. *Methylene Blue (simple stain).* Dissolve 1 g of methylene blue in 99 ml of distilled water.

10. Preparation of IAA Stock Solutions

A stock solution of 100 mg/liter indoleacetic acid is prepared by dissolving 100 mg of indoleacetic acid in 1 to 2 ml of 95% ethyl alcohol and adding approximately 900 ml of water. Warm the mixture gently on a *hot plate* or *steam bath* to evaporate the alcohol; then dilute it with distilled water to one liter. This solution should not be used after it is two weeks old. It should be stored in a refrigerator, preferably in a brown bottle or flask and covered with aluminum foil to exclude light.

For Investigation 30, dilutions are needed of 20.0, 2.0, 0.2, and 0.02 mg/liter of indoleacetic acid. The following procedure is an accurate way to make up these stock solutions. Take 200 ml of the 100 mg/liter stock solution and dilute it to 1 liter; this is the 20 mg/liter solution. Take 100 ml of the 20 mg/liter solution and dilute to 1 liter; this is the 2.0 mg/liter solution. Take 100 ml of the 2 mg/liter indoleacetic acid solution and dilute it to 1 liter; this is the 0.2 mg/liter solution. Take 100 ml of this solution and dilute it to 1 liter; this is the 0.02 mg/liter solution. It is not advisable to make all the solutions from the 100 mg/liter indoleacetic acid stock solution since it is very difficult to measure accurately the smaller quantities involved. Each class will need about 100 ml of each of the final solutions.

11. Gibberellic-Acid Stock Solution

Gibberellic acid	100 mg
Ethyl alcohol	3 ml
Distilled water	1000 ml

The gibberellic-acid stock solution (100 mg/liter) may be prepared by weighing out the gibberellic acid, dissolving it in the ethyl alcohol, and then diluting the resultant solution with distilled water to 1 liter.

E. SORGHUM-PLANT MINERAL REQUIREMENTS

Prepare the four mineral-culture solutions in which you will grow the plants (as shown in Table 19). Add each of the indicated stock solutions (preparations for which are shown in Table 20) to about 500 ml of distilled or deionized water and then add enough water to make

TABLE 19. MINERAL-CULTURE SOLUTIONS

Stock solution	Number of milliliters to be added for 1 liter of culture solution			
	Complete	Minus N	Minus P	Minus Fe
$Ca(NO_3)_2 \cdot 4\ H_2O$	10	—	10	10
KNO_3	10	—	10	10
$MgSO_4 \cdot 7\ H_2O$	10	10	10	10
KH_2PO_4	10	10	—	10
$CaCl_2$	—	10	—	—
KCl	—	10	10	—
Iron chelate	10	10	10	—
Trace-element stock	10	10	10	10
Water (distilled or deionized) enough to make:	1 liter	1 liter	1 liter	1 liter

TABLE 20. STOCK SOLUTIONS

Chemical	Amount	Amount of distilled or deionized water
1. $Ca(NO_3)_2 \cdot 4\ H_2O$	23.6 grams	200 ml
2. KNO_3	10.0 grams	200 ml
3. $MgSO_4 \cdot 7\ H_2O$	4.2 grams	300 ml
4. KH_2PO_4	4.2 grams	300 ml
5. $CaCl_2$	5.6 grams	100 ml
6. KCl	1.6 grams	200 ml
7. Iron chelate (iron ethylenediaminetetraacetate)	1.0 gram	100 ml

8. Trace elements (A stock solution should be made up to contain the following salts in the concentrations shown.)

a. $MnCl_2 \cdot 4\ H_2O$	1.8 grams
b. H_3BO_3	2.8 grams
c. $ZnSO_4 \cdot 7\ H_2O$	0.22 grams*
d. $CuSO_4 \cdot 5\ H_2O$	0.08 grams*
e. $Na_2MoO_4 \cdot 2\ H_2O$	0.025 grams*

   f. Distilled or deionized water to make 1,000 ml.

*The difficulty in weighing these small amounts of material directly can be avoided by using aliquots of more concentrated solutions of each. For example,

Weigh out:	Dissolve in:	Use:
2.2 grams of $ZnSO_4 \cdot 7\ H_2O$	100 ml of water	10 ml per liter
0.8 grams of $CuSO_4 \cdot 5\ H_2O$	100 ml of water	10 ml per liter
2.5 grams of $Na_2MoO_4 \cdot 2\ H_2O$	100 ml of water	1 ml per liter

a total of one liter. (If you mix the stock solutions without diluting them, precipitates, which are likely to form, may be difficult to redissolve.)

APPENDIX D

PREPARATION OF CULTURE MEDIA

Dehydrated growth media for microorganisms are commercially available. Commonly used media such as "nutrient agar," "nutrient broth," "potato dextrose agar," and others, can be purchased in this form. In addition, the components of many of these media may be purchased separately, including beef, yeast or malt extracts, peptone, and agar. Bouillon cubes are a dehydrated product, inexpensive and readily available in groceries. They need only be diluted three to four times more than called for on the label directions to serve as an excellent medium for many bacteria.

Although less convenient to prepare than dehydrated ones, "homemade" media are cheaper and sometimes better. Moreover, it is necessary on occasion to make media that are not available commercially. Detailed directions for the preparation of many media are provided in the references. Some simple, yet effective, media may be prepared from common food materials. Pieces of potato, turnip, and other bland tuber and root vegetables serve as a substrate upon which many bacteria can be grown. Boiled extracts of hay, lettuce, and other plant materials also furnish excellent media for growth of certain microorganisms. Sterilization of most media may be accomplished by autoclaving at 121°C for 15 to 20 minutes.

A. PROCEDURES FOR PREPARING MEDIA

Media may be made in less than liter quantities by proportioning volumes and weights. To prepare either slants or tubes of medium, make up the medium as directed and heat by placing the flask in boiling water or in an autoclave with steam on and door slightly ajar. The liquified medium can be dispensed with a pipette to tubes; the plugged tubes are then autoclaved for sterilization. If slants are needed, the tubes are tilted while the medium is still in liquid form after sterilization. Otherwise, the sterilized medium is allowed to harden in upright tubes.

Pipettes should be sterilized by dry heat in stainless-steel cans made for this purpose or wrapped in aluminum foil. It is possible to sterilize pipettes in the autoclave by wrapping each one in newspaper. The pipettes will be wet but usable.

The glass dally is made by cutting pieces of solid glass tubing into suitable lengths of about 6 inches long. Then heat each in a Bunsen burner to get an angle so that the horizontal segment will be about 1.5 inches long.

Pasteur pipettes are prepared by heating pieces of glass tubing and drawing the ends apart while hot. The drawn end must be long enough to reach into test tubes for sucking up and ejecting the suspension of conidia. Pasteur pipettes may be sterilized in very long test tubes.

Plastic petri plates are less expensive than glass plates and are sterile when they arrive. When large orders are placed, volume discounts bring the price down to an even lower level. Do not attempt to sterilize plastic plates because they melt at autoclave temperatures.

B. FORMULAS FOR ROUTINE MICROBIOLOGICAL MEDIA USED IN THIS COURSE

1. Brain-Heart Infusion

Use of commercial dehydrated medium is recommended. Follow instructions on container for preparation.

2. Nutrient Broth

Beef extract	3.0 g
Peptone	5.0 g
Distilled water	1000.0 ml

(Adjust pH from 6.8 to 7.2.)

Should beef extract or peptone be difficult to obtain, a clear soup made from beef or chicken bouillon cubes or a clear, canned soup (beef or chicken) is a suitable substitute.

3. Nutrient Agar

Beef extract	3.0 g
Peptone	5.0 g
Distilled water	1000.0 ml

(Adjust pH from 6.8 to 7.2.)

Add the above solution to 15 g of agar.

4. *Serratia marcescens* Agar

Add 0.05% glycerol to nutrient agar before sterilizing.

5. Soil Agar

> 1 gm peptone
> 1 gm yeast extract
> 1 gm glucose
> 0.5 gm K_2HPO_4
> 20 gm agar
> to 1 liter of tap water.

Sometimes soil extract (250 ml) is added, but it can usually be eliminated when using tap water. Soil extract is prepared by autoclaving 500 g fertile soil with 1500 ml tap water for 30 min, filtering, and adjusting the filtrate to 1000 ml.

6. Starch Agar

Agar	6.0 g
Soluble starch	3.5 g
Distilled water	500.0 ml

Bring the mixture to a boil; stir briskly to dissolve.

The pH of microbiological media should be adjusted to their specified values before they are used. This may be done by having on hand 1 N solutions of sodium hydroxide (NaOH) and sulfuric acid (H_2SO_4), plus some pH paper. After dissolving all the medium ingredients, take a sample of known volume from the medium (say 10 ml from a liter of medium) and check its pH. If no adjustment is necessary, proceed to sterilize the medium. If the sample is too acid, carefully pipette the necessary volume of 1 N sodium hydroxide to bring it to the proper pH. If it is too basic, add the needed amount of sulfuric acid. Now calculate the amount of acid or base needed to adjust the entire batch. Add this amount to the medium, recheck its pH, and if it is correct, proceed to sterilize. Some media may undergo pH changes during autoclaving. Check for this change by aseptically removing a sample of known size of the sterile medium and finding its pH. If pH adjustment is needed in sterile media, the acid or base used for this purpose must, of course, also be sterile.

APPENDIX E

MAINTENANCE OF LIVING ORGANISMS

A. MAINTAINING MICROORGANISMS IN CULTURE

An organism can be maintained in culture for extended periods of time once it has been provided with a good medium for growth. Before attempting any of the procedures outlined below, the organisms must be permitted to develop on the medium until there is evidence of vigorous growth. In most cases two or three days of growth after inoculation should suffice, but certain slow-growing organisms should be incubated longer. When the microbe has grown sufficiently, one of the following techniques can be used to preserve the culture.

1. Storage at Cold Temperatures

A temperature of 4°C will maintain many microbes in a viable form for months. Some, however, will not survive under these conditions, so their viability at cold temperatures should be checked.

2. Storage Under Reduced Oxygen Tension

Dip a cork which fits into the culture tube into alcohol and flame it so that it is sterilized. Push it down the test tube containing the culture until it rests about an inch above the culture. Replace the cotton plug. If the cork fits tightly enough, it will restrict evaporation and access of oxygen, and cultures will last for six months.

3. Pliofilm Coating

Push the cotton stopper into the neck of the test tube so that a small tuft remains for you to hold for its withdrawal. Wrap "Pliofilm" or the equivalent around the mouth of the test tube in order to prevent the drying-out of the agar.

4. Soil Preservation*

Autoclave about a quarter-inch of soil in a test tube for at least half an hour. Transfer a rich suspension of the organism you wish

---

* This technique is recommended as being most satisfactory.

to store and mix it with the soil. This technique is one of the most effective and easiest to use and works with a large number of organisms, including actinomycetes, fungi, and some algae.

5. Storage Under Mineral Oil

Autoclave a bottle of "Nujol," or some other fine-grade mineral oil, for an hour. Carefully pipette enough of this sterile oil over the surface of your culture so that about a half-inch of oil rests above the highest part of the agar. This technique, as well as the soil method, will preserve the viability of microorganisms—in some cases for years.

B. CULTURE OF MICROBES REQUIRED FOR THIS COURSE

1. *Serratia marcescens* (optional)

Wild type, D1; and colorless mutants, 933, and WCF.

2. *Saccharomyces cerevisiae*

As dried Fleischmann's active yeast or compressed yeast.

C. CARE OF FROG EMBRYOS

Examine the embryos daily and remove any dead ones. When the water becomes cloudy or develops a foul odor, change it by decanting it through a strainer or fine-mesh aquarium net which will catch embryos. Replace the old water with fresh water near the temperature of the original water. Replace any embryos which were caught in the strainer or net into the fresh water.

Frog embryos need no food until they have reached Shumway's Stage 25 of development. Once they reach this stage, follow the directions in "Maintaining Tadpoles in the Laboratory."

D. INDUCED HIBERNATION IN ADULT FROGS

Adult frogs can be maintained alive for several weeks in a moist environment in a refrigerator. Place about 6 frogs, along with plenty of wet cotton, in a gallon jar. Cover the jar with screen wire and place in a refrigerator. Examine the frogs periodically and remove any dead ones.

E. MAINTAINING TADPOLES IN THE LABORATORY

Tadpoles are vegetarians and are relatively easy to keep in the laboratory. They should be maintained in an aquarium or a similar container holding plenty of water. The water may be either pond water or tap water, but before using, tap water should be allowed to remain several days in the open so that the chlorine can escape.

Foods such as algae, boiled lettuce, or spinach leaves, or any of the several canned, strained, green, baby-food vegetables, are satisfactory for tadpoles. Algae have an advantage in that they do not cloud the water; therefore, the water can go as long as a week or more without changing. If the other foods are used, the water should be changed thirty minutes or so after each feeding. The animals should be fed three times a week, the amount depending on their size. Large tadpoles can eat as much as a teaspoon of vegetable matter per feeding.

F. MAINTAINING BABY CHICKS IN THE LABORATORY

A custom brooder is best for this work, but if one is not available, baby chicks can be kept in practically any warm enclosure. For the hormone work, a hair dryer with a brooder thermostat serves well as a heat source. (An electric-light bulb cannot be used since the chicks must not be exposed to continuous light.) The chicks must be fed and watered daily. Regular baby-chick feed from a feed store is best; however, finely ground cereals, such as corn meal, rolled oats, or dried baby-food cereal, are satisfactory substitutes. A commercial feeding tray and watering device are the best appliances for maintaining a clean environment.

APPENDIX F

MATH TABLES

COMMON LOGARITHMS

N	0	1	2	3	4	5	6	7	8	9
10	0000	0043	0086	0128	0170	0212	0253	0294	0334	0374
11	0414	0453	0492	0531	0569	0607	0645	0682	0719	0755
12	0792	0828	0864	0899	0934	0969	1004	1038	1072	1106
13	1139	1173	1206	1239	1271	1303	1335	1367	1399	1430
14	1461	1492	1523	1553	1584	1614	1644	1673	1703	1732
15	1761	1790	1818	1847	1875	1903	1931	1959	1987	2014
16	2041	2068	2095	2122	2148	2175	2201	2227	2253	2279
17	2304	2330	2355	2380	2405	2430	2455	2480	2504	2529
18	2553	2577	2601	2625	2648	2672	2695	2718	2742	2765
19	2788	2810	2833	2856	2878	2900	2923	2945	2967	2989
20	3010	3032	3054	3075	3096	3118	3139	3160	3181	3201
21	3222	3243	3263	3284	3304	3324	3345	3365	3385	3404
22	3424	3444	3464	3483	3502	3522	3541	3560	3579	3598
23	3617	3636	3655	3674	3692	3711	3729	3747	3766	3784
24	3802	3820	3838	3856	3874	3892	3909	3927	3945	3962
25	3979	3997	4014	4031	4048	4065	4082	4099	4116	4133
26	4150	4166	4183	4200	4216	4232	4249	4265	4281	4298
27	4314	4330	4346	4362	4378	4393	4409	4425	4440	4456
28	4472	4487	4502	4518	4533	4548	4564	4579	4594	4609
29	4624	4639	4654	4669	4683	4698	4713	4728	4742	4757
30	4771	4786	4800	4814	4829	4843	4857	4871	4886	4900
31	4914	4928	4942	4955	4969	4983	4997	5011	5024	5038
32	5051	5065	5079	5092	5105	5119	5132	5145	5159	5172
33	5185	5198	5211	5224	5237	5250	5263	5276	5289	5302
34	5315	5328	5340	5353	5366	5378	5391	5403	5416	5428
35	5411	5453	5465	5478	5490	5502	5514	5527	5539	5551
36	5563	5575	5587	5599	5611	5623	5635	5647	5658	5670
37	5682	5694	5705	5717	5729	5740	5752	5763	5775	5786
38	5798	5809	5821	5832	5843	5855	5866	5877	5888	5899
39	5911	5922	5933	5944	5955	5966	5977	5988	5999	6010
40	6021	6031	6042	6053	6064	6075	6085	6096	6107	6117
41	6128	6138	6149	6160	6170	6180	6191	6201	6212	6222
42	6232	6243	6253	6263	6274	6284	6294	6304	6314	6325
43	6335	6345	6355	6365	6375	6385	6395	6405	6415	6425
44	6435	6444	6454	6464	6474	6484	6493	6503	6513	6522

COMMON LOGARITHMS (continued)

N	0	1	2	3	4	5	6	7	8	9
45	6532	6542	6551	6561	6571	6580	6590	6599	6609	6618
46	6628	6637	6646	6656	6665	6675	6684	6693	6702	6712
47	6721	6730	6739	6749	6758	6767	6776	6785	6794	6803
48	6812	6821	6830	6839	6848	6857	6866	6875	6884	6893
49	6902	6911	6920	6928	6937	6946	6955	6964	6972	6981
50	6990	6998	7007	7016	7024	7033	7042	7050	7059	7067
51	7076	7084	7093	7101	7110	7118	7126	7135	7143	7152
52	7160	7168	7177	7185	7193	7202	7210	7218	7226	7235
53	7243	7251	7259	7267	7275	7284	7292	7300	7308	7316
54	7324	7332	7340	7348	7356	7364	7372	7380	7388	7396
55	7404	7412	7419	7427	7435	7443	7451	7459	7466	7474
56	7482	7490	7497	7505	7513	7520	7528	7536	7543	7551
57	7559	7566	7574	7582	7589	7597	7604	7612	7619	7627
58	7634	7642	7649	7657	7664	7672	7679	7686	7694	7701
59	7709	7716	7723	7731	7738	7745	7752	7760	7767	7774
60	7782	7789	7796	7803	7810	7818	7825	7832	7839	7846
61	7853	7860	7868	7875	7882	7889	7896	7903	7910	7917
62	7924	7931	7938	7945	7952	7959	7966	7973	7980	7987
63	7993	8000	8007	8014	8021	8028	8035	8041	8048	8055
64	8062	8069	8075	8082	8089	8096	8102	8109	8116	8122
65	8129	8136	8142	8149	8156	8162	8169	8176	8182	8189
66	8195	8202	8209	8215	8222	8228	8235	8241	8248	8254
67	8261	8267	8274	8280	8287	8293	8299	8306	8312	8319
68	8325	8331	8338	8344	8351	8357	8363	8370	8376	8382
69	8388	8395	8401	8407	8414	8420	8426	8432	8439	8445
70	8451	8457	8463	8470	8476	8482	8488	8494	8500	8506
71	8513	8519	8525	8531	8537	8543	8549	8555	8561	8567
72	8573	8579	8585	8591	8597	8603	8609	8615	8621	8627
73	8633	8639	8645	8651	8657	8663	8669	8675	8681	8686
74	8692	8698	8704	8710	8716	8722	8727	8733	8739	8745
75	8751	8756	8762	8768	8774	8779	8785	8791	8797	8802
76	8808	8814	8820	8825	8831	8837	8842	8848	8854	8859
77	8865	8871	8876	8882	8887	8893	8899	8904	8910	8915
78	8921	8927	8932	8938	8943	8949	8954	8960	8965	8971
79	8976	8982	8987	8993	8998	9004	9009	9015	9020	9025
80	9031	9036	9042	9047	9053	9058	9063	9069	9074	9079
81	9085	9090	9096	9101	9106	9112	9117	9122	9128	9133
82	9138	9143	9149	9154	9159	9165	9170	9175	9180	9186
83	9191	9196	9201	9206	9212	9217	9222	9227	9232	9238
84	9243	9248	9253	9258	9263	9269	9274	9279	9284	9289

COMMON LOGARITHMS (continued)

N	0	1	2	3	4	5	6	7	8	9
85	9294	9299	9304	9309	9315	9320	9325	9330	9335	9340
86	9345	9350	9355	9360	9365	9370	9375	9380	9385	9390
87	9395	9400	9405	9410	9415	9420	9425	9430	9435	9440
88	9445	9450	9455	9460	9465	9469	9474	9479	9484	9489
89	9494	9499	9504	9509	9513	9518	9523	9528	9533	9538
90	9542	9547	9552	9557	9562	9566	9571	9576	9581	9586
91	9590	9595	9600	9605	9609	9614	9619	9624	9628	9633
92	9638	9643	9647	9652	9657	9661	9666	9671	9675	9680
93	9685	9689	9694	9699	9703	9708	9713	9717	9722	9727
94	9731	9736	9741	9745	9750	9754	9759	9763	9768	9773
95	9777	9782	9786	9791	9795	9800	9805	9809	9814	9818
96	9823	9827	9832	9836	9841	9845	9850	9854	9859	9863
97	9868	9872	9877	9881	9886	9890	9894	9899	9903	9908
98	9912	9917	9921	9926	9930	9934	9939	9943	9948	9952
99	9956	9961	9965	9969	9974	9978	9983	9987	9991	9996

TABLE 10.   CRITICAL VALUES OF χ^2

Values of χ^2 equal to or greater than those tabulated occur by chance less frequently than the indicated level of p.

d.f.	p = 0.9	p = 0.5	p = 0.2	p = 0.05	p = 0.01	p = 0.001
1	.0158	.455	1.642	3.841	6.635	10.827
2	.211	1.386	3.219	5.991	9.210	13.815
3	.584	2.366	4.642	7.815	11.345	16.268
4	1.064	3.367	5.989	9.488	13.277	18.465
5	1.610	4.351	7.289	11.070	15.086	20.517
6	2.204	5.348	8.558	12.592	16.812	22.457
7	2.833	6.346	9.803	14.067	18.475	24.322
8	3.490	7.344	11.303	15.507	20.090	26.125
9	4.168	8.343	12.242	16.919	21.666	27.877
10	4.865	9.342	13.442	18.307	23.209	29.588

Note: The shading in Table 10 emphasizes the area of increasing probability that the null hypothesis should be rejected; that is, the area of increased confidence that the two samples represent two different populations. In evaluating data from the Investigations in this course, p values of less than 0.05 will be considered as reason for rejection.

Degrees of freedom. This expression refers to the value $n - 1$. It may be defined as the number of individuals or events, or sets of individuals or events, in a given sample which are free to vary.

In chi-square tests, the degrees of freedom (d.f.) are one less than the number of attributes being observed, or $n - 1$, where n equals the number of attributes. If there are two attributes, the d. f. is 1.

TABLE 8.   DISTRIBUTION OF τ PROBABILITY

Probability

Degrees of Freedom	0.1	0.05	0.01	0.001
1	6.314	12.706	63.657	636.619
2	2.920	4.303	9.925	31.598
3	2.353	3.182	5.841	12.941
4	2.132	2.776	4.604	8.610
5	2.015	2.571	4.032	6.859
6	1.943	2.447	3.707	5.959
7	1.895	2.365	3.499	5.405
8	1.860	2.306	3.355	5.041
9	1.833	2.262	3.250	4.781
10	1.812	2.228	3.169	4.587
11	1.796	2.201	3.106	4.437
12	1.782	2.179	3.055	4.318
13	1.771	2.160	3.012	4.221
14	1.761	2.145	2.977	4.140
15	1.753	2.131	2.947	4.073
16	1.746	2.120	2.921	4.015
17	1.740	2.110	2.898	3.965
18	1.734	2.101	2.878	3.922
19	1.729	2.093	2.861	3.883
20	1.725	2.086	2.845	3.850
21	1.721	2.080	2.831	3.819
22	1.717	2.074	2.819	3.792
23	1.714	2.069	2.807	3.767
24	1.711	2.064	2.797	3.745
25	1.708	2.060	2.787	3.725
26	1.706	2.056	2.779	3.707
27	1.703	2.052	2.771	3.690
28	1.701	2.048	2.763	3.674
29	1.699	2.045	2.756	3.659
30	1.697	2.042	2.750	3.646
40	1.684	2.025	2.704	3.551
60	1.671	2.000	2.660	3.460
120	1.658	1.980	2.617	3.373
∞	1.645	1.960	2.576	3.291

Note: *The shading in this table emphasizes the areas of increasing probability that the null hypothesis should be rejected—that is, the areas of increased confidence that the two samples represent two different populations. In evaluating data from the Investigations in this course,* p *values of less than 0.05 will generally be considered as adequate for rejection.*
Source: Abridged from Table III of R. A. Fisher and F. Yates: *Statistical Tables for Biological, Agricultural, and Medical Research,* published by Oliver and Boyd Ltd., Edinburgh, by permission of the authors and publisher.

APPENDIX G

BIBLIOGRAPHY BY PHASE

PHASE ONE Sections 1, 2, and 3

Bronowski, J. 1958. The creative process. Scientific American. September.

BSCS. 1968. Biological science: molecules to man. Blue version. Houghton Mifflin Co., Boston.

BSCS. 1968. High school biology. Green version. Rand McNally and Co., Chicago.

BSCS. 1968. Biological science: an inquiry into life. Yellow version. Harcourt, Brace and World, Inc., New York.

BSCS. 1963. The BSCS biology teacher's handbook. John Wiley and Sons, Inc., New York.

Conant, James B. 1951. Science and common sense. Yale University Press, New Haven, Conn.

Glass, Bentley. 1961. Revolution in biology. BSCS Newsletter No. 9, September.

Grobman, A. B. 1961. The threshold of a revolution in biological education. The Journal of Medical Education. October.

Holton, Gerald, and Duane H. D. Roller. 1958. Foundations of modern physical science. Addison-Wesley Publishing Co., Inc., Reading, Mass.

Hurd, P. D. 1961. Biological education in American secondary schools, 1890–1960. A BSCS Publication. Waverly Press, Inc., Baltimore.

McKeon, R. 1947. Introduction to Aristotle. Modern Library, Random House, New York.

Voss, Burton E. and Stanley B. Brown. 1968. Biology as inquiry: a book of teaching methods. C. V. Mosby, New York.

Wald, George. 1958. Innovation in biology. Scientific American. September.

PHASE TWO Sections 4, 5, 6, 7, and 8

Bailey, N. T. J. 1959. Statistical methods in biology. John Wiley and Sons, Inc., New York.

Beveridge, W. I. B. 1957. The art of scientific investigation. Vintage Books, New York.

Brock, Thomas H. 1961. Milestones in microbiology. Prentice-Hall, Inc., Englewood Cliffs, N. J.

Dixon, W. J., and F. J. Massey. 1957. Introduction to statistical analysis. McGraw-Hill Book Co., Inc., New York.

Downie, N. M., and R. W. Heath. 1965. Basic statistical methods. Second Edition. Harper and Brothers, New York.

Edwards, A. L. 1967. Statistical methods. Second Edition. Holt, Rinehart and Winston, Inc., New York.

Fisher, R. A., and F. Yates. 1953. Statistical tables for biological, agricultural, and medical research. Oliver and Boyd, Ltd., Edinburgh.

Freund, E. 1967. Modern elementary statistics. Third Edition. Prentice-Hall, Inc., Englewood Cliffs, N. J.

Gabriel, M. L., and S. Fogel (Editors). 1955. Great experiments in biology. Prentice-Hall, Inc., Englewood Cliffs, N. J.

Snedecor, G. W. 1967. Statistical methods. Sixth Edition. Iowa State College Press, Ames, Iowa.

Umbreit, W. W. 1964. Manometric techniques. Burgess Publishing Co., Minneapolis.

Van Norman, R. W. 1963. Experimental biology. Prentice-Hall, Inc., Englewood Cliffs, N. J.

PHASE THREE   Section 9

Braun, W. 1965. Bacterial genetics. Second Edition. W. B. Saunders Co., Philadelphia.

Bunting, M. I. 1940. A description of some color variants produced by *Serratia marcescens*, strain 274. Journal of Bacteriology. 59:241.

Cairns, John. 1966. The bacterial chromosome. Scientific American. January.

Clifton, C. I. 1957. Introduction to bacterial physiology. McGraw-Hill Book Co., Inc., New York.

Dulbecco, Renato. 1967. The induction of cancer by viruses. Scientific American. April.

Edgar, R. S. and R. H. Epstein. 1965. The genetics of a bacterial virus. Scientific American. February.

Fraenkel-Conrat, Heinz. 1964. The genetic code of a virus. Scientific American. October.

Fraenkel-Conrat, Heinz. 1956. Rebuilding a virus. Scientific American. June.

Frobisher, M. 1968. Fundamentals of microbiology. Eighth Edition. W. B. Saunders Co., Philadelphia.

Gabriel, M. L., and S. Fogel (Editors). 1955. Great experiments in biology. Prentice-Hall, Inc., Englewood Cliffs, N.J.

Hanawalt, Phillip C. and Robert H. Haynes. 1967. The repair of DNA. Scientific American. February.

Hayes, W. 1964. The genetics of bacteria and their viruses. John Wiley and Sons, Inc., New York.

Hoagland, Mahlon B. 1959. Nucleic acids and proteins. Scientific American. December.

Horne, R. W. 1963. The structure of viruses. Scientific American. January.

Hotchkiss, Rollin D. and Esther Weiss. 1956. Transformed bacteria. Scientific American. November.

Hurwitz, Jerard and J. J. Furth. 1962. Messenger RNA. Scientific American. February.

Jacob, Francois. 1966. The genetics of the bacterial cell. Science, Vol. 152, p. 1470.

Jacob, Francois and Elie L. Wollman. 1961. Viruses and genes. Scientific American. June.

Kellenberger, Edouard. 1966. The genetic control of the shape of a virus. Scientific American. December.

Lamanna, C. and M. F. Mallette. 1965. Basic bacteriology. Third Edition. Williams and Wilkins Co., Baltimore.

Pelczar, M. J. and R. D. Reid. 1965. Microbiology. Second Edition. McGraw-Hill Book Co., Inc., New York.

Phillips, David C. 1966. The three-dimensional structure of an enzyme molecule. Scientific American. November.

Salle, A. J. 1961. Fundamental principles of bacteriology. Fifth Edition. McGraw-Hill Book Co., Inc., New York.

Santer, U. V., and H. J. Vogel. 1956. Prodigiosin synthesis in *Serratia marcescens:* isolation of a pyrrole-containing precursor. Biochimica et Biophysica Acta. 19:578.

Smith, F. E. 1954. Quantitative aspects of population growth. Princeton University Press, Princeton, N.J.

Spiegelman, S. 1964. Hybrid nucleic acids. Scientific American. May.

Stanier, R. Y., M. Duodoroff, and E. A. Adelberg. 1963. The microbial world. Second Edition. Prentice-Hall, Inc., Englewood Cliffs, N.J.

Stearman, R. L. 1955. Statistical concepts in microbiology. Bacteriological Reviews. 19:160–215.

Umbreit, W. W. 1962. Modern microbiology. W. H. Freeman, San Francisco, California.

Wood, William B. and R. S. Edgar. 1967. Building a bacterial virus. Scientific American. July.

PHASE FOUR   Sections 10, 11, 12, 13, 14

Section 10

Butler, W. L., and R. J. Downs. 1960. Light and plant development. Scientific American. December.

Galston, A. W. 1964. The life of the green plant. Second Edition. Prentice-Hall, Inc., Englewood Cliffs, N.J.

Jacobs, W. P. 1959. What substance normally controls a given biological process? 1. Formulation of some rules. Developmental Biology. 1:527–533.

Jensen, C. O., W. Sacks, and F. A. Baldavski. 1951. The reduction of triphenyltetrazolium chloride by dehydrogenases of corn embryos. Science. 113 (2925):65–66.

Koller, D. 1959. Germination. Scientific American. April.

Mattson, A. M., C. O. Jensen, and R. A. Dutcher. 1947. Triphenyltetrazolium chloride as a dye for vital tissues. Science. 106 (2752):294–295.

Miller, E. C. 1938. Plant physiology. McGraw-Hill Book Co., Inc., New York.

Smith, F. E. 1951. Tetrazolium salt. Science. 113 (2948):751–754.

Steward, F. C. 1963. The control of growth in plant cells. Scientific American. October.

Went, F. W. 1957. The experimental control of plant growth. Chronica Botanica Co., Waltham, Mass.

Section 11

Allen, R. D. 1953. The moment of fertilization. Scientific American. June.

Asimov, Isaac. 1963. The genetic code. New American Library, Inc., New York.

Beermann, Wolfgang and Ulrich Clever. 1964. Chromosome puffs. Scientific American. April.

Benzer, Seymour. 1962. The fine structure of the gene. Scientific American. January.

Bonner, D., and S. Mills. 1964. Heredity. Second Edition. Prentice-Hall, Inc., Englewood Cliffs, N.J.

Changeux, Jean-Pierre. 1965. The control of biochemical reactions. Scientific American. April.

Crick, F. H. C. 1962. The genetic code. Scientific American. October.

———. 1966. The genetic code: III. Scientific American. October.

———. 1957. Nucleic acids. Scientific American. September.

———. 1966. Of molecules and men. Univ. of Wash. Press, Seattle, Wash.

———. 1954. The structure of the hereditary material. Scientific American. October.

Davidson, Eric H. 1965. Hormones and genes. Scientific American. June.

Gray, George W. 1953. Human growth. Scientific American. October.

Kendrew, John C. 1966. The thread of life. Harvard University Press, Cambridge, Mass.

Nirenberg, Marshall W. 1963. The genetic code: II. Scientific American. March.

Patten, B. M. 1964. Foundations of embryology. Second Edition. McGraw-Hill Book Co., Inc., New York.

Rugh, R. 1953. The frog: its reproduction and development. McGraw-Hill Book Co., Inc., New York.

Shumway, Waldo. 1940. The Anatomical Record, Vol. 78, No. 2, October.

Singer, M. 1958. The regeneration of body parts. Scientific American. October.

Sinsheimer, Robert L. 1962. Single-stranded DNA. Scientific American. July.

Stein, William H. and Stanford Moore. 1961. The chemical structure of proteins. Scientific American. February.

Taylor, J. Herbert. 1958. The duplication of chromosomes. Scientific American. June.

Trembley, Abraham. 1744. *Histoire des polypes*. Verbeek, Leiden, Holland.

Waddington, C. H. 1953. How do cells differentiate? Scientific American. September.

Watson, James D. 1965. Molecular biology of the gene. W. A. Benjamin, Inc., New York.

Weissmann, August. 1893. The germ-plasm: a theory of heredity. (Translated from the German.) Walter Scott, Ltd., London.

Yanofsky, Charles. 1967. Gene structure and protein structure. Scientific American. May.

Section 12

Jacobs, W. P. and F. N. Rossetter. 1953. Studies on abscission: the stimulating role of nearby leaves. American Journal of Botany. 40:276–280.

Jacobs, W. P., and R. H. Wetmore. 1953. Studies on abscission: the inhibiting effect of auxin. American Journal of Botany. 40:272–275.

LaRue, C. D. 1936. The effect of auxin on the abscission of petioles. Proceedings of the National Academy of Science (Washington). 22:254–259.

Wiggleworth, V. B. 1959. Metamorphosis, polymorphism, and differentiation. Scientific American. February.

Williams, C. M. 1950. The metamorphosis of insects. Scientific American. April.

Section 13

Armstrong, E. A. 1963. A study of bird song. Oxford University Press, New York.

Brower, L. P. and J. V. Z. Brower. 1962. Investigations into mimicry. Natural History. 71:8–19 (April).

Brown, F. 1962. Biological clocks. BSCS Pamphlet No. 2. D. C. Heath and Co., Boston.

Brown, M. E. (Editor). 1957. The physiology of fishes. 2 vols. Academic Press, New York.

Bullough, W. S. 1961. Vertebrate reproductive cycles. Second Edition. John Wiley and Sons, Inc., New York.

Carr, A. 1962. Guideposts of animal navigation. BSCS Pamphlet No. 1. D. C. Heath and Co., Boston.

Carthy, J. D. 1962. An introduction to the behavior of invertebrates. The Macmillan Co., New York.

Cloudsley-Thompson, J. L. 1961. Rhythmic activity in animal physiology and behavior. Academic Press, New York.

Darling, F. F. 1937. A herd of red deer. Oxford University Press, New York.

Dethier, V. G. and E. Stellar. 1964. Animal behavior. Second Edition. Prentice-Hall, Inc., Englewood Cliffs, N.J.

Dorst, J. 1963. The migration of birds. Houghton Mifflin Co., Boston.

Etkin, William (Editor). 1964. Social behavior and organization among vertebrates. University of Chicago Press, Chicago.

Evans, Howard E. 1963. Predatory wasps. Scientific American. April.

————. 1963. Wasp farm. Natural History Press, Garden City, New York.

Fabre, J. Henri. 1961. Insect world. Apollo Editions, Inc., New York.

Farner, D. 1964. Photoperiodism in animals. BSCS Pamphlet No. 15. D. C. Heath and Co., Boston.

Ford, Clellan S. and Frank A. Beach. 1951. Patterns of sexual behavior (K128S). Ace Books, New York.

Frisch, Karl von. 1965. The dancing bees. Harcourt, Brace and World, Inc., New York.

Fuller, John L. and W. R. Thompson. 1960. Behavior genetics. John Wiley and Sons, Inc., New York.

Gans, C. 1961. A bullfrog and its prey. Natural History. 70:26–37, (February).

Goetsch, Wilhelm. 1957. Ants. University of Michigan Press (Ann Arbor Science), Ann Arbor, Mich.

Gray, Sir James. 1968. Animal locomotion. Norton Publishers, New York.

Griffin, Donald R. 1948. The navigation of birds. Scientific American. December.

———. 1950. The navigation of bats. Scientific American. August.

———. 1958. Listening in the dark. Yale University Press, New Haven, Conn.

———. 1958. More about bat radar. Scientific American. July.

———. 1959. Echoes of bats and men. Doubleday and Co., Inc., (Anchor Books), Garden City, New York.

Hafez, E. (Editor). 1968. The behavior of domestic animals. Second Edition. The Williams and Wilkins Co., Baltimore.

Harlow, H. F. and M. K. Harlow. 1961. A study of animal affection. Natural History. December.

Hasler, Arthur D. and James A. Larsen. 1955. The homing salmon. Scientific American. August.

Hediger, H. 1964. Wild animals in captivity. Peter Smith Publishers, London.

Hess, Eckhard. 1958. Imprinting in animals. Scientific American. March.

Kellog, W. N. 1962. Dolphins and hearing. Natural History. February.

Klopfer, P. 1962. Behavioral aspects of ecology. Prentice-Hall, Inc., Englewood Cliffs, N.J.

Krogh, August. 1948. The language of the bees. Scientific American. August.

Lindauer, M. 1967. Communication among social bees. Atheneum Press, New York.

Lissman, H. W. 1963. Electricity location by fish. Scientific American. March.

Lorenz, Konrad Z. 1958. The evolution of behavior. Scientific American. December.

Meyerriecks, A. J. 1962. Courtship in animals. BSCS Pamphlet No. 3. D. C. Heath and Co., Boston.

Pfeiffer, J. 1962. Vision in frogs. Natural History. November.

Portmann, Adolf. 1961. Animals as social beings. Viking Press, New York.

Pringle, J. 1961. The flight of the bumblebee. Natural History. August.

Rheingold, H. (Editor). 1963. Maternal behavior in mammals. John Wiley and Sons, Inc., New York.

Roe, A. and G. G. Simpson. 1958. Behavior and evolution. Yale University Press, New Haven, Conn.

Roeder, K. 1967. Nerve cells and insect behavior. Revised Edition. Harvard University Press, Cambridge, Mass.

Schaller, G. 1963. The mountain gorilla. University of Chicago Press, Chicago.

———. 1964. The year of the gorilla. University of Chicago Press, Chicago.

Schneirla, T. C. and Gerard Piel. 1948. The army ant. Scientific American. June.

Thorpe, W. H. 1963. Learning and instinct in animals. Revised Edition. Harvard University Press, Cambridge, Mass.

Thorpe, W. H. and O. Zangwill. 1961. Current problems in animal behavior. Cambridge University Press, New York.

Tinbergen, N. 1952. The curious behavior of the stickleback. Scientific American. December.

———. 1968. Curious naturalists. Anchor-Doubleday, New York.

———. 1961. The herring gull's world. Revised Edition. Basic Books, Inc., New York.

———. 1965. Animal behavior. Time-Life Books, Chicago.

Wells, M. J. 1962. Brain and behavior in *Cephalopoda*. Stanford University Press, Palo Alto, California.

Wenner, A. M. 1962. Buzzing the queen. Scientific American. December.

Wilson, Edward O. 1963. Pheromones. Scientific American. May.

Wynne-Edwards, V. C. 1962. Animal dispersion in relation to social behavior. Hafner Publishing Co., Inc., New York.

Section 14

Buddenbrock, Wolfgang von. 1958. The senses. University of Michigan Press (Ann Arbor Science), Ann Arbor, Mich.

Fraenkel, Gottfried S. and Donald L. Gunn. 1961. The orientation of animals. Dover Publications, Inc., New York.

Frisch, Karl von. 1965. The dancing bees: an account of the life and senses of the honeybee. Harcourt, Brace and World, Inc., New York.

Heinroth, Oskar and Katharina. 1958. Birds. University of Michigan Press (Ann Arbor Science), Ann Arbor, Mich.

Klüver, Heinrich. 1957. Behavior mechanisms in monkeys. University of Chicago Press (Phoenix), Chicago.

Köhler, Wolfgang. 1959. Mentality of apes. Vintage, New York.

Lockley, Ronald M. 1954. Shearwaters. The Devin-Adair Co., New York.

Lorenz, Konrad. 1961. King Solomon's ring. Apollo Books, New York.

Maxwell, Gavin. 1962. Ring of bright water. Fawcett, New York.

Pavlov, Ivan P. 1927. Conditioned reflexes. Dover Publications, Inc., New York.

Portmann, Adolf. 1959. Animal camouflage. University of Michigan Press (Ann Arbor Science), Ann Arbor, Mich.

Richards, O. W. 1961. The social insects. Peter Smith, New York.

Scott, J. P. 1958. Animal behavior. University of Chicago Press, Chicago.

Sinnott, Edmund W. 1955. The biology of the spirit. Viking Press, New York.

Southwick, Charles H. 1963. Primate social behavior. D. Van Nostrand Co., Inc., Princeton, N.J.

Tinbergen, N. 1965. Social behavior in animals. Second Edition. Methuen Company, New York.

van Bergeijk, Willem A. M., J. R. Pierce, and E. E. David, Jr. 1960. Waves and the ear. Doubleday and Co., Inc. (Anchor Books), Garden City, New York.

Young, John. 1960. Doubt and certainty in science: a biologist's reflections on the brain. Oxford University Press, New York.

PHASE FIVE Sections 15, 16

Carson, Rachel. 1962. Silent spring. Houghton Mifflin Co., Boston.

Deason, H. J. and W. Blacklow (Editors). 1957. A guide to science reading. The New American Library of World Literature, Inc., New York.

Dubos, R. 1962. The torch of life. Pocket Books, Inc., New York.

Ehrlich, Paul R., and Anne H. Ehrlich. 1970. Population resources environment. W. H. Freeman and Co., San Francisco.

Huxley, Julian. 1963. The human crisis. University of Washington Press, Seattle.

Northrop, F. S. C. 1963. Man, nature and god. Pocket Books, Inc., New York.

Snow, C. P. 1964. The two cultures and the scientific revolution. Mentor Books, New York.

INDEX

Numbers in **boldface** represent pages with figures.

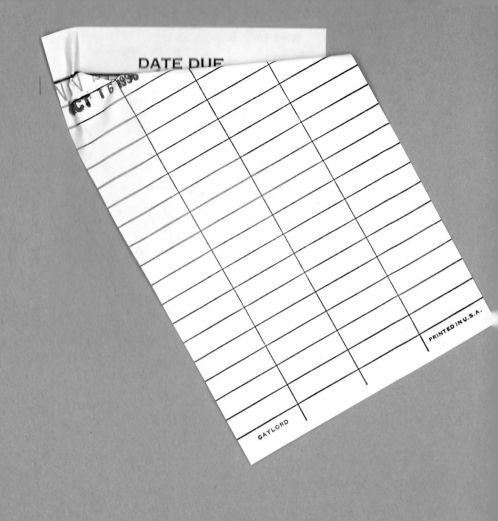